Bacterial Adaptation to Co-resistance

Santi M. Mandal • Debarati Paul
Editors

Bacterial Adaptation to Co-resistance

Springer

Editors
Santi M. Mandal
Central Research Facility
Indian Institute of Technology Kharagpur
Kharagpur, India

Debarati Paul
Amity Institute of Biotechnology
Amity University
Noida, Uttar Pradesh, India

ISBN 978-981-13-8502-5 ISBN 978-981-13-8503-2 (eBook)
https://doi.org/10.1007/978-981-13-8503-2

This Springer imprint is published by the registered company Springer Nature Singapore Pte Ltd.
The registered company address is: 152 Beach Road, #21-01/04 Gateway East, Singapore 189721, Singapore

Contents

1 **Plasmids: The Necessary Knowledge Wealth for Encountering Antibiotic-Resistance Menace** . . . 1
Shriparna Mukherjee and Ranadhir Chakraborty

2 **Disinfectants Amend the Expression of Membrane Bound Efflux Transporters to Augment Antimicrobial Resistance** . . . 19
Govindan Rajamohan and Vijaya Bharathi Srinivasan

3 **Knowledge Gaps and Research Needs in Bacterial Co-Resistance in the Environment** . . . 39
Paola Grenni and Gianluca Corno

4 **Microbial Resistance to Antibiotics** . . . 61
Martha Premlatha

5 **Do Non-medical Uses of Antibiotics Develop Cross-Resistance in Clinical Pathogens?** . . . 81
Lalitha Cheepurupalli, Sudarshan Singh Rathore, Thiagarajan Raman, Niranjana Sri Sundaramoorthy, and Jayapradha Ramakrishnan

6 **Biofilms in Antimicrobial Activity and Drug Resistance** . . . 109
Timsy Bhando, Vineet Dubey, and Ranjana Pathania

7 **Antimicrobial Resistance in Microbes: Mode of Action of TolC Like Protein and Their Mechanism of Regulating Stress Resistance and Physiology** . . . 141
Vijaya Bharathi Srinivasan and Govindan Rajamohan

8 **Efflux Mediated Co-resistance** . . . 161
Amit Gaurav, Atin Sharma, and Ranjana Pathania

9 **Biofilm and Antibiotic Resistance in *Acinetobacter baumannii*** . . . 181
Rajagopalan Saranathan, Sudhakar Pagal, and K. Prashanth

10 **Mechanism of Bacterial Co-resistance** . . . 191
Piyush Baindara

11 **Antibiotics and Microbial Antibiotic Resistance in Soil** 211
Ali-Akbar Safari-Sinegani, Mehdi Rashtbari, Nayereh Younessi, and Babak Mashkoori

12 **Microbial Adaptation and Resistance to Pesticides** 233
Debarati Paul and Santi M. Mandal

13 **Antimicrobial Agents Used in Food Preservation or as Agricides and Effect on Microbes in Developing Antimicrobial Resistance** 251
Soma Mukherjee, Nitin Dhowlaghar, and Wes Schilling

14 **Molecular Mechanisms of Action and Resistance of Antimalarial Drugs** 267
Juveria Khan, Monika Kaushik, and Shailja Singh

15 **Management and Control of Antimalarial Drug Resistance** 297
Amrita Chakrabarti, Vigyasa Singh, and Shailja Singh

About the Editors

Santi M. Mandal obtained his PhD in the field of Molecular Microbiology and continuing research with major focus on Antimicrobial Chemotherapy. He visited UTMB-USA and NUS-Singapore for his postdoctoral training. He worked as an Assistant Professor of Microbiology at Vidyasagar University, India. He has published more than 110 research papers in reputed journals and conferred upon several prestigious awards for his research contribution. Currently, he is engaged at Central Research Facility, Indian Institute of Technology Kharagpur, Kharagpur, India.

Debarati Paul (PhD), currently Associate Professor in the Department of Biotechnology, Amity University, has completed her PhD from the Institute of Microbial Technology-CSIR (Chandigarh), and thereafter, she carried our research work abroad in EAWAG (Switzerland) and Mississippi State University (USA) for about 5 years, before joining Amity University, Uttar Pradesh, Noida, as Assistant Professor. Her main areas of research are environmental biotechnology, encompassing biodegradation, biofuels, genomics and proteomics, and also antimicrobial resistance. She has published over 70 papers including research articles, reviews, conference proceedings and posters, and book chapters. She has actively participated in national and international projects as Research Fellow and also as Principal Investigator (PI) or as co-PI. She has received the prestigious Indo-US Science and Technology Forum Award and Bharat Gaurav Award along with DST travel grants, poster award, etc.

Plasmids: The Necessary Knowledge Wealth for Encountering Antibiotic-Resistance Menace

1

Shriparna Mukherjee and Ranadhir Chakraborty

Abstract

Infections caused by multidrug resistant bacteria have been identified as an emergent global public health problem as it significantly increases the rate of morbidity and mortality and simultaneously health care costs. It has already been identified that the foremost cause of mortality by the mid of twenty-first century would be drug-resistant disease having direct implication in annual direct costs amounting to 2–3% of the global output GDP. European Center for Disease Prevention and Control (ECDC) and the Center for Disease control and Prevention (CDC) have jointly made the definitions for MDR (multidrug resistant), XDR (extensively or extremely drug resistant) and PDR (Pan drug resistance). Only acquired antimicrobial resistance but not intrinsic resistance was considered for defining MDR, XDR and PDR types. The consecutive development of drug resistance in a population occurs stepwise and organisms with low-level resistance may form the genetic platform for the development of higher resistance levels. Molecular analyses have revealed that widespread multi resistance has commonly been achieved by the acquisition of preexisting determinants followed by amplification in response to selection (acquired resistance). Plasmids, being an important carrier of mobile genetic elements bearing antibiotic resistance genes, play a fundamental role in the dissemination of antimicrobial resistance genes within the gene pool. Certain connections between antibiotic-resistant plasmids and pathogenic bacterial clones are hugely extensive. Plasmids, mainly the conjugative ones are undoubtedly the most significant

S. Mukherjee
Prasannadeb Women's College, Jalpaiguri, West Bengal, India

R. Chakraborty (✉)
OMICS Laboratory, Department of Biotechnology, University of North Bengal, Siliguri, West Bengal, India
e-mail: rcnbusiliguri@gmail.com

S. M. Mandal, D. Paul (eds.), *Bacterial Adaptation to Co-resistance*,
https://doi.org/10.1007/978-981-13-8503-2_1

drivers of antibiotic-resistance dissemination in the bacterial families such as Enterobacteriaceae and Enterococcaceae, which comprise some of the most important hospital-borne- pathogens. Role of plasmids in developing antibiotic resistance and also endemic and epidemic behavior in certain important pathogenic *bacterial genera* like *Pseudomonas*, *Acinetobacter baumanii*, *Staphylococcus aureus* have been evidenced from several studies. Survival of these bacterial pathogens in different challenging environments is mediated mainly by plasmids. In the near future, scientists will be able to blend the technological and theoretical knowledge in clubbing genomics, transcriptomics, proteomics and metabolomics to delineate the molecular basis of the fitness-cost and compensation in plasmid–bacterium relationships

Keywords
Plasmids · MDR · XDR · PDR · Co-resistance · Cross-resistance

1.1 Introduction

The antibiotic era began with the discovery of the first three significant antibiotics-tyrothricin, penicillin and actinomycin, in 1939 and 1940. Within the first 18 years of the antibiotic era, about 30 antimicrobial agents had come into use (Swartz 2000). Since their discovery, the use of manufactured antibiotics to control diseases has revolutionized medicine. It has also greatly reduced the threat of many once fatal illnesses. The use of these wonder drugs, combined with improvements in sanitation, housing and nutrition and the advent of widespread immunization programme, has led to a dramatic drop in deaths from diseases that were previously widespread, untreatable and frequently fatal. By helping to bring many infectious diseases under control, these drugs have also contributed to the major gains in life expectancy experienced during the latter part of the last century. These gains were soon jeopardized by the emergence and spread of antibiotic resistant microbes. The organizations such as US Centres for Disease Control and Prevention (CDC), the European Centre for Disease Prevention and Control (ECDC) and World Health Organization (WHO) have identified the infections caused by multidrug resistant (MDR) bacteria as an emergent global disease and a major public health problem as it significantly increases the rate of morbidity and mortality and simultaneously health care costs.

The successive development of antibiotic resistant bacterial populations follows the Darwinian principle of 'Survival of the fittest' (White and McDermott 2001). Many genes determining resistance had been present in nature and predate the clinical use of antimicrobial drugs. Antibiotic resistant bacteria, estimated at over 2000 year old have been isolated from deep within glaciers in Canada's high arctic regions (Dancer et al. 1997). Resistant microorganisms have also been found among historic bacterial cultures collected before the beginning of antimicrobial era (Smith

1967). Over millions of years, bacteria have evolved a number of strategies to coexist peacefully, including the capacity to produce antibiotics to fight off competitors. The last five decades have seen an outstanding ability of bacterial populations to develop and share resistance to every antibiotics that has been developed, often by quite amazing mechanisms and much more readily than was formerly predicted. Dramatic increase in both the proportion and absolute number of multidrug resistant bacterial pathogens are posing worldwide threat (Roca et al. 2015). The foremost cause of mortality by the mid of twenty-first century would be drug-resistant disease having direct implication in annual direct costs amounting to 2–3% of the global output GDP.

1.2 Who Will Be Regarded as Multidrug Resistant Ones?

1.2.1 Definitions for MDR, XDR and PDR

Standard terminology for defining different types of drug resistant microbes is of vital importance. This will help to get consistent reporting of comparable data that will track trends of antimicrobial resistance both locally and globally. The database would therefore enable in promoting the sensible use of antimicrobials and adoption of other public health measures by the authorities concerned with public health (Carmeli et al. 2010; Jones and Masterton 2001). A joint initiative was taken by the European Center for Disease Prevention and Control (ECDC) and the Center for Disease control and Prevention (CDC) to define bacterial pathogens exhibiting resistance toward a significant number of antimicrobial agents with suitable standardized international terminology like MDR (multidrug resistant), XDR (extensively or extremely drug resistant) and PDR (Pan drug resistance) (Magiorakos et al. 2012). Final proposed definitions were presented to the ECDC advisory forum on 30th September, 2010. Only acquired antimicrobial resistance but not intrinsic resistance was considered for defining MDR, XDR and PDR types. A bacterial isolate is considered to be resistant to an 'antimicrobial class' when it exerts/ exhibits non susceptibility to one or more antimicrobial agents within that class (Hidron et al. 2008; Kallen et al. 2010).

Bacterial isolate resisting at least one agent in three or more antimicrobial categories is considered as MDR. XDR exhibits non-susceptibility to at least one agent in all but two or fewer antimicrobial categories (i.e. bacterial isolates remain susceptible to only one or two categories). PDR is resistant against all agents in every antimicrobial category (i.e. no agents tested as susceptible for that organism). Thus, a bacterial isolate which was previously characterized as XDR, will also be characterized as MDR. Similarly, the prerequisite of a bacterial isolate to be branded as PDR, it has to be primarily be a XDR. As per set theory, a subset of MDR is XDR and subset of XDR is PDR. The serious implication of bacteria being PDR is that no approved antibiotic agents are left to be used against them.

1.2.2 Other Terminologies for MDR Phenotypes: Co-Resistance, Cross-Resistance and Pleiotropic Resistance

A number of additional terms are also used to refer multidrug-resistant phenotypes. These are co-resistance, cross-resistance and pleiotropic resistance. Co- resistance involves presence of different resistance mechanisms encoded by either mutated or acquired genes leading to development of complex drug resistance phenotypes affecting different antimicrobial classes, selection for one gene encourages the safeguarding of another resistance gene, one that does not essentially tender a selective advantage to the chemical in question (Johnson et al. 2016). Cross-resistance is produced due to the presence of mutated or acquired resistance genes affecting different antimicrobial agents from the same categories, one resistance gene can proffer defense against several toxic chemicals (Curiao et al. 2016). Pleiotropic resistance affects several antimicrobials from different categories following mutation or acquisition of resistance genes (Canton and Morosini 2011).

1.3 Mobilization of Resistance Factors

The interplay of many mechanistic and epidemiological factors brings about the progress and spread of antibiotic resistance. Emergence of resistance traits in nature occur at random, at anywhere at any time. These traits are often associated with some fitness cost. The consecutive development of drug resistance in a population occurs stepwise and organisms with low-level resistance may form the genetic platform for the development of higher resistance levels (Normark and Normark 2002). Once resistance is developed, resistant bacteria appear to acquire a 'life of their own' (Barbosa and Levy 2000).

In contrast to clinical microorganisms, environmental microorganisms that produce antibiotics express a considerable degree of intrinsic resistance that emerges to be independent of the selective pressure–a trait that has been proposed to be primordial and existing even before the commercial production and therapeutic use of antibiotics (D'Costa et al. 2011; Cox and Wright 2013). The momentous international concern is the danger of intrinsic resistances (originating in environmental microorganisms) being transferred to pathogens (Forsberg et al. 2012; Cox and Wright 2013). Further augmentation of intrinsic resistance can occur via modification or over-expression of cellular constituents like penicillin-binding protein 5 (PBP 5) that guard against the effects of the antibiotic (Sifaoui et al. 2001); such modifications are then regarded as acquired resistance.

Molecular analyses have revealed that widespread multi resistance has commonly been achieved by the acquisition of preexisting determinants followed by amplification in response to selection (acquired resistance). The capture, accumulation, and dissemination of resistance genes are largely due to the actions of conjugative and mobilizable genetic elements (MGE). Conjugative elements (plasmids) can transfer themselves from one bacterium to another whilst MGEs use the conjugation functions of co-resident conjugative elements (conjugative plasmids or conjugative transposons) to transfer to another host. Horizontal gene transfer of these genetic elements is the major contributor to emergence, recombination and dissemination of multidrug resistance among bacterial pathogens. Mobile genetic elements are discrete DNA molecules capable of translocation from one part of the genome to another part or between the genomes. MGEs can also be differentiated by their capacity to liberally pass on from cell to cell (i.e., intracellular MGEs) and those which after integrating into intracellular MGEs are capable of dissemination from cell to cell (Siguier et al. 2014). The significance of all these MGEs is that a lot of them have 'cargo genes' that are able to present phenotypes such as antibiotic resistance to the recipients (Roberts and Mullany 2011; Johnson and Grossman 2015).

1.4 Plasmids in Dissemination of Resistance

Plasmids often play an important role as carriers of mobile genetic elements bearing antibiotic resistance genes to serve as vectors for nonconjugative and conjugative resistance-gene(s)- bearing transposons. Horizontal gene transfer via plasmids could be narrow or broad; narrow when horizontal gene transfer occurs between members of the same species or broad when takes place between bacteria belonging to diverse species and genera. Plasmids are highly supple DNA elements, and by the possession of various mechanisms they can recombine, form co-integrates, or become integrated in part or whole into the host chromosomal DNA or into other plasmids. Plasmids, per se, participate in playing a fundamental role in the dissemination of antimicrobial resistance genes within the gene pool. They are important cargo vessels for carrying MGE and acquired antimicrobial resistance genes in both Gram-negative and Gram-positive genera.

Plasmids that mediate horizontal movement of resistance- genes are responsible for global spread of resistance (Carattoli 2013). Resistance plasmids (feature resistance to frequently used antibiotics) are mostly conjugative; additional are mobilizable. With the aid of conjugation, relocation of genes, acquired through homologous recombination, integration and excision from the host chromosome, takes place from donor to the recipient cells. Conjugative plasmid-encoded complexes help the donor by attaching to capable recipient thereby escorting processes of secured association, required earlier for the transfer of DNA. Plasmids that are unsuccessful to get transferred by this approach are relocated by conjugative elements following the progress of transitory or steady fusion called co-integrates. Plasmids also promote cell contact development in the course of production of pheromone manipulated micro-fibrillar peripheral casing materials. Mobilizable plasmids bear DNA transfer genes necessary for structure of all or element of the relaxasome, but are lacking genes essential for the formation of mating pores. Mobilizable plasmids have the ability to exploit conjugative plasmids bearing two classes of transposons, composite transposons (Class I; holding a range of resistance genes which possess identical structural and functional attributes, but less DNA homology) and complex transposons (class II; constituting three dissimilar but interrelated families; Tn3, Tn21 and

Tn2501; Schmitt 1986; Wiedemann et al. 1986; Lafond et al. 1989). Some of the composite transposons in gram- negative bacteria are Tn5, Tn9, Tn10, Tn903, Tn1525, and Tn2350 and among gram positive bacteria are Tn4001 and Tn4003. These components hold the capability to interact both intra and inter-molecularly so that they can jump within a DNA molecule or from one DNA molecule to another (Bennett 2008). The most well studied Tn21, bear OXA (possess oxacillinase activity) and PSE (β -lactamase gene, *Pseudomonas* specific enzyme), genes that render antibiotic resistance. Tn21 also carry resistance genes towards mercury compounds and trimethoprim (Brown et al. 1986; Sundström et al. 1988). Retro-transposons or class I function by copying RNA from DNA by transcription and RNA to DNA by reverse transcription; in that way get inserted into the genome at a different location (Kapitonov and Jurka 2008). Class II transposons bypass RNA intermediate and acts by cut and paste mechanism (Wicker et al. 2007). Transposases generate staggered incision at specific site, producing sticky ends and after transposition to the aimed site normally pursue by target site duplication forming short direct repeat at the sites of insertion. Nonetheless, transposons endow with antibiotic resistance owing to the existence of an additional gene on a plasmid there are possibilities that transposons can hop from chromosomal DNA to plasmid DNA and vice versa for development of resistance (Wagner 2006). Transposons contain insertion sequences (ISs; size <2.5 kb) at both ends. ISs are fundamental form of mobile genetic elements disseminated in bacteria and considered as non-complex bacterial mobile DNA as per their structure (El Salabi et al. 2013). There are more than 19 families of ISs, having dissimilar size (Wagner et al. 2007). ISs contain an open reading frame that codes for a transposase enzyme, flanked by inverted repeat sequences of 10–40 base pairs at both ends. The transposase enzyme severs target DNA and interleaves the IS due to potential association with the inverted repeat sequences. IS often show keenness toward AT-rich area of DNA that lead to elevated chances of undergoing homologous recombination, creates variety of possibilities such as deletions, inversions and duplications. There are evidences that when two identical IS elements surround a region of DNA, a composite transposon is produced, and the total interceded DNA flanked by the terminal inverted repeats get mobilized by one or both of the IS coded transposases (Ochi et al. 2009; Gyles and Boerlin 2014).

1.5 Resistance Plasmids in Enterobacteriaceae

1.5.1 Plasmids Carrying ESBLs

Enterobacteriaceae producing expanded-spectrum β-lactamases (ESBLs), those of the CTX-M type in particular, are a major problem worldwide (Paterson and Bonomo 2005). It was observed that pandemic dissemination of the CTX-M-15 enzyme took place due to the presence of highly virulent *Escherichia coli* O25:H4-ST131 (Leflon-Guibout et al. 2008; Woodford 2008). Plasmids belonging to the IncF group carry the *bla*CTX-M-15 gene along with the *bla*TEM-1, *bla*OXA-1, and *aac*(6)-Ib-cr resistance genes. IncF plasmids carrying the blaCTX-M-15 gene were found in *E. coli* sequence types ST405, ST354, ST28, and ST695, in a *Shigella sonnei* strain, in *Salmonella enterica* serovar *Enteritidis* in the United Kingdom, and in a *Klebsiella pneumoniae* strain in Spain (Coque et al. 2008; Diestra et al. 2009; Hrabak et al. 2008). The blaVEB-1 gene in Enterobacteriaceae isolated from Canada, France, Thailand, and Turkey was located on one single plasmid type in the IncA/C group (Poirel et al. 2007). Klebsiella producers of SHV-5 largely prevailed in the United States during the 1993–2000 period, carrying the blaSHV-5 gene located on IncL/M plasmids that were also reported in *Salmonella* isolated from children in Albania (Tosini et al. 1998). The blaSHV-12 gene variant largely prevailed in K. pneumoniae isolates from Europe, and it was located on plasmids from different families. Recently, the *bla*CTX-M-1 gene was associated with IncI1 in *E. coli* isolated from human patients in different parts of France, suggesting a potential link between animals and humans for the dissemination of this gene variant in this country (Marcade et al. 2009).

1.5.2 Plasmids Carrying AmpC -Lactamases

The family of AmpC -lactamases includes the chromosomal enzymes of E. coli, Enterobacter spp., Citrobacter spp., Morganella spp., Proteus spp., P. aeruginosa, and other species, in addition to a growing number of plasmid-mediated β-lactamases related to the above-mentioned enzymes. Dissemination of organisms that produce CMY, an AmpC that originated from the chromosome of Citrobacter freundii, has been linked to specific plasmid families that are recurrent in isolates from animal sources (Andrysiak et al. 2008; Welch et al. 2007). The majority of the blaCMY-2 plasmids identified in E. coli and Salmonella spp. in the United States were categorized in the IncA/C group. IncA/C-positive strains were isolated from beef, chicken, turkey, and pork and were found in samples from different regions of the United States, revealing that this common plasmid backbone is broadly disseminated among resistant zoonotic pathogens associated with agriculture in this country (Mulvey et al. 2009; Winokur et al. 2001). Several blaCMY gene variants were also associated with the IncI1 plasmid family. IncI1 plasmids are widespread in E. coli animal strains (17.4% and 41% in avian commensal and pathogenic E. coli strains, respectively), again suggesting that the dissemination of this gene could occur in the intestinal tract of animals (Johnson et al. 2007).

1.5.3 Plasmids Carrying Carbapenem Resistance

Carbapenemases that hydrolyze most -lactams, including carbapenems, are classified in four molecular classes, and most of the class A carbapenemases are chromosomally encoded (IMI, NMC-A, and SME) with the exception of KPC enzymes identified in Enterobacteriaceae (and rarely in Pseudomonas aeruginosa) and the GES-type enzymes identified in Enterobacteriaceae and P. aeruginosa. The class B enzymes are the most clinically significant carbapenemases; they are metallo--lactamases (MBLs), mostly of the IMP and the VIM series. They have been reported worldwide, and their genes are carried by plasmids and integrons, hydrolyzing all -lactams with the exception of aztreonam. The 1998–2004 global SENTRY survey found only rare examples of MBL genes (blaIMP-1, blaIMP-11, and blaVIM-1) among Enterobacteriaceae isolates (Deshpande et al. 2006).The blaIMP-4 gene was recognized in Australian Enterobacteriaceae from Sydney in 2003 to 2006 and caused outbreaks in Melbourne in 2004 and 2005. IncL/M plasmids were identified in 22 of 23 Sydney isolates over 3 years, while IncA/C plasmids were detected in all Melbourne isolates. Thus, in Australia, distinct broad-host-range plasmids carrying identical cassette arrays in different contexts simultaneously emerged in two cities, with no apparent mixing over several years (Espedido et al. 2008). VIM enzymes have been found mainly in nonfermenting gram-negative bacteria, but their numbers are increasing in Enterobacteriaceae. K. pneumoniae isolates carrying the blaVIM-1 gene and E. coli isolates carrying blaVIM-1 and blaCMY-13 genes, randomly collected from five different hospitals in Athens and Piraeus from 2001 to 2003 and representative of the VIM-1-producing isolates circulating in Greece, were all assigned to the IncN group, indicating the spread of an epidemic plasmid associated with the emergence of the blaVIM-1 gene in that country (Carattoli et al. 2006). Moreover, recent identification of blaVIM-1, often associated with blaSHV-5 within the same cell but located on a different plasmid, confirmed the transfer via self-transmissible IncN plasmids in K. pneumoniae (Miriagou et al. 2008). The blaVIM-1 gene was also recently identified on IncW broad-host-range plasmids in Serratia liquefaciens and Klebsiella oxytoca from Greece, suggesting a novel vehicle for a larger dissemination of this resistance threat in this country. Four MBL-producing species (K. pneumoniae, K. oxytoca, Enterobacter cloacae, and E. coli) have been described in Spain. The strains showed different blaVIM-1 genetic environments, and the gene was located on different plasmid scaffolds. A 60-kb conjugative plasmid belonging to the IncI1 group was observed in the K. pneumoniae clone and in E. coli, while plasmids belonging to the IncH12 group were found among E. cloacae isolates. IncA/C plasmids carrying both the blaVIM-4 and blaCMY-4 genes were identified in Italy in clinical isolates of K. pneumoniae and E. cloacae. The scaffolds of these plasmids were similar to those of the IncA/C plasmids carrying blaCMY-2 or blaCMY-4 from S. enterica isolated in the United States and the United Kingdom, but the carbapenemase gene was not present on these Salmonella plasmids and likely represents a novel acquisition for the IncA/C plasmids (Colinon et al. 2007; Hopkins et al. 2006). IncA/C-IncP multireplicon plasmids carrying the blaIMP-13 gene were also identified in Salmonella found in food sources in Colombia (O'Mahony et al. 2006).

1.6 Resistance Plasmids in *Pseudomonas aeruginosa*

Opportunistic pathogen like *Pseudomonas* is genetically equipped with outstanding intrinsic antibiotic resistance machinery (Cabot et al. 2016). Role of resistance plasmids in acquired antibiotic resistance is also detected in *Pseudomonas aeruginosa* (Strateva and Yordanov 2009). Antibiotic resistance in *Pseudomonas aeruginosa* is commonly associated with conjugative plasmids of IncP group (Korfhagen et al. 1976; Pansegrau et al. 1994). Several studies have shown the role of different mobile genetic elements like transposons, integrons, IS elements and also the genomic island encoded virulence (Di Pilato et al. 2014; Odumosu et al. 2013; Kobayashi et al. 2013; Xiong et al. 2013; Naas et al. 2013; Jovcic et al. 2013; Li et al. 2008) as carriers of antibiotic resistance genes other than plasmid mediated pathogenecity (Harrison et al. 2010). Plasmids from *P. aeruginosa* often found to carry the same metallo-β-lactamases found on mobile genetic elements from Enterobacteriaceae. The routes of dissemination of these genes between Pseudomonads and Enterobacteria are not known. A study reported the presence of two new plasmids, pAMBL1 and pAMBL2, coding for VIM-1 carbapenemase from *P. aeruginosa* and demonstrated that the presence of three copies of blaVIM-1 in pAMBL2 produces high-level resistance to carbapenems (Millan et al. 2015). Bi et al. (2016) sequenced two novel, native plasmids from a clinical *P. aeruginosa* isolate HS87. One of these plasmids (pHS87a) with the size of 26 kb carries mobile genetic elements and an aminoglycoside-resistant gene. An imipenem-resistant *Pseudomonas aeruginosa* strain HN39 harbors a blaSIM-2-carrying novel 282-kb megaplasmid pHN39-SIM that carries a novel 38.8-kb multidrug resistance region. This multidrug resistance region is composed of Tn6284 generated from the insertion of an In4-family integron In1208 into Tn5046, and a Tn4662a-derived element with the insertion of ΔTn512 connected with two other genes. In1208 carries besides blaSIM-2 several additional genes that mediate resistance to erythromycin, chloramphenicol, rifampicin, streptomycin, quaternary ammonium compounds, sulphonamides and mercury (Sun et al. 2016). Another study reported characterization of a novel multidrug resistance megaplasmid carrying qnrVC6 and bla_{IMP-45} from *Pseudomonas aeruginosa*. This is the first ever finding of a qnrVC6 gene in P. aeruginosa The qnrVC6 gene is flanked by two copies of insertion sequence (IS) elements ISCR1, a multiresistance class 1 integron In786 containing aacA4-bla_{IMP-45}-bla_{OXA-1}-catB3 cassettes, an armA gene, and an aphA7 gene flanked by two copies of IS26 (Liu et al. 2018). The work of Shi et al. (2018) presented the first two fully sequenced resistance plasmids, bla_{KPC-2}-carrying p14057A and *tetA*(A) -carrying p14057B, coexisting in *P. aeruginosa*, which is a significant cause of nosocomial infection. The findings of their work offer a deeper insight into how *P. aeruginosa* becomes drug-resistant by diversified mechanisms.

1.7 Resistance Plasmids in *Acinetobacter baumannii*

Acinetobacter baumannii is a nosocomial human pathogen. Mobile genetic elements like plasmids and transposons help to develop multidrug resistance in *Acinetobacter* like other bacterial genera. Palsmid profiling has revealed the presence of multiple indigenous plasmids of varying molecular sizes in more than 80% of the Acinetobacter isolates (Roca et al. 2012). Study conducted by Poirel and Nordmann (2006) conclusively inferred that plasmids/integrons play an important role in development of antibiotic resistance and also endemic and epidemic behavior of *Acinetobacter baumannii.* Plasmid pIP1841 (63 kb) that confers resistance to aminoglycosides obtained from *A. baumannii* is self-transferable to other *Acinetobacter baumannii*, *Acinetobacter haemolyticus* and *Acinetobacter lwoffii* but not to *Escherichia coli* (Lambert et al. 1988). Several studies have shown the presence of antibiotic resistance genes like β-lactams ($bla_{\mathrm{GES\text{-}11}}$), carbapenems ($bla_{\mathrm{IMP}}$, bla_{VIM}, $bla_{\mathrm{OXA\text{-}23}}$, $bla_{\mathrm{OXA\text{-}24}}$, $bla_{\mathrm{OXA\text{-}58}}$, and $bla_{\mathrm{NDM\text{-}1}}$), sulfonamide (*sul*2), and streptomycin (*str*AB) on plasmids in *A. baumannii* (Da Silva and Domingues 2016; Hamidian et al. 2016). Leungtongkam et al. (2018) demonstrated conjugation-mediated gene transfer from the extensively drug-resistant *Acinetobacter baumannii* (XDR-AB) and New Delhi Metallo-β-lactamase-1-producing *Acinetobacter baumannii* (NDM-AB) to environmental isolates of *Acinetobacter* spp. Resistance to ticarcillin and kanamycin was transferred from four donors to two sodium azide-resistant *A. baumannii* strains, namely, NU013R and NU015R. It has been observed that donor AB364 strain can transfer the $bla_{\mathrm{OXA\text{-}23}}$ and $bla_{\mathrm{PER\text{-}1}}$ genes residing on a 240 kb plasmid to both recipients in association with *int1*. The *aphA6* and $bla_{\mathrm{PER\text{-}1}}$ genes were too co-transferred with the *int1* gene from the donor strains AB352 and AB405. The transfer of a 220-kb plasmid from the donors to recipient was detected. The GR6 plasmid containing the kanamycin resistance gene (*aphA6*) was successfully transferred from the donor strain AB140 to both recipient strains. This study for the first time has shown co-transfer of antimicrobial resistance elements with integron cassettes or with the plasmid group GR6 in *in vitro* conjugation. Plasmid pNDM-BJ01carrying blaNDM-1 in Tn125 is found in both *A. baumannii* and other *Acinetobacter* spp. This plasmid was also reported later from *Enterobacter aerogenes* (Chen et al. 2015) suggesting their horizontal transfer. Plasmid pAb-G7-2 and pACICU2 carry aphA6 (encoding kanamycin and amikacin resistance) in Tn*aph*A6 (Hamidian et al. 2014). Recently, presence of small plasmids usually ranged from 2 to 10 kb in size, has been observed in majority of *A. baumannii* genomes. Some of them are found to harbor antibiotic resistance genes and homologs of plasmid mobilization genes. Some of these plasmids, such as the Rep-3 superfamily group and the pRAY-type, are quite widespread among diverse *A. baumannii* clinical isolates worldwide (Lean and Yeo 2017).

1.8 Resistance Plasmids in Staphylococci

Survival of *Staphylococcus aureus* in different challenging environments is mediated mainly by plasmids. 15–20% of the Staphylococcal genome comprises of MGEs, including bacteriophages (phages), pathogenicity islands, plasmids, transposons, integrative conjugative elements (ICEs), integrons, and staphylococcal chromosome cassettes (SCCs). All of these MGEs except the phages are potential carriers of antibiotic-resistance genes (ARGs) (Lindsay 2010; Planet et al. 2017; Alibayov et al. 2014). Size range of the plasmids falls between 1–60 kbp in most clinical isolates of *Staphylococcus*. Small plasmids are found to resist erythromycin, chloramphenicol, or tetracycline, whereas the larger plasmids exert resistance to β-lactams, aminoglycosides, and macrolides (McCarthy and Lindsay 2012). Conjugative plasmids often carry transposons that can provide resistance to aminoglycosides, erythromycin, spectinomycin, tetracycline, trimethoprim, vancomycin, or β-lactams (Haaber et al. 2017). Of the three plasmid borne tetracycline resistance genes, *tet*K are most widespread followed by *tet*M but *tet*L is less widely disseminated among *Staphylococcus* population. Plasmid pSWS47 from methicillin resistant *Staphylococcus epidermidis* carries both *tet*M and *tet*L along with three other resistance genes (Weiss et al. 2014). The gene *cfr*, initially identified on multiresistance plasmid pSCFS1 from S. sciuri, was further detected in various other staphylococcal species (Kehrenberg et al. 2004). *cfr* gene confers resistance to phenicols, lincosamides, pleuromutilins, streptogramin A antibiotics, selected 16-membered macrolides and also to the oxazolidinone linezolid. The transferable linezolid resistance conferred by the multiresistance gene *cfr* poses a significant and interdisciplinary public health challenge (Shen et al. 2013). Several plasmid borne genes conferring resistance to aminoglycosides, aminocyclitols, or streptothricins are found as parts of transposons that can integrate into plasmids and also to the chromosomal DNA. Transposon Tn5405 found in *S. aureus* is a composite transposon with a central resistance gene region encompassing the streptomycin resistance gene *aad*E, the streptothricin acetyltransferase gene *sat*4, and the kanamycin/neomycin/amikacin resistance gene *aph*A3.

Initially, complete Tn5405 was discovered on plasmid pIP1085B. This composite transposon along with its gene cluster has also been detected on pSERP from *S. epidermidis* (Gill et al. 2005), p18806-P03 and related plasmids from the *S. aureus* clone USA300 (Kennedy et al. 2010). The *aad*E gene was shown to be part of the aforesaid chromosomal or plasmidic multi-resistance gene cluster in MRSA ST398, MSSA ST9, and MRSA ST9 (Lozano et al. 2012; Wendlandt et al. 2013; Li et al. 2013).

Recently, spectinomycin resistance gene, *spd*, has been found on plasmid pDJ91S from MRSA ST398 of chicken origin. Plasmid pSWS2889 carrying the same *spd* gene has been detected among MRSA CC398 from diverse sources. The gene *apm*A that confers resistance to the aminocyclitol antibiotic apramycin was identified on the multi-resistance plasmid pAFS11 from a bovine MRSA CC398 isolate (Feβler et al. 2011). Resistance genes *erm*(B), *aad*D, *tet*(L), and dfrK are also present on this plasmid. The conjugative plasmid pKM01 obtained from a canine *Staphylococcus pseudintermedius* has shown to carry the *mup*A gene along with the *aac*A-*aph*D gene (Matanovic et al. 2013). In 2002 first clinical vancomycin resistant *S. aureus* was detected. The isolate was found to carry conjugative multiresistance plasmid pLW043. This plasmid consisted of a pSK41-like *S. aureus* resistance plasmid (*dfr*A, *aac*A-*aph*D, *bla*Z, and *qac*C) with an insertion of a Tn1546-like element carrying the *van*A gene cluster (Weigel et al. 2003). In staphylococci, resistance to penicillins is conferred by the *bla*Z-encoded β-lactamase. The transposon Tn552, carrier of the *bla*Z-*bla*I-*bla*R1 operon (Rowland and Dyke 1989; Rowland and Dyke 1990), has been detected on plasmids as well as in the chromosomal DNA (Lyon and Skurray 1987; Jensen and Lyon 2009; Rowland and Dyke 1990), but for the first time it was detected on on plasmid pI258 (Rowland and Dyke 1990). These β-lactamase plasmids also carry additional antimicrobial resistance genes like *erm*(B) (Novick et al. 1979), *fus*B (O'Brien et al. 2002), and *aac*A *aph*D (Gillespie and Skurray 1986). More recently, *bla*Z-carrying multiresistance plasmids that harbored the enterotoxin genes *sed*, *sej*, and *ser* (Shearer et al. 2011) or the gene for the exfoliative toxin B (Olsen et al. 2006) have been described. Of the four different trimethoprim resistance genes (*dfr*A, *dfr*D, *dfr*G, and *dfr*K) of staphylococci, only three (*dfr*A, *dfr*D, and *dfr*K) are transferable and often located on plasmids. Gene *dfr*D, was found on plasmid pABU17 from *S. haemolyticus* (Dale et al. 1995). Gene *dfr*K was initially identified on plasmid pKKS2187 from porcine MRSA CC398 (Kadlec and Schwarz 2009).

1.9 Conclusion

At present humanity is encountering one of the biggest threats and that is antibiotic resistance. In this context, plasmids have a crucial function in the propagation of antibiotic resistance among important human pathogens. Certain connections between antibiotic-resistant plasmids and pathogenic bacterial clones are hugely extensive. There is still a void in understanding the contribution of the fitness cost of plasmids and recompense of evolution in the increase and broadening of booming plasmid–bacterium associations. Plasmids play a key role in the evolution of antibiotic-resistance in bacteria and broadcasting resistance genes among the most troublesome clinical pathogens. Conjugative plasmids are undoubtedly the most significant drivers of antibiotic-resistance dissemination in the bacterial families such as *Enterobacteriaceae* and *Enterococcaceae*, which comprise some of the most important hospital-borne- pathogens. Remarkably, several of these plasmid–bacterium associations turn out to be particularly triumphant, generating 'superbugs' that spread frenziedly in the hospital set-ups. These explicit alliances between plasmids and bacterial clones are all-encompassing, with certain resistance-plasmids sturdily allied to specific bacterial lineages. Studies twining genomics and transcriptomics are in demand to reveal the convolution of plasmid–bacterium interactions. In the near future, scientists will be able to blend the technological and theoretical knowledge in clubbing genomics, transcriptomics, proteomics and metabolomics to delineate the molecular basis of the fitness-cost and compensation in plasmid–bacterium relationships.

References

Alibayov B et al (2014) *Staphylococcus aureus* mobile genetic elements. Mol Biol Rep 41:5005–5018

Andrysiak AK, Olson AB, Tracz DM, Dore K, Irwin R, Ng LK, Gilmour MW (2008) Genetic characterization of clinical and agri-food isolates of multi drug resistant *Salmonella enterica* serovar Heidelberg from Canada. BMC Microbiol 8:89

Barbosa TM, Levy SB (2000) The impact of antibiotic use on resistance development and persistence. Drug Resist Updat 3(5):03–311

Bennett PM (2008) Plasmid encoded antibiotic resistance: acquisition and transfer of antibiotic resistance genes in bacteria. Br J Pharmacol 153(suppl 1):S347–S357

Bi D, Xie Y, Tai C, Jiang X, Zhang J, Harrison EM et al (2016) A site-specific integrative plasmid found in *Pseudomonas aeruginosa* clinical isolate HS87 along with a plasmid carrying an aminoglycoside-resistant gene. PLoS One 11(2):e0148367

Brown NL, Misra TK, Winnie JN, Schmidt A, Seiff M, Silver S (1986) The nucleotide sequence of the mercuric resistance operons of plasmid R100 and transposon TnSOZ; further evidence for mer genes which enhance the activity of the mercuric ion detoxification system. Mol Gen Genet 202:143–151

Cabot G, Zamorano L, Moyà B, Juan C, Navas A, Blázquez J, Oliver A (2016) Evolution of *Pseudomonas aeruginosa* antimicrobial resistance and fitness under low and high mutation rates. Antimicrob Agents Chemother 60:1767–1778

Canton R, Morosini MI (2011) Emergence and spread of antibiotic resistance following exposure to antibiotics. FEMS Microbiol Rev 35:977–991

Carattoli A (2013) Plasmids and the spread of resistance. Int J Med Microbiol 303(6–7):298–304

Carattoli A, Miriagou V, Bertini A, Loli A, Colinon C, Villa L, Whichard JM, Rossolini GM (2006) Replicon typing of plasmids encoding resistance to newer β-lactams. Emerg Infect Dis 12:1145–1148

Carmeli Y, Akova M, Cornaglia G et al (2010) Controlling the spread of carbapenemase-producing Gram-negatives: therapeutic approach and infection control. Clin Microbiol Infect 16:102–111

Chen Z, Li H, Feng J, Li Y, Chen X, Guo X, Chen W, Wang L, Lin L, Yang H, Yang W, Wang J, Zhou D, Liu C, Yin Z (2015) NDM-1 encoded by a pNDM-BJ01-like plasmid p3SP-NDM in clinical *Enterobacter aerogenes*. Front Microbiol 6:294

Colinon C, Miriagou V, Carattoli A, Luzzaro F, Rossolini GM (2007) Characterization of the IncA/C plasmid pCC416 encoding VIM-4 and CMY-4-lactamases. J Antimicrob Chemother 60:258–262

Coque TM, Novais A, Carattoli A, Poirel L, Pitout J et al (2008) Dissemination of clonally related *Escherichia coli* strains expressing extended-spectrum β-lactamase CTX-M-15. Emerg Infect Dis 14:195–200

Cox G, Wright GD (2013) Intrinsic antibiotic resistance: mechanisms, origins, challenges and solutions. Int J Med Microbiol 303(6–7):287–292

Curiao T, Marchi E, Grandgirard D, León-Sampedro R, Viti C, Leib SL et al (2016) Multiple adaptive routes of *Salmonella enterica* Typhimurium to biocide and antibiotic exposure. BMC Genomics 17:491

D'Costa VM, King CE, Kalan L, Morar M, Wilson WL et al (2011) Antibiotic resistance is ancient. Nature 477:457–461

Da Silva GJ, Domingues S (2016) Insights on the horizontal gene transfer of carbapenemase determinants in the opportunistic pathogen *Acinetobacter baumannii*. Microorganisms 4(3):29

Dale GE, Langen H, Page MG, Then RL, Stüber D (1995) Cloning and characterization of a novel, plasmid-encoded trimethoprim-resistant dihydrofolate reductase from *Staphylococcus haemolyticus* MUR313. Antimicrob Agents Chemother 39:1920–1924

Dancer SJ, Shears P, Platt DJ (1997) Isolation and characterization of coliforms from glacial ice and waters in Canadas High Arctic. J Appl Bacteriol 82:597–609

Deshpande LM, Jones RN, Fritsche TR, Sader HS (2006) Occurrence and characterization of carbapenemase-producing *Enterobacteriaceae*: report from the SENTRY Antimicrobial Surveillance Program (2000–2004). Microb Drug Resist 12:223–230

Di Pilato V, Pollini S, Rossolini GM (2014) Characterization of plasmid pAX22, encoding VIM-1 metallo-betalactamase, reveals a new putative mechanism of In70 integron mobilization. J Antimicrob Chemother 69(1):67–71

Diestra K, Juan C, Curiao T, Moya B, Miro E et al (2009) Characterization of plasmids encoding *bla*ESBL and surrounding genes in Spanish clinical isolates of *Escherichia coli* and *Klebsiella pneumoniae*. J Antimicrob Chemother 63:60–66

El Salabi A, Walsh TR, Chouchani C (2013) Extended spectrum beta lactamases, carbapenemases and mobile genetic elements responsible for antibiotics resistance in gram negative bacteria. Crit Rev Microbiol 39:113–122

Espedido BA, Partridge SR, Iredell JR (2008) *bla*IMP-4 in different genetic contexts in *Enterobacteriaceae* isolates from Australia. Antimicrob Agents Chemother 52:2984–2987

Feβler AT, Kadlec K, Schwarz S (2011) Novel apramycin resistance gene apmA in bovine and porcine methicillin-resistant *Staphylococcus aureus* ST398 isolates. Antimicrob Agents Chemother 55:373–375

Forsberg KJ, Reyes A, Wang B, Selleck EM, Sommer MOA, Dantas G (2012) The shared antibiotic resistome of soil bacteria and human pathogens. Science 337(6098):1107–1111

Gill SR, Fouts DE, Archer GL, Mongodin EF, Deboy RT et al (2005) Insights on evolution of virulence and resistance from the complete genome analysis of an early methicillin-resistant Staphylococcus aureus strain and a biofilm-producing methicillin-resistant *Staphylococcus epidermidis* strain. J Bacteriol 187:2426–2438

Gillespie MT, Skurray RA (1986) Plasmids in multiresistant *Staphylococcus aureus*. Microbiol Sci 3:53–58

Gyles C, Boerlin P (2014) Horizontally transferred genetic elements and their role in pathogenesis of bacterial disease. Vet Pathol 51:328–340

Haaber J, Penadés JR, Ingmer H (2017) Transfer of antibiotic resistance in *Staphylococcus aureus*. Trends Microbiol 25(11):893–905

Hamidian M, Holt KE, Pickard D, Dougan G, Hall RM (2014) A GC1 *Acinetobacter baumannii* isolate carrying AbaR3 and the aminoglycoside resistance transposon Tn*aphA6* in a conjugative plasmid. J Antimicrob Chemother 69:955–958

Hamidian M, Ambrose SJ, Hall RM (2016) A large conjugative *Acinetobacter baumannii* plasmid carrying the *sul2* sulphonamide and *strAB* streptomycin resistance genes. Plasmid 87:43–50

Harrison EM, Carter ME, Luck S, Ou HY, He X, Deng Z et al (2010) Pathogenicity islands PAPI-1 and PAPI-2 contribute individually and synergistically to the virulence of *Pseudomonas aeruginosa* strain PA14. Infect Immun 78(4):1437–1446

Hidron AI, Edwards JR, Patel J et al (2008) NHSN annual update: antimicrobial-resistant pathogens associated with healthcare-associated infections: annual summary of data reported to the National Healthcare Safety Network at the Centers for Disease Control and Prevention, 2006–2007. Infect Control Hosp Epidemiol 29:996–1011

Hopkins KL, Liebana E, Villa L, Batchelor M, Threlfall EJ, Carattoli A (2006) Replicon typing of plasmids carrying CTX-M or CMY β-lactamases circulating among *Salmonella* and *Escherichia coli* isolates. Antimicrob Agents Chemother 50:3203–3206

Hrabak J, Empel J, Gniadkowski M, Halbhuber Z, Rebl K, Urbaskova P (2008) CTX-M-15-producing *Shigella sonnei* strain from a Czech patient who traveled in Asia. J Clin Microbiol 46:2147–2148

Jensen SO, Lyon BR (2009) Genetics of antimicrobial resistance in *Staphylococcus aureus*. Future Microbiol 4:565–582

Johnson CM, Grossman AD (2015) Integrative and conjugative elements (ICEs): what they do and how they work. Ann Rev Genetics 49:577–601

Johnson TJ, Wannemuehler YM, Johnson SJ, Logue CM, White DG, Doetkott C, Nolan LK (2007) Plasmid replicon typing of commensal and pathogenic *Escherichia coli* isolates. Appl Environ Microbiol 73:1976–1983

Johnson TA, Stedtfeld RD, Wang Q, Cole JR, Hashsham SA, Looft T et al (2016) Clusters of antibiotic resistance genes enriched together stay together in swine agriculture. MBio 7:e02214–e02215

Jones RN, Masterton R (2001) Determining the value of antimicrobial surveillance programs. Diagn Microbiol Infect Dis 41:171–175

Jovcic B, Lepsanovic Z, Begovic J, Rakonjac B, Perovanovic J, Topisirovic L et al (2013) The clinical isolate *Pseudomonas aeruginosa* MMA83 carries two copies of the blaNDM-1 gene in a novel genetic context. Antimicrob Agents Chemother 57(7):3405–3407

Kadlec K, Schwarz S (2009) Identification of a novel trimethoprim resistance gene, *dfr*K, in a methicillin-resistant *Staphylococcus aureus* ST398 strain and its physical linkage to the tetracycline resistance gene tet(L). Antimicrob Agents Chemother 53:776–778

Kallen AJ, Hidron AI, Patel J, Srinivasan A (2010) Multidrug resistance among gram-negative pathogens that caused healthcare-associated infections reported to the National Healthcare Safety Network, 2006–2008. Infect Control Hosp Epidemiol 31:528–531

Kapitonov VV, Jurka J (2008) A universal classification of eukaryotic transposable elements implemented in Repbase. Nat Rev Gent 9:411–412

Kehrenberg C, Ojo KK, Schwarz S (2004) Nucleotide sequence and organization of the multiresistance plasmid pSCFS1 from *Staphylococcus sciuri.* J Antimicrob Chemother 54:936–939

Kennedy AD, Porcella SF, Martens C, Whitney AR, Braughton KR et al (2010) Complete nucleotide sequence analysis of plasmids in strains of Staphylococcus aureus clone USA300 reveals a high level of identity among isolates with closely related core genome sequences. J Clin Microbiol 48:4504–4511

Kobayashi K, Hayashi I, Kouda S, Kato F, Fujiwara T, Kayama S et al (2013) Identification and characterization of a novel aac(6′)-Iag associated with the blaIMP-1-integron in a multidrug-resistant *Pseudomonas aeruginosa.* PLoS One 8(8):e70557

Korfhagen TR, Ferrel JA, Menefee CL, Loper JC (1976) Resistance plasmids of *Pseudomonas aeruginosa*:change from conjugative to nonconjugative in a hospital population. Antimicrob Agents Chemother 9(5):810–816

Lafond M, Couture F, Vezina G, Levesque RC (1989) Evolutionary perspectives on multiresistance beta lactam transposons. J Bactriol 171:6423–6429

Lambert T, Gerbaud G, Courvalin P (1988) Transferable amikacin resistance in *Acinetobacter* spp. due to a new type of 3′-aminoglycoside phosphotransferase. Antimicrob Agents Chemother 32:15–19

Lean SS, Yeo CC (2017) Small, enigmatic plasmids of the nosocomial pathogen, *Acinetobacter baumannii*: good, bad, who knows? Front Microbiol 8:1547

Leflon-Guibout V, Blanco J, Amaqdouf K, Mora A, Guize L, Nicolas-Chanoine MH (2008) Absence of CTX-M enzymes but a high prevalence of clones, including clone ST131, among the fecal *Escherichia coli* isolates of healthy subjects living in the Paris area. J Clin Microbiol 46:3900–3905

Leungtongkam U, Thummeepak R, Tasanapak K, Sitthisak S (2018) Acquisition and transfer of antibiotic resistance genes in association with conjugative plasmid or class 1 integrons of *Acinetobacter baumannii.* PLoS One 13(12):e0208468

Li H, Toleman MA, Bennett PM, Jones RN, Walsh TR (2008) Complete sequence of p07-406, a 24,179-basepair plasmid harboring the blaVIM-7 metallo-beta-lactamase gene in a *Pseudomonas aeruginosa* isolate from the United States. Antimicrob Agents Chemother 52(9):3099–3105

Li B, Wendlandt S, Yao J, Liu Y, Zhang Q, Shi Z, Wei J, Shao D, Schwarz S, Wang S, Ma Z (2013) Detection and new genetic environment of the pleuromutilin-lincosamide-streptogramin A resistance gene lsa(E) in methicillin-resistant Staphylococcus aureus of swine origin. J Antimicrob Chemother 68:1251–1255

Lindsay JA (2010) Genomic variation and evolution of *Staphylococcus aureus*. Int J Med Microbiol 300:98–103

Liu J, Yang L, Chen D, Peters BM, Li L, Li B, Xu Z, Shirtliff ME (2018) Complete sequence of pBM413, a novel multidrug resistance megaplasmid carrying *qnrVC6* and *bla*IMP-45 from *Pseudomonas aeruginosa*. Int J Antimicrob Agents 51:145–150

Lozano C, Aspiroz C, Sáenz Y, Ruiz-García M, Royo-García G et al (2012) Genetic environment and location of the lnu(A) and lnu(B) genes in methicillin resistant *Staphylococcus aureus* and other staphylococci of animal and human origin. J Antimicrob Chemother 67:2804–2808

Lyon BR, Skurray R (1987) Antimicrobial resistance of *Staphylococcus aureus*: genetic basis. Microbiol Rev 51:88–134

Magiorakos AP, Srinivasan A, Carey RB, Carmeli Y, Falagas ME et al (2012) Multidrug- resistant, extensively drug- resistant and pan drug-resistant bacteria: an international expert proposal for interim standard definitions for acquired resistance. Clin Microbiol Infect 18:268–281

Marcade G, Deschamps C, Boyd A, Gautier V, Picard B, Branger C, Denamur E, Arlet G (2009) Replicon typing of plasmids in *Escherichia coli* producing extended-spectrum β-lactamases. J Antimicrob Chemother 63:67–71

Matanovic K, Pérez-Roth E, Pintarić S, Šeol Martinec B (2013) Molecular characterization of high-level mupirocin resistance in *Staphylococcus pseudintermedius*. J Clin Microbiol 51:1005–1007

McCarthy AJ, Lindsay JA (2012) The distribution of plasmids that carry virulence and resistance genes in *Staphylococcus aureus* is lineage associated. BMC Microbiol 12:104

Millan AS, Toll-Riera M, Escudero JA, Canto'n R, Coque TM, MacLean RC (2015) Sequencing of plasmids pAMBL1 and pAMBL2 from *Pseudomonas aeruginosa* reveals a blaVIM-1 amplification causing high-level carbapenem resistance. J Antimicrob Chemother 70:3000–3003

Miriagou V, Douzinas EE, Papagiannitsis CC, Piperaki E, Legakis NJ, Tzouvelekis LS (2008) Emergence of *Serratia liquefaciens* and *Klebsiella oxytoca* with metallo-beta-lactamase-encoding IncW plasmids: further spread of the *bla*VIM-1-carrying integron in-e541. Int J Antimicrob Agents 32:540–541

Mulvey MR, Susky E, McCracken M, Morck DW, Read RR (2009) Similar cefoxitin-resistance plasmids circulating in *Escherichia coli* from human and animal sources. Vet Microbiol 134:279–287

Naas T, Bonnin RA, Cuzon G, Villegas MV, Nordmann P (2013) Complete sequence of two KPC-harbouring plasmids from *Pseudomonas aeruginosa*. J Antimicrob Chemother 68(8):1757–1762

Normark BH, Normark S (2002) Evolution and spread of antibiotic resistance. J Intern Med 252(2):91–106

Novick RP, Edelman I, Schwesinger MD, Gruss AD, Swanson EC, Pattee PA (1979) Genetic translocation in *Staphylococcus aureus*. Proc Natl Acad Sci U S A 76:400–404

O'Brien FG, Price C, Grubb WB, Gustafson JE (2002) Genetic characterization of the fusidic acid and cadmium resistance determinants of *Staphylococcus aureus* plasmid pUB101. J Antimicrob Chemother 50:313–321

O'Mahony R, Quinn T, Drudy D, Walsh C, Whyte P, Mattar S, Fanning S (2006) Antimicrobial resistance in nontyphoidal *Salmonella* from food sources in Colombia: evidence for an unusual plasmid-localized class1 integron in serotypes Typhimurium and Anatum. Microb Drug Resist 12:269–277

Ochi S, Shimizu T, Ohtani K (2009) Nucleotide sequence analysis of the enterotoxigenic Escherichia coli Ent plasmid. DNA Res 16:299–309

Odumosu BT, Adeniyi BA, Chandra R (2013) Analysis of integrons and associated gene cassettes in clinical isolates of multidrug resistant *Pseudomonas aeruginosa* from Southwest Nigeria. Ann Clin Microbiol Antimicrob 12:29

Olsen JE, Christensen H, Aarestrup FM (2006) Diversity and evolution of blaZ from Staphylococcus aureus and coagulase-negative staphylococci. J Antimicrob Chemother 57:450–460

Pansegrau W, Lanka E, Barth PT, Figurski DH, Guiney DG, Haas D et al (1994) Complete nucleotide sequence of Birmingham IncP alpha plasmids. Compilation and comparative analysis. J Mol Biol 239(5):623–663

Paterson DL, Bonomo RA (2005) Extended-spectrum β-lactamases: a clinical update. Clin Microbiol Rev 18:657–686
Planet PJ et al (2017) Architecture of a species: phylogenomics of *Staphylococcus aureus*. Trends Microbiol 25:153–166
Poirel L, Nordmann P (2006) Carbapenem resistance in *Acinetobacter baumannii*: mechanisms and epidemiology. Clin Microbiol Infect 12:826–836
Poirel L, Villa L, Bertini A, Pitout JD, Nordmann P, Carattoli A (2007) Expanded spectrum β-lactamase and plasmid-mediated quinolone resistance. Emerg Infect Dis 13:803–805
Roberts AP, Mullany P (2011) Tn916-like genetic elements: a diverse group of modular mobile elements conferring antibiotic resistance. FEMS Microbiol Rev 35(5):856–871
Roca I, Espinal P, Vila-Farrés X, Vila J (2012) The *Acinetobacter baumannii* oxymoron: commensal hospital dweller turned pan-drug-resistant menace. Front Microbiol 3:148
Roca I, Akova M, Baquero F, Carlet J, Cavaleri M et al (2015) The global threat of antimicrobial resistance: science for intervention. New Microbes New Infect 6:22–29
Rowland SJ, Dyke KG (1989) Characterization of the staphylococcal beta-lactamase transposon Tn552. EMBO J 8:2761–2773
Rowland SJ, Dyke KG (1990) Tn552, a novel transposable element from *Staphylococcus aureus*. Mol Microbiol 4:961–975
Schmitt R (1986) Molecular biology of transposable elements. J Antimicrob Chemother 18:25–34
Shearer JE, Wireman J, Hostetler J, Forberger H, Borman J et al (2011) Major families of multiresistant plasmids from geographically and epidemiologically diverse staphylococci. G3 (Bethesda) 1:581–591
Shen J, Wang Y, Schwarz S (2013) Presence and dissemination of the multiresistance gene cfr in Gram-positive and Gram-negative bacteria. J Antimicrob Chemother 68:1697–1706
Shi L, Liang Q, Feng J, Zhan Z, Zhao Y, Yang W, Yang H, Chen Y, Huang M, Tong Y, Li X, Yin Z, Wang J, Zhou D (2018) Coexistence of two novel resistance plasmids, *bla*KPC-2-carrying p14057A and *tetA*(A)-carrying p14057B, in *Pseudomonas aeruginosa*. Virulence 9(1):306–311
Sifaoui F, Arthur M, Rice LB, Gutmann L (2001) Role of penicillin-binding protein 5 in expression of ampicillin resistance and peptidoglycan structure in *Enterococcus faecium*. Antimicrobial Agents Chemother 45:2594–2597
Siguier P, Gourbeyre E, Chandler M (2014) Bacterial insertion sequences: their genomic impact and diversity. FEMS Microbiol Rev 38(5):865–891
Smith DH (1967) R factor infection of *Escherichia coli* lyophilized in 1946. J Bacteriol 94:2071–2072
Strateva T, Yordanov D (2009) *Pseudomonas aeruginosa* – a phenomenon of bacterial resistance. J Med Microbiol 58(9):1133–1148
Sun F, Zhou D, Wang Q, Feng J, Feng W, Luo W, Zhang D, Liu Y, Qiu X, Yin Z, Chen W, Xia P (2016) The first report of detecting the *bla*SIM-2 gene and determining the complete sequence of the SIM-encoding plasmid. Clin Microbiol Infect 22:347–351
Sundstrom L, Radstrom P, Swedberg G, Skold O (1988) Site specific recombination promotes linkage between trimethoprim and sulfonamide – resistance genes. Sequence characterization of dhfrV and sulI and a recombination active locus of Tn21. Mol Gen Genet 213:191–201
Swartz MN (2000) Impact of antimicrobial agents and chemotherapy from 1972 to 1998. Antimicrob Agents Chemother 44:2009–2016
Tosini F, Visca P, Luzzi I, Dionisi AM, Pezzella C, Petrucca A, Carattoli A (1998) Class 1 integron-borne multiple-antibiotic resistance carried by IncFI and IncL/M plasmids in *Salmonella enterica* serotype typhimurium. Antimicrob Agents Chemother 42:3053–3058
Wagner A (2006) Periodic extinctions of transposable elements in bacterial lineages: evidence from intragenomic variation in multiple genomes. Mol Biol Evol 23:723–733
Wagner A, Lewis C, Bichsel M (2007) A survey of bacterial insertion sequences using IScan. Nucleic Acids Res 35:5284–5293
Weigel LM, Clewell DB, Gill SR, Clark NC, McDougal LK et al (2003) Genetic analysis of a high-level vancomycin-resistant isolate of *Staphylococcus aureus*. Science 302:1569–1571

Weiss S, Kadlec K, Feβler AT, Schwarz S (2014) Complete sequence of a multiresistance plasmid from a methicillin-resistant *Staphylococcus epidermidis* ST5 isolated in a small animal clinic. J Antimicrob Chemother 69:847–849

Welch TJ, Fricke WF, McDermott PF, White DG, Rosso ML et al (2007) Multiple antimicrobial resistance in plague: an emerging public health risk. PLoS One 2:e309

Wendlandt S, Lozano C, Kadlec K, Gómez-Sanz E, Zarazaga M, Torres C, Schwarz S (2013) The enterococcal ABC transporter gene lsa(E) confers combined resistance to lincosamides, pleuromutilins and streptogramin A antibiotics in methicillin-susceptible and methicillin-resistant Staphylococcus aureus. J Antimicrob Chemother 68:473–475

White DG, McDermott PF (2001) Emergence and transfer of antibacterial resistance. J Dairy Sci 84:E151–E155

Wicker T, Sabot F, Hua-Van A, Bennetzen JL, Capy P, Chalhoub B et al (2007) A unified classification system for eukaryotic transposable elements. Nat Rev Genet 8:973–982

Wiedemann B, Meyer JF, Ziihlsdorf MT (1986) Insertions of resistance genes in to Tn21 like transposons. J Antimicrob Chemother 18:85–92

Winokur PL, Vonstein DL, Hoffman LJ, Uhlenhopp EK, Doern GV (2001) Evidence for transfer of CMY-2 AmpC β-lactamase plasmids between *Escherichia coli* and *Salmonella* isolates from food animals and humans. Antimicrob Agents Chemother 45:2716–2722

Woodford N (2008) Successful, multiresistant bacterial clones. J Antimicrob Chemother 61:233–234

Xiong J, Alexander DC, Ma JH, Deraspe M, Low DE, Jamieson FB et al (2013) Complete sequence of pOZ176, a 500-kilobase IncP-2 plasmid encoding IMP-9-mediated carbapenem resistance, from outbreak isolate *Pseudomonas aeruginosa* 96. Antimicrob Agents Chemother 57(8):3775–3782

Disinfectants Amend the Expression of Membrane Bound Efflux Transporters to Augment Antimicrobial Resistance

2

Govindan Rajamohan and Vijaya Bharathi Srinivasan

Abstract

Disinfectants are extensively used in health care settings and these active agents demonstrate broad-spectrum antimicrobial activity; however, little is known about their role as inducers/selectors of multidrug resistance in Gram-negative bacteria. To investigate the disinfectant susceptibilities and its role in conferring multidrug resistance in *Acinetobacter baumannii*, various molecular techniques were performed using standard procedures. The expression of drug specific efflux transporters and non-specific porins were detected by quantitative real time PCR. The susceptibility testing and phenotypic assays demonstrated the role of active efflux in mediating decreased susceptibility to hospital based disinfectants. Interestingly, different disinfectants increased the minimum inhibition concentrations for some of the clinically relevant antibiotics. Additionally, disinfectants also enhanced the expression of drug specific transporters *adeB, adeJ, abeM, macB*, and non specific transporters that function as porins *ompA* and *carO* to augment antimicrobial resistance in *A. baumannii*. Overall, the study provides experimental evidence highlighting the role of disinfectants as inducers of multidrug resistance in *A. baumannii*. With such a concern that disinfectant exposure can select for a multidrug resistant population, it is important to implement strict intervention to control and prevent.

Keywords

Biofilm · Motility · Drug resistance · Biocide tolerance · Hospital acquired pathogen

G. Rajamohan (✉) · V. B. Srinivasan
Bacterial Signaling and Drug Resistance Laboratory, Council of Scientific Industrial Research, Institute of Microbial Technology, Chandigarh, India
e-mail: rmohan@imtech.res.in

S. M. Mandal, D. Paul (eds.), *Bacterial Adaptation to Co-resistance*,
https://doi.org/10.1007/978-981-13-8503-2_2

2.1 Introduction

According to the Biocides Directive (98/8/EC), active substances and preparations comprised of more than one active substances to destroy, or exert a controlling effect on any harmful organism by chemical or biological means is called biocides (Baker 2000). Examples include disinfectants, pesticides, herbicides, fungicides, insecticides, preservatives, and antiseptics (Patryn et al. 2011). Disinfectants are used on inanimate objects or intact skin to reduce the number of micro-organisms (Rutala and Weber 2001). Disinfectants are classified as low, medium or high-level disinfectants, depending on how many types of micro-organisms they kill (Rutala and Weber 2001). High-level disinfectants are called chemical sterilants. The antimicrobial modes of action of disinfectants include denaturation of bacterial proteins, damage to bacterial membrane causing leakage of intracellular molecules (phenols, surfactants) and interference of bacterial enzyme and metabolism (oxidants, alkylating agents) (Maris 1995). Disinfectants are derivatives of different functional group such as acridine derivatives for *e.g.* (ethacridine lactate), biguanides and amidines (chlorhexidine), phenol and derivatives (phenol, triclosan), nitrofuran derivatives (nitrofurazone), quinoline derivatives (dequalinium), iodine products (iodine/octylphenoxypolyglycolether), quaternary ammonium compounds (benzalkonium), mercurial products (mercuric amidochloride), silver compounds (silver nitrate), alcohols (propanol) and other (potassium permanganate) (Maris 1995). Disinfectants have wide applications for *e.g.* for benzalkonium chloride (N-Alkyl-N-benzyl-N,N-dimethylammonium chloride) the usage is extremely wide ranging, such as being an active ingredient in dettol and lysol brand products, to microbial corrosion inhibition in the oilfield sector, and a multi-surface mould, algae and moss remover. It is also used as skin antiseptics, such as bactine, to protect scrapes and cuts, in pharmaceuticals such as throat lozenges & various leave-on skin antiseptics (Graf 2001). Similarly, chlorhexidine (N′,N″-hexane-1,6-diylbis [N-(4-chlorophenyl) (imidodicarbonimidic diamide)]) is often used as an active ingredient in mouthwash designed to reduce dental plaque and oral bacteria (Robinson 1995). Triclosan (2, 4, 4′-trichloro-2′-hydroxydiphenyl ether) is a nonionic, broad spectrum, antimicrobial agent that, because of its approving safety outline, has been added into diverse personal care products, including deodorant soaps, underarm deodorants, shower gels, and health care personnel hand washes (Davies 2007).

Benzalkonium chloride, chlorhexidine and triclosan are commonly used chemicals in stringent sterilization procedures that are followed in clinical settings (Smith and Hunter 2008). Previous studies have reported the incidence of disinfectant resistance in Gram-negative as well as Gram-positive pathogens. Soumet et al. has previously demonstrated that adaptation to quaternary ammonium compounds in *E. coli* renders them resistant to phenicol compounds (Soumet et al. 2012). Several reports have documented the biocide tolerance behavior in *Burkholderia*, *Pseudomonas*, *Salmonella*, including *Staphylococcus* (Rose et al. 2009; Ferreira et al. 2011; Beier et al. 2011; Skovgaard et al. 2013). Resistance mechanisms altering the activity of biocide molecules include alteration in structure of bacterial envelope, modifications in outer membrane proteins and phospholipids, charges on exposed cell surface which leads to reduction in membrane permeability (Russell 1985; Rushton et al. 2013; Coenye et al. 2011). Efflux pumps fall into the ATP-binding cassette (ABC), major facilitator super family (MFS), resistance/nodulation/cell division (RND), multi drug and toxic compound extrusion (MATE) and the small multidrug resistance (SMR) groups and utilize the energy of the proton motive force or ATP to export antibiotics from the cell. Efflux pumps are involved in disinfectant resistance such as QACs, triclosan, phenolic compounds and *etc*. The efflux systems AcrAB-TolC, AcrEF-TolC, EmrE in *E. coli,* MexAB-OprM, MexCD-OprJ, MexEF-OprN and MexJK in *P. aeruginosa*, QacA-D, Smr, QacG and QacH in *S. aureus* have been shown to be involved in conferring biocide tolerance (Holdsworth and Law 2013; Gotoh et al. 1998; Köhler et al. 1997; Masuda et al. 2000; Sekiguchi et al. 2004). Plasmid mediated resistance to biocides is well demonstrated in many bacterial species by specific genes (*qacA*, B, C, D and E) (Russell 1997).

These prominent mechanisms that render bacteria disinfectant resistant are also responsible for conferring multidrug resistance (MDR) in both Gram-negative and positive bacteria (Russell and Day 1996). The involvement of disinfectants as selectors and/or inducers of mechanisms involved in the emergence of highly MDR strains have been highlighted in previous laboratory studies. It has been reported that sense mutation Gly95Ser in enoyl-acyl carrier protein reductase FabI, an enzyme involved in fatty acid biosynthetic pathway lead to high level triclosan resistance in bacteria (Zhu et al. 2010). The production of efflux pumps RND-type transporter *acrB*, transcriptional activator *marA* and up regulation of biofilms has been the underlying mechanism to induce MDR in biocide tolerant pathogens (Russell and Day 1996). One such highly disinfectant tolerant clinically significant pathogen is *Acinetobacter baumannii* (Pendleton et al. 2013).

Acinetobacter is an aerobic Gram-negative bacterium, widely known for causing serious infections in hospital intensive care units which are being reported worldwide (Pendleton et al. 2013). Most frequently encountered species is *A. baumannii* and it is commonly associated with simple to serious infections with high mortality rate of 30–75% in hospitalized patients (García-Quintanilla et al. 2013). Recently, significant increase in *A.baumannii* strains that are resistant to carbapenems, cephalosporins, aminoglycosides and fluoroquinolones have been reported from hospitals in Pennsylvania, Houston, Chicago, Walter Reed Army Medical Center (WRAMC), Washington, D.C., Los Angeles and multiple resistance determinants have been identified to explain their MDR phenotype (Valentine et al. 2008; Hujer et al. 2006). In our preliminary study, we demonstrated that β-lactamases, aminoglycoside-modifying enzymes, target gene mutations and efflux pumps *adeABC, abeS, amvA* were responsible for the MDR/biocide resistant behavior in the clinical strains isolated during 2005–2007 from tertiary care hospitals in central Ohio, USA (Srinivasan et al. 2009a, b). In Indian scenario 13.5% of nosocomial infections were caused by *Acinetobacter spp.* and the isolates were completely resistance to amikacin, piperacillin, tazobactum and streptomycin (Sinha et al. 2013). Bacterial MDR is now a global healthcare problem and requires urgent attention due to lack of new antibiotics. The development of resistance in dreaded human pathogens has highlighted the need for the rapid discovery of new families of drugs with novel targets and/or modes of action.

In our recent study we elucidated the functions of *rstAB* two component signal transduction system in biofilm formation, stress response, tolerance to antibiotics, disinfectants and virulence for the first time (unpublished observations). Though there is plethora of information available, however studies pertaining to reveal the direct underlying genetic mechanisms responsible for mediating decreased susceptibility to disinfectants in *A. baumannii* have never been explored before. The scope of the study has been expanded to identify the role of hospital based disinfectants in augmenting MDR in clonally distinct clinical isolates of *A. baumannii* isolated from India.

2.2 Materials and Methods

2.2.1 Study Population and the Bacterial Identification

A total of three *A.baumannii* strains were obtained as part of routine care were gifted for this study by doctors from medical center in Chandigarh and all sample identifiers had been previously removed and data were analyzed anonymously. Further the isolates were reconfirmed by 16sRNA and *gyrA* sequencing. The isolated were sub cultured on tryptic soy agar (Fisher Scientific, Suwanee, GA, USA) slants for further study. The selection criteria of the strains were based on the heterogeneity in their properties, levels of resistance and excluding multiple isolates of the same strain from one locality.

2.2.2 Enterobacterial Repetitive Intergenic Consensus Sequences (ERIC-PCR)

The ERIC-PCR was performed as described previously by Versalovic et al (Hulton et al. 1991). Using 5 μL from extracted DNA, 1.5 mM of $MgCl_2$, 10 pmol from primers ERIC1 and ERIC2 primers, 1.0 U of Taq DNA polymerase (NEB, USA), 1 X PCR buffer, and water to complete the total volume of 50 μL. The PCR reaction was conducted with 35 cycles, consisting of denaturation for 4 min at 94 °C, annealing for 60 s at 50 °C, and extension for 2.5 min at 72 °C, which increased with a final extension for 20 min at 72 °C, as described by Rafiee et al. The amplification products were detected through electrophoresis at 35 V for 20 h in 2% agarose.

2.2.3 Antibiotic Susceptibility Testing

Antibiotic susceptibility for *A. baumannii* isolates included in this study were examined using commercial discs such as ampicillin, AMP (10 μg/ml); carbenicillin, CAR (10 μg/ml); cefepime, CPM (30 μg/ml); ceftazidime, CAZ (30 μg/ml); ceftriaxone, CTR (30 μg/ml); chloramphenicol, CHL (10 μg/ml); ciprofloxacin, CIP (5 μg/ml); doxycycline, DOX (10 μg/ml); ertapenem, ETP (30 μg/ml); gentamicin,

GEN (10 μg/ml); imipenem, IPM (10 μg/ml); kanamycin, KAN (30 μg/ml); linezolid, LZD (10 μg/ml); minocycline, MIN (10 μg/ml); nalidixic acid, NAL (30 μg/ml); norfloxacin, NOR (30 μg/ml); piperacillin, PIP (30 μg/ml); rifampicin, RIF (5 μg/ml); spectinomycin, SPT (100 μg/ml); streptomycin, STR (100 μg/ml); tetracycline, TET (30 μg/ml); ticarcillin, TIC (30 μg/ml); tigecycline, TGC (30 μg/ml); tobramycin, TOB (10 μg/ml); and trimethoprim TMP (10 μg/ml) (Hi Media, Bombay, India) as described previously and interpreted according to the interpretation criteria recommended by CLSI (Clinical and Laboratory Standards Institute 2006). Where required disinfectants (3.2 μg/ml of benzalkonium chloride or chlorhexidine or triclosan) were included in the agar plate and the experiment was performed as described above. MDR was defined in this analysis as resistance to three or more members of the following classes of antibiotics: quinolones, extended-spectrum cephalosporins, aminoglycosides and carbapenems (Srinivasan et al. 2009b).

2.2.4 Minimum Inhibitory Concentration Determination for Different Compounds

Minimum inhibitory concentration (MIC) of the strains was determined using commercially available E-strips and agar double dilution method (Srinivasan et al. 2012a). In agar dilution method plates were prepared with antibiotic such as amikacin, ampicillin, carbenecillin, cefotaxime, ceftazidime, chloramphenicol, ciprofloxacin, erythromycin, gentamicin, kanamycin, neomycin, norfloxacin, oxacillin, polymyxin B, streptomycin, tetracycline and trimethoprim at different concentrations (0.5 μg/ml, 4 μg/ml, 16 μg/ml, 64 μg/ml, 256 μg/ml, 1024 μg/ml), structurally related compounds such as acridine orange, acriflavine, ethidium bromide, indole, rhodamine, saffrannin, sodium salicylate and sodium tungstate (16 μg/ml, 64 μg/ml, 256 μg/ml, 1024 μg/ml, 4096 μg/ml and 16,834 μg/ml), and biocides such as benzalkonium chloride, chlorhexidine, triclosan (16 μg/ml, 64 μg/ml, 256 μg/ml, 1024 μg/ml, 4096 μg/ml and 16,834 μg/ml) and bacteria at different dilutions were spotted and incubated at 37 °C and later were visually inspected, MIC were determined as per their growth pattern.

2.2.5 Bacterial Growth Curves

The growth inactivation assay to assess the impact on drug efflux capacity was performed as mentioned before with slight modifications (Srinivasan et al. 2012b). The *A. baumannii* cultures at mid log phase (OD_{600nm} = 0.2) were inoculated into LB broth containing disinfectant such as benzalkonium chloride, chlorhexidine or triclosan in independent experiments either alone or with efflux pump inhibitor carbonyl cyanide 3-chlorophenylhydrazone; CCCP 10 μg/ml (Sigma, St. Louis, MO). The growth profile of *A. baumannii* cultures thereafter at 37 °C was analyzed by measuring the absorbance at OD_{600nm} periodically. Optical densities were measured for 10 h at 37 °C shaking using spectrophotometer at 600 nm and recorded after every 1 h. The experiment was performed with freshly autoclaved medium in triplicates at least three independent times. Efflux pump inhibitors had no intrinsic antibacterial activity against strains at the concentration used in the experiments.

2.2.6 Motility Assay

The motility assays were performed as reported before (Peleg et al. 2008), where LB grown *A. baumannii* cultures (OD_{600nm} = 1.0) were inoculated with a toothpick on LB plates with 0.25%, 0.45% and 0.7% agar (in the presence and absence of disinfectants) and incubated for 14 h at 37 °C. In this assay the bacteria can swim through the soft agar and produce a halo. The diameter of the halo is a measure of the ability to swim.

2.2.7 Crystal Violet Binding Assay

The crystal violet binding assay was performed as described before (Peleg et al. 2008). Bacterial cultures were grown in LB and diluted to OD_{600nm} = 0.2; 150 μL of each strain was then placed into glass tubes (in the presence and absence of disinfectants). The tubes were incubated at 37 °C shaking for 24 h. After incubation, they were washed three times with PBS to remove planktonic growth. The remaining biofilm was fixed with methanol for 15 min. Once methanol was removed and tubes were dried, biofilms were stained with 1% crystal violet for 5 min. The stain was removed by washing with water and plates were dried. Biofilm thickness was measured by adding 33% glacial acetic acid and taking a reading at OD_{580nm} using an automated plate reader.

2.2.8 RNA Isolation and Real-Time Reverse Transcription PCR (RT-PCR)

Total RNA was extracted from the log-phase cultures using the RNeasy Mini Kit (Qiagen) according to the manufacturer's instructions (Srinivasan et al. 2012b). Aliquots of 500 ng of DNase I treated total RNA served as template for complementary DNA (cDNA) synthesis using superscript III reverse transcriptase (Invitrogen). The quantitative PCR reactions for different genes were performed using gene specific primers (Table 2.1). Gene expression levels were monitored by real time RT-PCR using Maxima SYBR Green qPCR master mix (Fermentas) in an iCycler thermal cycler (Bio-Rad) and the melting curve analysis were carried out to confirm amplification of a single product. Total RNA was isolated from at least two separately grown replicate cultures. All real-time RT-PCR experiments were performed more than three times with *gyrB as* an internal control.

Table 2.1 Primers used in this study

Primer name	Primer sequences (5′-3′)
*adeB*RT-F	AACGGACGACCATCTTTGAGTATT
*adeB*RT-R	CAGTTGTTCCATTTCACGCATT
*adeJ*RT-F	AGCGTATTGACAGCGGTATT
*adeJ*RT-R	ATAGAGCACGCCAGAGAAGA
abeM RT-F	GGTAGGTGTAGGCTTATGGA
abeM RT-R	CTTCGGCAACTAATGGTGT
*carO*RTF	TGCCGATGGTGTCAAAATTA
*carO*RTR	CTGCAGAACCTGACGAAACA
*ompA*RTF	TGCCGATGGTGTCAAAATTA
*ompA*RTR	CTGCAGAACCTGACGAAACA
macB-RT-F	CAACAATATTGACGTTGGCC
macB-RT-R	CGAACATCGACTGCTTTTCA
*gyrB*RT-F	GAGCATGTGTATGACTATGAAGGTG
*gyrB*RT-R	CAGCGAACATTTTGTTGGTAACTAT

2.2.9 Bioinformatic Analysis and Statistical Analysis

All data are presented as means ± the standard error of the mean. Plotting and calculation of the standard deviation was performed in Microsoft Excel. Statistical analysis was performed on crude data by using a paired Student t test. P values of <0.05 were considered significant.

2.3 Results

2.3.1 Characteristics of Study Population and Clonality

The isolates studied were a part of random sampling of specimens obtained from Chandigarh Medical Centers. The *A. baumannii* were isolated from different specimen types. All *A. baumannii* isolates are independent *i.e.* were recovered one per patient. The isolates were obtained from patients belonging to different age groups: 60–65 years (n = 2) and one isolate from an infant baby (3 day old baby, source being endotracheal secretion system). Occurrences of rapidly growing *A. baumannii* infections are increasingly being reported globally. Information about clonality of clinical isolates obtained during outbreaks can give opportunities for fast treatment. ERIC PCR is known to be a simple, high throughput, affordable, reproducible, and discriminatory molecular typing method for inference of genetic relatedness among Gram-negative bacteria. To understand the genetic relatedness, a total of three isolates collected during 2013 were selected and characterized. In a preliminary series of experiments using ERIC-R1/R2 primers, the *A. baumannii* strains yielded 4–9 bands ranging from 0.5 Kb to 5 Kb with different band intensities (data not shown). CAB2 strain was found identical to that of French strain and CAB1 and CAB3 were genetically unique and distinct.

2.3.2 MIC for Structurally Related Compounds and Hospital Based Disinfectants

The MIC values for different compounds were determined by agar dilution method. For deoxycholate (is a bile salt with mild detergent property) the strains had following MIC; CAB-1: 8 μg/ml, CAB-2: 256 μg/ml; CAB-3: 16 μg/ml; SDS (an anionic surfactants that possess biocidal properties) CAB-1: 8 μg/ml, CAB-2: 128 μg/ml; CAB-3: 8 μg/ml; safranine (biological stain) CAB-1: <16 μg/ml, CAB-2: 512 μg/ml; CAB-3: <16 μg/ml; acridine orange (dye used as an intercalating agent) CAB-1: 16 μg/ml, CAB-2: 256 μg/ml; CAB-3: 16 μg/ml; acriflavine (a dye used as an antiseptic) CAB-1: 8 μg/ml, CAB-2: 256 μg/ml; CAB-3: 8 μg/ml; EtBr (intercalating agent) CAB-1: 8 μg/ml, CAB-2: 256 μg/ml; CAB-3: 8 μg/ml. The MIC values for benzalkonium chloride (N-Alkyl-N-benzyl-N,N-dimethylammonium chloride) were CAB-1: 6.4 μg/ml, CAB-2: 6.4 μg/ml and CAB-3: 3.2 μg/ml. The MIC values for chlorhexidine (N′,N″-hexane-1,6-diylbis [N-(4-chlorophenyl) (imidodicarbonimidic diamide)]) were CAB-1: 6.4 μg/ml, CAB-2: 25.6 μg/ml and CAB-3: 3.2 μg/ml. The MIC values for triclosan (2, 4, 4′-trichloro-2′-hydroxydiphenyl ether) were CAB-1: <3.2 μg/ml, CAB-2: 51.2 μg/ml and CAB-3: <3.2 μg/ml.

2.3.3 Involvement of Active Drug Efflux in Conferring Disinfectant Resistance

It has been reported before that strains exhibiting high level biocide resistance might exhibit the involvement of efflux pumps to mediate the resistance. To examine if membrane transporters are involved in conferring antimicrobial resistance in *A. baumannii*, the accumulation profile of fluorescent substrate *EtBr* was examined, data analysis indicated that the *EtBr* efflux was most efficient in the highly biocide resistant strain CAB-2, whereas it was less efficient in other low level resistant strain CAB-2 and CAB-3. The addition of CCCP increased the *EtBr* levels which eventually reached a plateau in all strains. These data suggested that efflux is very much operative in these strains (Data not shown).

Thus, in order to determine the involvement of efflux in disinfectant resistance we performed the growth inactivation assay. The effect on growth due to the presence of benzalkonium chloride, chlorhexidine and triclosan in the absence and the presence of the efflux pump inhibitors, namely, CCCP (an inhibitor of the proton motive force [PMF]) and reserpine (a plant alkaloid known to inhibit the ATP binding cassette [ABC transporters]), was measured. It was seen that irrespective of the MICs, CCCP reduced the growth of bacteria (0.7 fold in CAB-1, 20.65 fold in CAB-2 and 0.9 fold in CBA-3; $p = 0.013$) in the presence of chlorhexidine in all strains, which emphasis that efflux was more proficient in CAB-2. We observed similar results on using benzalkonium chloride and triclosan (laboratory observations). Overall, these preliminary results suggest that efflux pumps were active in these strains and were involved in mediating disinfectant resistance. To further gain an insight on the role of disinfectants as inducers of antibiotic resistance in *A. baumannii*, the Kirby Bauer assay was performed in the absence and presence of either 3.2 μg/ml of benzalkonium chloride, 3.2 μg/ml of chlorhexidine and 3.2 μg/ml of triclosan.

2.3.4 Impact of Disinfectants on Antibiotic Resistance in Clinical Strains of *A. baumannii*

The antibiogram patterns of CAB-1, CAB-2 and CAB-3 in the presence and absence of disinfectants were monitored by Kirby Bauer disc diffusion assay. Precisely, 100% of the isolates were resistant to ampicillin; 100% of the isolates were resistant to azithromycin; 66.66% of the isolates were resistant to ceftazidime; 66.66% of the isolates were resistant to carbenicillin; 33.33% of the isolates were resistant to chloramphenicol; 33.33% of the isolates were resistant to ciprofloxacin; 66.66% of the isolates were resistant to cefepime; 33.33% of the isolates were resistant to doxycycline; 66.66% of the isolates were resistant to erythromycin; 66.66% of the isolates were resistant to ertapenem; 100% of the isolates were resistant to gentamicin; 33.33% of the isolates were resistant to imipenem; 100% of the isolates were resistant to methicillin; 33.33% of the isolates were resistant to meropenem; 100% of the isolates were resistant to neomycin; 33.33% of the isolates were resistant to norfloxacin; 33.33% of the isolates were resistant to ofloxacin; 100% of the isolates were resistant to penicillin; 66.66% of the isolates were resistant to rifampicin; 100% of the isolates were resistant to streptomycin; 33.33% of the isolates were resistant to tetracycline; 0% of the isolates were resistant to tigecycline; 100% of the isolates were resistant to tobramycin; 100% of the isolates were resistant to vancomycin (Fig. 2.1). These strains exhibited an only a slightly altered antibiotic profile in the presence of disinfectants benzalkonium chloride, chlorhexidine and triclosan (Fig. 2.1).

In order to get a true picture of the antibiotic susceptibility profile and understand the role of different disinfectants as inducers of MDR, the MIC values for different antibiotics namely ceftriaxone, chloramphenicol, erythromycin, gentamicin, linezolid, nalidixic acid, neomycin, norfloxacin, ofloxacin, polymyxin B, sparfloxacin, tetracycline, ticarcillin, tobramycin, trimethoprim and vancomycin were determined using by agar dilution method and E-strips for *A. baumannii* strains CAB-1, CAB-2 and CAB-3. In the presence of disinfectants the MIC values for different antibiotics were altered in the clinical strains.

With 3.2 μg/ml of benzalkonium chloride in the assay plate, it altered the MICs for CAB-1 (for ceftriaxone by 11.7 fold); CAB-2 (for trimethoprim by 20 fold); CAB-3 (for neomycin by 2 fold) respectively.

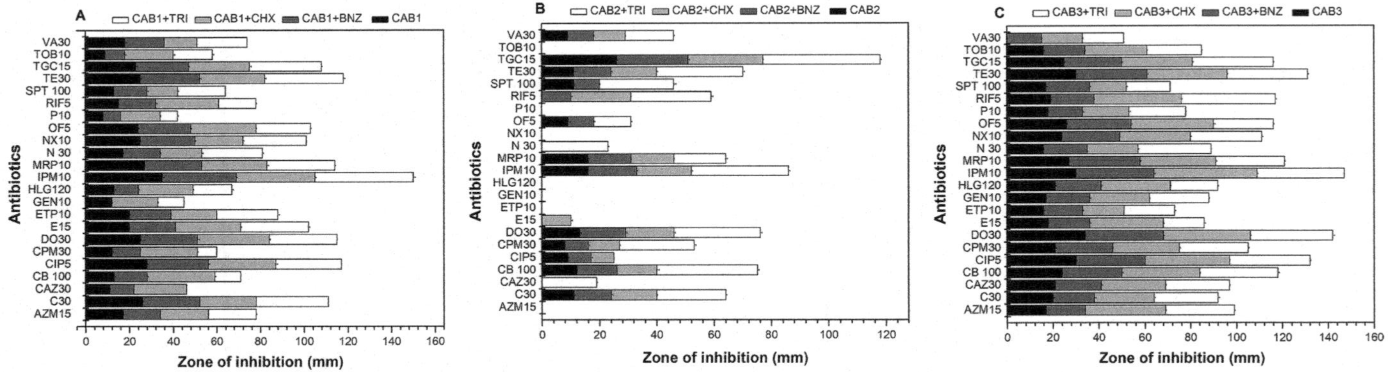

Fig. 2.1 Contributions of disinfectants in *A. baumannii* antimicrobial resistance. The Kirby Bauer disc diffusion assay was performed with different antibiotics using commercial discs in the presence and absence of disinfectants in the absence and presence of either of 3.2 μg/ml of benzalkonium chloride (BNZ), 3.2 μg/ml of chlorhexidine (CHX), 3.2 μg/ml of triclosan (TRI). Amoxicillin, AMX (30 mg/L); azithromycin, AZM (15 mg/L); cefepime, CPM (30 mg/L); ceftazidime, CAZ (30 mg/L); ceftriaxone, CRO (30 mg/L); ciprofloxacin, CIP (5 mg/L); colistin, CST (10 mg/L); doxycycline, DOX (30 mg/L); enrofloxacin, ELX (10 mg/L); ertapenem, ETP (10 mg/L); erythromycin, ERY (30 mg/L); gentamicin, GEN (10 mg/L); imipenem, IPM (10 mg/L); levofloxacin, LVX (5 mg/L); norfloxacin, NOR (10 mg/L); novobiocin, NOV (30 mg/L); piperacillin, PIP (100 mg/L); polymyxin B, PMB (300 mg/L); rifampicin, RIF (30 mg/L); spectinomycin, SPT (100 mg/L); streptomycin, STR (10 mg/L); tetracycline, TET (30 mg/L); ticarcillin, TIC (75 mg/L); tigecycline, TGC (15 mg/L); tobramycin, TOB (10 mg/L); trimethoprim, TMP (5 mg/L). Data for representative drugs have been shown here

With 3.2 μg/ml of chlorhexidine in the assay plate altered the MIC for CAB-2 (for trimethoprim by 40 fold).

With 3.2 μg/ml of triclosan in the assay plate altered the MICs for CAB-1 (for cefepime by 2 fold, ceftriaxone by 2 fold, and ticarcillin by 4.8 fold); CAB-2 (for chloramphenicol by 2 fold); CAB-3 (for norfloxacin by 50 fold) (Table 2.2). To expand our study the impact of disinfectants on motility behavior in *A. baumannii*, was studied in the absence and presence of either 3.2 μg/ml of benzalkonium chloride, 3.2 μg/ml of chlorhexidine and 3.2 μg/ml of triclosan.

2.3.5 Effect of Disinfectants on the Motility Behavior of Clinical Strains in *A. baumannii*

Bacterial motility is a complex process that employs mechanisms like swimming, swarming, twitching with the help of surface appendages such as flagella and pili. Therefore, we wanted to investigate whether the disinfectants affected *A. baumannii* motility. To test this, 1 μl drop of an overnight culture of wild type strain was placed on LB plates with following agar concentrations 0.25%, 0.45% and 0.7% respectively, and evaluated their ability to migrate away from the point of inoculation.

Approximately 4–5 h after inoculation, outward motility was evident and after 14 h of growth, cells were motile and formed concentric motility rings around the point of inoculation and distance of migration was bigger indicating robust motility (0.25%: CAB-1 = 85 mm ±0.011, CAB-2 = 10 mm ± 0.007, CAB-3 = 20 mm ± 0.011; 0.45%: CAB-1 = 85 mm ± 0.007, CAB-2 = 13 mm ± 0.0032, CAB-3 = 10 mm ± 0.054; 0.7%: CAB-1 = 75 mm ± 0.012, CAB-2 = 10 mm ± 0.013, CAB-3 = 11 mm ± 0.025;).

In stark contrast, in the presence of disinfectants for *e.g.* chlorhexidine (0.25%: CAB-1 = 31 mm ± 0.005, CAB-2 = 8 mm ± 0.034, CAB-3 = 5 mm ± 0.078; 0.45%: CAB-1 = 20 mm ± 0.032, CAB-2 = 10 mm ± 0.017, CAB-3 = 9 mm ± 0.018; 0.7%: CAB-1 = 13 mm ± 0.011, CAB-2 = 9 mm ± 0.009, CAB-3 = 9 mm ± 0.012) the cells had impaired motility and distance of migration was decreased indicating complete loss in motility (Fig. 2.2).

We did not observe any further motility upon prolonged incubation, although some cell growth was observed. These primary evidences strongly suggest that disinfectant influences motility which is known to help *A. baumannii* to establish colonization during development of infection.

2.3.6 Impact on Biofilm Forming Ability of Clinical Strains in *A. baumannii*

To further gain an insight on the role of disinfectants in influencing biofilm formation in *A. baumannii*, the biofilm assay was performed in the absence and presence of either 3.2 μg/ml of benzalkonium chloride, 3.2 μg/ml of chlorhexidine and 3.2 μg/ml of triclosan. A bacterial biofilm is a community-like structure that comprises bacterial cells embedded in an exopolysaccharide matrix and is attached to a living solid surface.

Table 2.2 MIC of *A. baumannii* clinical strains in presence of different hospital based disinfectants

	MIC (mg/ml)											
	CAB1				CAB2				CAB3			
Antibiotics	LB	BNZ	CHX	TRI	LB	BNZ	CHX	TRI	LB	BNZ	CHX	TRI
Ceftriaxone	0.256	3	0.016	0.512	>2.048	>2.048	0.512	>2.048	1	0.128	0.032	<0.016
Chloramphenicol	0.01	0.1	0.1	0.1	5	5	0.1	10	5	0.1	0.1	0.1
Erythromycin	0.1	0.1	0.1	0.01	>240	>240	>240	>240	0.1	0.1	0.01	0.1
Gentamicin	10	5	0.1	5	>240	>240	>240	>240	<8	<8	0.01	0.01
Linezolid	10	10	5	0.01	>240	>240	10	>240	10	10	0.1	10
Nalidixic acid	0.1	0.1	0.01	0.1	>240	60	30	>240	0.01	0.01	0.01	0.01
Neomycin	0.1	0.1	0.1	0.01	>240	>240	>240	>240	5	10	0.1	0.1
Norfloxacin	0.1	0.1	0.1	0.01	>240	>240	>240	>240	0.1	0.1	0.1	5
Ofloxacin	0.15	0.01	0.04	0.15	>8	2	>8	>4	0.004	0.004	0.004	0.004
Polymyxin B	60	60	0.1	0.1	0.01	<0.01	0.01	<0.01	<0.01	<0.01	<0.01	<0.01
Sparfloxacin	0.01	<0.01	<0.01	0.01	2	2	1	0.01	<0.01	<0.01	<0.01	<0.01
Tetracycline	0.01	<0.01	<0.01	<0.01	10	0.1	0.01	0.1	0.01	<0.01	<0.01	<0.01
Ticarcillin	50	10	<0.01	>240	60	60	60	60	0.1	0.01	0.01	0.01
Tobramycin	2	2	0.1	2	>16	>16	8	>16	0.1	0.1	0.1	0.1
Trimethoprim	>32	>32	>32	>32	0.1	2	4	0.1	0.1	0.1	0.1	0.1
Vancomycin	0.1	0.1	<0.01	0.01	10	10	0.1	10	32	2	<0.01	<0.01

MIC values for different antibiotics were determined using E-strips for *A. baumannii* CAB-1, CAB-2 and CBA-3 in the absence and presence of either of 3.2 μg/ml of benzalkonium chloride (BNZ), 3.2 μg/ml of chlorhexidine (CHX), 3.2 μg/ml of triclosan (TRI)

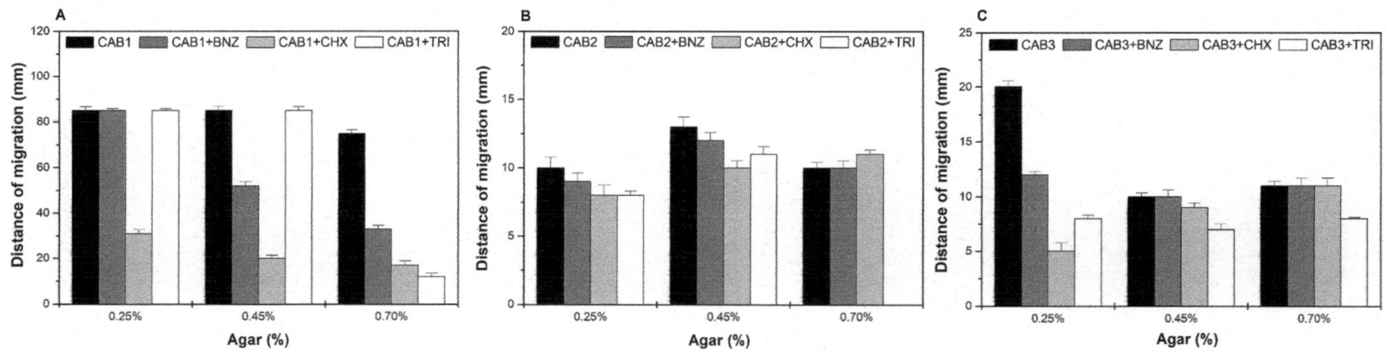

Fig. 2.2 Alterations in motility behavior. The motility behavior was estimated in the presence and absence of disinfectants in the absence and presence of either of 3.2 μg/ml of benzalkonium chloride (BNZ), 3.2 μg/ml of chlorhexidine (CHX), 3.2 μg/ml of triclosan (TRI). The average diameter of swimming halos from three different experiments is plotted with standard deviations. P value for the differences between CAB-1, CAB-2, CAB-3 strains were < 0.01. Results are given in mm

Formation of a biofilm increases bacterial survival in environments with limited nutrients or harmful chemicals. Many studies have found that biofilm can protect bacterial cells from the attacks of bacteriophages, and toxic substances. The bacteria in biofilm are more tolerant to drugs and hamper therapy. Therefore, it was imperative to elucidate the role of disinfectants in influencing biofilm formation by using standard method: quantifying adhesion to polyvinyl chloride tubes by semi-quantitative colorimetric assay.

We found that *in-vitro* biofilm forming ability of *A. baumannii* strains in presence of disinfectants were impaired by 2-fold compared to the control condition (laboratory observations).

2.3.7 Expression Levels of Efflux Transporters and Membrane Bound Porins in *A. baumannii*

We grew the cultures in presence of different disinfectants, at following concentrations 3.2 μg/ml of benzalkonium chloride, 3.2 μg/ml of chlorhexidine and 3.2 μg/ml of triclosan. RNA was isolated from these cultures and expression levels of different transporters and porins were tested.

Upon growing the cultures with 3.2 μg/ml of benzalkonium chloride, we found increased expression levels for homolog of ABAYE1822, membrane bound RND-efflux transporter *adeB* (0.14 fold ± 0.018, 2.64 fold ± 0.002, 0.20 fold ± 0.004), homolog of ABAYE0747, membrane bound RND-efflux transporter *adeJ* (2 fold ± 0.007, 3.03 fold ± 0.006, 1.87 fold ± 0.003), homolog of ABAYE3381, MATE family efflux pump *abeM* (0.27 fold ± 0.002, 0.41 fold ± 0.043, 0.05 fold ± 0.034), homolog of ABAYE3248 ABC family efflux transporter *macB* (0.06 fold ± 0.023, 0.71 fold ± 0.012, 0.31 fold ± 0.065), homolog of ABAYE0649 membrane porin *ompA* (24.25 fold ± 0.034, 36.76 fold ± 0.044, 1.23 fold ± 0.008) and homolog of ABAYE0924 membrane porin *carO* (14.93 fold ± 0.028, 0.08 fold ± 0.056, 0.20 fold ± 0.034) for CAB-1, CAB-2 and CAB-3 respectively.

The levels of expression in presence of 3.2 μg/ml of chlorhexidine were increased for *adeB* (0.5 fold ± 0.012, 2.46 fold ± 0.011, 0.18 fold ± 0.008), *adeJ* (1.87 fold ± 0.045, 1.52 fold ± 0.008, 0.81 fold ± 0.003), *abeM* (0.41 fold ± 0.01, 0.71 fold ± 0.005, 0.07 fold ± 0.003), *macB* (0.50 fold ± 0.032, 0.93 fold ± 0.054, 0.06 fold ± 0.064), *ompA* (1.32 fold ± 0.054, 2.14 fold ± 0.076, 7.46 fold ± 0.056) and *carO* (2.30 fold ± 0.032, 0.04 fold ± 0.033, 0.020 fold ± 0.089) for CAB-1, CAB-2 and CAB-3 respectively.

The levels of expression in presence of 3.2 μg/ml of triclosan were increased for *adeB* (1.74 fold ± 0.054, 2.83 fold ± 0.034, 3.25 fold ± 0.012), *adeJ* (0.76 fold ± 0.011, 1.07 fold ± 0.023, 0.62 fold ± 0.018), *abeM* (1.41 fold ± 0.028, 1.62 fold ± 0.043, 0.29 fold ± 0.076), *macB* (0.16 fold ± 0.045, 0.15 fold ± 0.023, 0.81 fold ± 0.012), *ompA* (8 fold ± 0.089, 9.19 fold±0.011, 21.11 fold ± 0.001) and *carO* (2.14 fold ± 0.024, 0.41 fold ± 0.002, 0.12 fold ± 0.004) $p = <0.05$ for CAB-1, CAB-2 and CAB-3 respectively (Table 2.3).

Table 2.3 Relative expression levels of efflux transporters and cell envelope porins in *A. baumannii*

	adeB	*adeJ*	*abeM*	*macB*	*ompA*	*carO*
CAB-1 BNZ	0.14 ± 0.018	2 ± 0.007	0.27 ± 0.002	0.06 ± 0.023	24.25 ± 0.034	14.93 ± 0.028
CAB-2 BNZ	2.64 ± 0.002	3.03 ± 0.006	0.41 ± 0.043	0.71 ± 0.012	36.76 ± 0.044	0.08 ± 0.056
CAB-3 BNZ	0.20 ± 0.004	1.87 ± 0.003	0.05 ± 0.034	0.31 ± 0.065	1.23 ± 0.008	0.20 ± 0.034
CAB-1 CHX	0.5 ± 0.012	1.87 ± 0.045	0.41 ± 0.01	0.50 ± 0.032	1.32 ± 0.054	2.30 ± 0.032
CAB-2 CHX	2.46 ± 0.011	1.52 ± 0.008	0.71 ± 0.005	0.93 ± 0.054	2.14 ± 0.076	0.04 ± 0.033
CAB-3 CHX	0.18 ± 0.008	0.81 ± 0.003	0.07 ± 0.003	0.06 ± 0.064	7.46 ± 0.056	0.020 ± 0.089
CAB-1 TRI	1.74 ± 0.054	0.76 ± 0.011	1.41 ± 0.028	0.16 ± 0.045	8 ± 0.089	2.14 ± 0.024
CAB-2 TRI	2.83 ± 0.034	1.07 ± 0.023	1.62 ± 0.043	0.15 ± 0.023	9.19 ± 0.011	0.41 ± 0.002
CAB-3 TRI	3.25 ± 0.012	0.62 ± 0.018	0.29 ± 0.076	0.81 ± 0.012	21.11 ± 0.001	0.12 ± 0.004

The expression levels for membrane bound RND-efflux transporter *adeB*, membrane bound RND-efflux transporter *adeJ,* MATE family efflux pump *abeM,* ABC family efflux transporter *macB*, membrane porin *ompA* and membrane porin *carO* has been quantified from RNA that was isolated in the absence and presence of either of 3.2 μg/ml of benzalkonium chloride (BNZ), 3.2 μg/ml of chlorhexidine (CHX), 3.2 μg/ml of triclosan (TRI)

2.4 Discussion

Over the millennia, bacteria have evolved evasion strategies to overcome arrays of chemical and environmental assails including antimicrobial drugs (Peleg et al. 2008). The occurrence of nosocomial infections in hospital intensive care units due to *A.baumannii* currently ranges from 2% to 10% of all Gram-negative bacterial infections in Europe and account for about 2.5% of them in the United States (Fournier et al. 2006). An increased incidence of MDR *A. baumannii* bloodstream infections in US Army servicemen injured during Afghanistan and Iraq/Kuwait military operations was reported previously (Dijkshoorn et al. 2007). MDR isolates belonging to the different ribotypes endemic to Los Angeles, New York, Washington, Chicago, Houston and Pennsylvania has been described very recently (Valentine et al. 2008; Hujer et al. 2006; Srinivasan et al. 2009a). Currently this bacterium ranks among the most dreaded nosocomial pathogens. Till date few MDR *A. baumannii* strains have been sequenced which includes *A. baumannii* ACICU (an epidemic MDR clinical isolate responsible for an outbreak in Rome (Italy) in 2005, Accession number CP000863), *A. baumannii* AB307-0294 (obtained from the blood of a patient hospitalized in Buffalo, NY, in 1994, Accession number CP001172), *A. baumannii* AB900 (perineal isolate obtained in 2003 from an active duty military patient at WRAMC, Accession number ABXK00000000), *A. baumannii* AB0057 (MDR bloodstream isolate collected in 2004 from a patient at WRAMC, Accession number ABJM01000001) and *A. baumannii* AYE (an epidemic MDR clinical isolate responsible for a nationwide outbreak in France in 2001, Accession number CU459141), however the precise molecular mechanism involved in conferring disinfectant resistance has never been examined especially on Indian isolates.

Considering the importance of disinfection in the prevention of nosocomial infection, the aim of this study was to evaluate the biocide susceptibilities of a set of MDR *A. baumannii* clinical isolates and to delineate the role of disinfectants as inducers of antibiotic resistance. The current set of clonally distinct *A. baumannii* isolates was found to be disinfectant tolerant.

Drug efflux indicates to be the important strategic mechanism exploited by bacteria to tolerate antibiotic and environmental assails . Efflux genes represents 6–18% of membrane proteins in bacterial genomes. Particularly interesting among mediators of MDR in *A. baumannii* are the efflux pumps belonging to MFS, RND, ATP binding cassette, small multidrug resistance, multidrug and toxic compound extrusion families (Piddock 2006). We have previously shown that *amvA*, *abeS*, *adeABC* has role in biocide and drug resistance (Srinivasan et al. 2009b; Rajamohan et al. 2010a; Rajamohan et al. 2010b). Loss of outer membrane porins are also considered as prominent mechanism of drug resistance (Mussi et al. 2007). In these clinically distinct strains efflux transporters were responsible for mediating antimicrobial resistance.

The role of disinfectants as inducers of antibiotic resistance has been characterized in many Gram-negative pathogens. Silvia *et al* had previously demonstrated that a *P. aeruginosa* strain obtained from epidemic displayed high level triclosan resistance of MIC-2125 mg/L, but when adapted to MIC of 3400 mg/L, strain exhibited an increased MIC for tetracycline, ciprofloxacin, amikacin, levofloxacin, carbenecillin and chloramphenicol (D'Arezzo et al. 2012). A similar study done in *Salmonella* serovar Virchow showed that strains adapted to resist benzalkonium chloride displayed resistance to amoxicillin, amoxicillin-clauvalinic acid, chloramphenicol, imipenem and trimethoprim. In a study with *E. coli* 0157, strains which were adapted to resist benzalkonium chloride also displayed resistance to amoxicillin, amoxicillin-clauvalinic acid, chloramphenicol, imipenem, tetracycline and trimethoprim (Braoudaki and Hilton 2004).

In our study we found that presence of disinfectants as selective pressure made cells to exhibit increased MIC for chloramphenicol, cefepime, ceftriaxone, norfloxacin and neomycin. It is worthy to state that such a study has been done for the first time in *A. baumannii* from India. We found a decrease in the ability of the strains to exhibit motility and biofilm formation.

Resistance to antiseptics and disinfectants associated with integrons carrying efflux-related transporters such as QacE has been well studied (Sekiguchi et al. 2004). Various studies in Gram-negative bacteria have demonstrated that efflux pumps play an important role in intrinsic resistance to disinfectants including quaternary ammonium compounds. Biocide resistance can result from the increased expression of efflux pumps, particularly in Enterobacteriaceae the AcrAB-TolC system. AcrAB-TolC is the major multidrug efflux pump in many Gram-negative bacteria, including *Salmonella* and *E. coli* (Russell and Day 1996). Studies done previously have demonstrated that over expression of multidrug efflux transporters AcrAB–TolC with broad substrate specificity can confer decreased resistance to antibiotics, dyes, detergents including disinfectants (Piddock 2006). Huet *et al* has shown that over expression of *mepA*, *mdeA*, *norA* and *norC* was responsible for mediating resistance to biocides and dyes in *S. aureus* (Huet et al. 2008). In our recent study we have shown the role of *kpnO*, an outer membrane porin in mediating antimicrobial and disinfectant resistance in *K. pneumoniae* for the very first time (Srinivasan et al. 2012b). The authors have previously shown that RND type efflux systems *adeABC* and *adeIJK* transport systems can extrude not only structurally unrelated compounds, detergents, but also disinfectants, providing experimental evidence for the broad substrate specificity of efflux pumps in *A. baumannii* (Rajamohan et al. 2010b).

Here in to elucidate the role of disinfectants as inducers of drug resistance, we found that in the presence of disinfectants the clonally distinct strains exhibited an altered expression for membrane transporters *adeB*, *adeJ*, *abeM*, *macB*, and porins *ompA* and *carO*. Overall this study provided direct experimental evidence for the role of disinfectants as inducers of MDR in *A. baumannii*. With such a concern that disinfectant exposure can select for a multidrug resistant population, it is important to determine the susceptibility of clinical *A. baumannii* to various disinfectants on a large scale and systematic manner and to implement strict intervention of control and preventive measures.

Acknowledgement We are highly thankful to Director, CSIR-Institute of Microbial Technology, Chandigarh, for providing excellent facility to carry out this work. VBS remains thankful to DST (SB/YS/LS-177/2014) for the fellowship. The funding from CSIR, DST (SB/YS/LS-177/2014) and DBT (BT/PR14304/BRB/10/822/2010, BT/HRD/NBA/34/01/2012) are highly acknowledged.

References

Baker S (2000) Biocidal Products Directive 98/8/EC. Complement Ther Nurs Midwifery 6:48–49

Beier RC, Anderson PN, Hume ME et al (2011) Characterization of *Salmonella enterica* isolates from turkeys in commercial processing plants for resistance to antibiotics, disinfectants, and a growth promoter. Foodborne Pathog Dis 8:593–600

Braoudaki M, Hilton AC (2004) Adaptive resistance to biocides in *Salmonella enterica* and *Escherichia coli* O157 and cross-resistance to antimicrobial agents. J Clin Microbiol 42:73–78

Clinical and Laboratory Standards Institute (2006) Methods for dilution antimicrobial susceptibility tests for bacteria that grow aerobically; approved standard. M7-A7, vol 26-2. Clinical and Laboratory Standards Institute, Wayne

Coenye T, Van Acker H, Peeters E et al (2011) Molecular mechanisms of chlorhexidine tolerance in *Burkholderia cenocepacia* biofilms. Antimicrob Agents Chemother 55:1912–1929

D'Arezzo S, Lanini S, Puro V et al (2012) High-level tolerance to triclosan may play a role in *Pseudomonas aeruginosa* antibiotic resistance in immunocompromised hosts: evidence from outbreak investigation. BMC Res Notes 5:43

Davies RM (2007) The clinical efficacy of triclosan/copolymer and other common therapeutic approaches to periodontal health. Clin Microbiol Infect 4:25–29

Dijkshoorn L, Nemec A, Seifert H (2007) An increasing threat in hospitals: multidrug-resistant *Acinetobacter baumannii*. Nat Rev Microbiol 5:939–951

Ferreira C, Pereira AM, Pereira MC et al (2011) Physiological changes induced by the quaternary ammonium compound benzyldimethyldodecylammonium chloride on *Pseudomonas fluorescens*. J Antimicrob Chemother 66:1036–1043

Fournier PE, Vallenet D, Barbe V, Audic S et al (2006) Comparative genomics of multidrug resistance in *Acinetobacter baumannii*. PLoS Genet 2:e7

García-Quintanilla M, Pulido MR, López-Rojas R et al (2013) Emerging therapies for multidrug resistant *Acinetobacter baumannii*. Trends Microbiol 21:157–163

Gotoh NH, Tsujimoto A, Nomura K et al (1998) Functional replacement of OprJ by OprM in the MexCD-OprJ multidrug efflux system of *Pseudomonas aeruginosa*. FEMS Microbiol Lett 165:21–27

Graf P (2001) Benzalkonium chloride as a preservative in nasal solutions: re-examining the data. Respir Med 95:728–733

Holdsworth SR, Law CJ (2013) The major facilitator super family transporter MdtM contributes to the intrinsic resistance of *Escherichia coli* to quaternary ammonium compounds. J Antimicrob Chemother 68:831–839

Huet AA, Raygada JL, Mendiratta K et al (2008) Multidrug efflux pump overexpression in *Staphylococcus aureus* after single and multiple in vitro exposures to biocides and dyes. Microbiology 154:3144–3153

Hujer KM, Hujer AM, Hulten EA et al (2006) Analysis of antibiotic resistance genes in multidrug-resistant *Acinetobacter* sp. isolates from military and civilian patients treated at the Walter Reed Army Medical Center. Antimicrob Agents Chemother 50:4114–4123

Hulton CS, Higgins CF, Sharp PM (1991) ERIC sequences: a novel family of repetitive elements in the genomes of *Escherichia coli*, *Salmonella* Typhimurium and other enterobacteria. Mol Microbiol 5:825–834

Köhler TM, Michea-Hamzehpour U, Henze N et al (1997) Characterization of MexE-MexF-OprN, a positively regulated multidrug efflux system of *Pseudomonas aeruginosa*. Mol Microbiol 23:345–354

Maris P (1995) Modes of action of disinfectants. Rev Sci Tech 14:47–55

Masuda N, Sagagawa E, Ohya S et al (2000) Contribution of the MexX-MexY-OprM efflux system to intrinsic resistance in *Pseudomonas aeruginosa*. Antimicrob Agents Chemother 44:2242–2246

Mussi MA, Relling VM, Limansky AS et al (2007) CarO, an *Acinetobacter baumannii* outer membrane protein involved in carbapenem resistance, is essential for L-ornithine uptake. FEBS Lett 581:5573–5578

Patryn R, Jarosz MJ, Włoszczak-Szubzda A et al (2011) Considerations on Directive 98/8 of the European Commission – the biocide directive. Ann Agric Environ Med 18:452–458

Peleg AY, Seifert H, Paterson DL (2008) *Acinetobacter baumannii*: emergence of a successful pathogen. Clin Microbiol Rev 2:538–582

Pendleton JN, Gorman SP, Gilmore BF (2013) Clinical relevance of the ESKAPE pathogens. Expert Rev Anti-Infect Ther 11:297–308

Piddock LJ (2006) Clinically relevant chromosomally encoded multidrug resistance efflux pumps in bacteria. Clin Microbiol Rev 19:382–402

Rajamohan G, Srinivasan VB, Gebreyes WA (2010a) Molecular and functional characterization of a novel efflux pump, AmvA, mediating antimicrobial and disinfectant resistance in *Acinetobacter baumannii*. J Antimicrob Chemother 65:1919–1925

Rajamohan G, Srinivasan VB, Gebreyes WA (2010b) Novel role of *Acinetobacter baumannii* RND efflux transporters in mediating decreased susceptibility to biocides. J Antimicrob Chemother 65:228–232

Robinson JG (1995) Chlorhexidine gluconate-the solution for dental problems. J Vet Dent 12:29–31

Rose H, Baldwin A, Dowson CG et al (2009) Biocide susceptibility of the *Burkholderia cepacia* complex. J Antimicrob Chemother 63:502–510

Rushton L, Sass A, Baldwin A et al (2013) Key role for efflux in the preservative susceptibility and adaptive resistance of *Burkholderia cepacia* complex bacteria. Antimicrob Agents Chemother 57:2972–2980

Russell AD (1985) The role of plasmids in bacterial resistance to antiseptics, disinfectants and preservatives. J Hosp Infect 6:9–19

Russell AD (1997) Plasmids and bacterial resistance to biocides. J Appl Microbiol 83:155–165

Russell AD, Day MJ (1996) Antibiotic and biocide resistance in bacteria. Microbios 85:45–65

Rutala WA, Weber DJ (2001) New disinfection and sterilization methods. Emerg Infect Dis 7:348–353

Sekiguchi J, Hama T, Fujino T et al (2004) Detection of the antiseptic- and disinfectant-resistance genes *qacA*, *qacB*, and *qacC* in methicillin-resistant *Staphylococcus aureus* isolated in a Tokyo hospital. J Infect Dis 57:288–291

Sinha N, Agarwal J, Srivastava S et al (2013) Analysis of carbapenem-resistant *Acinetobacter* from a tertiary care setting in North India. Indian J Med Microbiol 31(1):60–63

Skovgaard S, Nielsen LN, Larsen MH et al (2013) *Staphylococcus epidermidis* isolated in 1965 are more susceptible to triclosan than current isolates. PLoS One 8:e62197

Smith K, Hunter IS (2008) Efficacy of common hospital biocides with biofilms of multi-drug resistant clinical isolates. J Med Microbiol 57:966–973

Soumet C, Fourreau E, Legrandois P (2012) Resistance to phenicol compounds following adaptation to quaternary ammonium compounds in *Escherichia coli*. Vet Microbiol 158:147–152

Srinivasan VB, Rajamohan G, Preeti P et al (2009a) Genetic relatedness and molecular characterization of resistance determinants in multidrug resistant *Acinetobacter baumannii* isolated in central OH, USA. Ann Clin Microbiol Antimicrob 8:21

Srinivasan VB, Rajamohan G, Gebreyes WA (2009b) Role of AbeS, a novel efflux pump of the SMR family of transporters in resistance to antimicrobial agents in *Acinetobacter baumannii*. Antimicrob Agents Chemother 53:5312–5316

Srinivasan VB, Vaidyanathan V, Mondal A et al (2012a) Role of the two component signal transduction system CpxAR in conferring cefepime and chloramphenicol resistance in *Klebsiella pneumoniae* NTUH-K2044. PLoS One 7:e33777

Srinivasan VB, Venkataramaiah M, Mondal A et al (2012b) Functional characterization of a novel outer membrane porin KpnO, regulated by PhoBR two-component system in *Klebsiella pneumoniae* NTUH-K2044. PLoS One 7:e41505

Valentine SC, Contreras D, Tan S et al (2008) Phenotypic and molecular characterization of *Acinetobacter baumannii* clinical isolates from nosocomial outbreaks in Los Angeles County, California. J Clin Microbiol 46:2499–2507

Zhu L, Lin J, Ma J et al (2010) Triclosan resistance of *Pseudomonas aeruginosa* PAO1 is due to FabV, a triclosan-resistant enoyl-acyl carrier protein reductase. Antimicrob Agents Chemother 54:689–698

Knowledge Gaps and Research Needs in Bacterial Co-Resistance in the Environment

3

Paola Grenni and Gianluca Corno

Abstract

This chapter describes the different factors that increase or stimulate the presence of resistance to antibiotics in bacteria in the environment. Particular factors accentuating the spread and evolution of antibiotic resistance are various pollutants such as heavy metals, disinfectant products, other organic pollutants and nutrients.

In particular, co-factors inducing resistance in water and sediments are explained. A particular emphasis is placed on the co-selection of antibiotic resistance genes due to the presence of metals, both in soil and in waters. Moreover, the role of nutrients and other organic contaminants in improving antibiotic resistance in bacteria is also highlighted. Finally, the role of the disinfection of waters and wastewaters in abatement of antibiotic resistance genes is reported.

Keywords

Antibiotic resistance · AMR · Co-resistance · Cross-resistance · Metals · Disinfectant

P. Grenni (✉)
Water Research Institute (IRSA), National Research Council (CNR), Area della Ricerca di Roma 1 Montelibretti (Roma), Monterotondo, Rome, Italy
e-mail: grenni@irsa.cnr.it

G. Corno
Water Research Institute, CNR – National Research Council of Italy, Verbania, Italy
e-mail: gianluca.corno@cnr.it

S. M. Mandal, D. Paul (eds.), *Bacterial Adaptation to Co-resistance*,
https://doi.org/10.1007/978-981-13-8503-2_3

3.1 Introduction

Antibiotics are considered in the recent year environmental contaminants. In fact, they can reach the environment from several sources. For example, human and animal urine and faeces after pharmaceutical treatment can contain active residues or their metabolites, which finally reach sewage water and sludge. The latter can be spread on farmland to recycle nutrients were can exercise a selective pressure on natural microorganisms (Ju et al. 2018).

Three main processes are responsible for acquiring resistance to antibiotics in bacteria: spontaneous mutation, recombination and horizontal gene transfer through plasmids and transposons (mainly by conjugation; Berendonk et al. 2015; von Wintersdorff et al. 2016). Spontaneous mutation can be caused by continuous exposure to antibiotics, which develops selection and produce genetic modification in microbes by mutation. This gives rise to a new resistant phenotype with a stable genetic recombination guaranteeing the survival of microorganisms (see for example Zaman et al. 2017). The developed gene, that confers resistance towards the antibiotic, could be transferred to the next generation or to other microorganisms (even in some cases genetically distant) through horizontal gene transfer (HGT), also known as horizontal evolution. HGT uses transferable genetic elements causing a wide diffusion of the resistance trait. HGT is a process permitting resistance genes to spread between, possibly very distant, bacterial lineages (environmental but also pathogenic bacteria) which may have been sensitive to antibiotics.

Several different resistance mechanisms have been identified (e.g. antibiotic-altering or degrading enzymes, genetic variation of the antibiotic target action, efflux pumps; Grenni et al. 2018).

In many cases, the same class of antimicrobial agents are used in both human and veterinary medication. In some countries (not in Europe), veterinary antibiotics are used in prevention and treatment of animal diseases and as growth promoters, far outweighing their use as animal therapeutics (Ben et al. 2019).

The acquisition of antimicrobial resistance (AMR) increase the difficulties to treat infections. AMR can therefore be viewed as a common challenge threatening the sustainability of health care in both medical and veterinary practice. Many infectious bacteria are common in humans and animals and it is very well established that there is a frequent transfer of these species and acquired antimicrobial resistance genes (AMRGs) among humans, animals and the environment (see for example Perry and Wright 2013; Yamaji et al. 2018).

Consequently, the prevention of AMR requires a strong response in human, veterinary and environmental fields. The 'One Health' concept highlights the interconnections among human, animal and environmental health, and promotes the importance of collaborative efforts to achieve optimal health for all three. This is also true for combatting AMR, in order to avoid terrible episodes, becoming ever more frequent in human and animal medicine, where antibiotics are ineffective. The difficulties of infection disease treatment are described in some reports as the start of the post-antibiotic era (Alanis 2005). Unfortunately, a decrease in antibiotic use would not necessarily prevent or avoid the resistance in clinical or natural environments (Salyers and Amábile-Cuevas 1997) because the contagion, that is the spread of resistant strains and resistance genes, appears to be the dominant contributing factor.

We therefore need to discovery additional methods to reduce this phenomenon and to fully comprehend the evolution and dissemination of antibiotic resistance mechanisms (Aminov 2010).

Some authors assume that a bacterial strain is resistant to an antibiotic if the minimum concentration with an effect on microorganism (minimal inhibitory concentration, MIC) is higher than for the equivalent wild-type strain. Hence, a 'resistance gene' permits a bacterium to survive at higher antibiotic concentration or decreases its susceptibility to antibiotics (Martinez 2014). Resistance genes, have, in any case, fitness costs (Melnyk et al. 2015). If the latter are negligible, these genes can be kept and mobilized anywhere but need a selection pressure to be maintained on a mobile genetic element until they have evolved. The concentrations of antibiotics (or other contaminants that can confer resistance to the bacteria) required to maintain these genes (called minimal selective concentrations, MSCs), were in all cases below (almost up to 140-fold) the MIC of the plasmid-free susceptible bacteria. This finding indicates that the very low antibiotic (or other pollutant) levels found in the environments and in treated humans and animals might be sufficiently high to maintain multi-resistance plasmids.

More recently, it has been found that the selection for naturally occurring resistance genes in a microbial community remains constant at antibiotic concentrations from low sub-inhibitory to above clinically important quantities (Murray et al. 2018).

To determine the amount of antibiotic pollution, it is very useful to quantify the antibiotic resistance genes (ARG) using the quantitative PCR (qPCR), that give an approximation of the prevalence of known ARGs in environmental samples. Although qPCR can give a good evaluation of the presence of ARGs, a careful standardization of gene copy numbers is needed (Berendonk et al. 2015).

A factor making this problem worse is that the spread and evolution of antibiotic resistance is also triggered or catalysed by the presence of different pollutants in the environment such as heavy metals, disinfectant products (e.g. quaternary ammonium compounds and other biocides), other organic pollutants and nutrients (Fig. 3.1). It has been showed that, for example, antimicrobial agents different from antibiotics (e.g. disinfectants like quaternary ammonium compounds) have the capability to stimulate a co-selection, promoting an indirect selection for antibiotic resistance (Pal et al. 2014; Wales and Davies 2015). Another example is heavy metal contaminations, widespread in agriculture (Han et al. 2001) and aquaculture (Burridge et al. 2010), which contribute to this environmental burden. Disinfectants and their by-products, detergents, biocides and organic solvents can also induce bacterial expression of multidrug resistance determinants (Baker-Austin et al. 2006).

Co-selection can be realised in two modes: (1) co-resistance, in which the selection for one gene promotes the maintenance of a resistance gene, one that does not necessarily offer a selective advantage to the substance in question (Johnson et al. 2016); and (2) cross-resistance, in which one resistance gene can be efficient to protect from multiple toxic chemicals (Curiao et al. 2015), Fig. 3.2. Co-resistance is

Fig. 3.1 Natural microbial communities are subject to different pollutants: antibiotics, metals, PAHs, PCBs, disinfectant products. This fact can promote a co-selection process, indirectly selecting for antibiotic resistance

equivalent to have toolbox; a microorganism might only need one or two tools (genes) from the toolbox at any one time, but there are many tools "casually" available if their presence could be useful in some situation. The microbial genes as a whole (including ARGs in plasmids, integrons and transposons) is the toolbox and the resistance genes present in the genome are the tools. Cross-resistance could be similar to a tool capable of multiple functions, that can be used in different ways depending on the needs.

The ARGs, that recently are considered as emerging environmental contaminant (Grenni et al. 2018) can be present in the environment depending on several aspects other than antibiotics alone (Andersson and Hughes 2010).

In this chapter different co-factors that induce or promote resistance to antibiotic in different environments are reported.

3.1.1 Co-Factors That Induce Resistance in Water and Sediments

Antibiotic Resistant Bacteria (ARB) and genes (ARGs) providing resistance against specific antibiotics are common in natural settings, representing a significant proportion of the resident microbial communities and of the microbiomes of seas, lakes

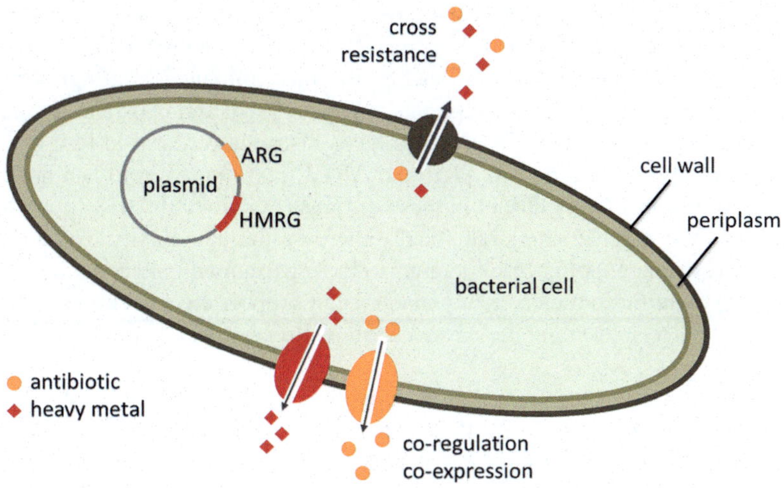

Fig. 3.2 Different ways for bacteria to acquire resistance against antibiotics and heavy metals: Co-resistance: ARG and HMRG located on the same genetic element, with high possibilities of co-selection under specific stress. Cross-resistance: the same system ensure resistance to heavy metals and antibiotics (e.g. general efflux pumps). Co-regulation and co-expression: the same regulator is controlling the expression of resistance systems to heavy metals and antibiotics. (Figure modified from Baker-Austin et al. 2006)

and rivers and of their sediments (Eckert et al. 2018). The origins of these cells and genes (further referred as determinants of resistance) are various: they can be allochthonous, introduced in natural environments by point or diffuse sources of pollutions (e.g. wastewater effluents, agricultural soils washout) (Di Cesare et al. 2017), or they can be autochthonous, developed and selected within the resident bacterial communities as a response to a specific stress (Czekalski et al. 2014; Di Cesare et al. 2015). In the second case, the promoting factor can be the release of natural antibiotics, by either fungi or other bacteria, as confirmed by the presence of resistant bacteria even in pristine environments, such as Antarctic ice (D'Costa et al. 2011). Nevertheless, the magnitude of ARB and ARGs in waters and sediments correlates very well to the degree of human activities affecting the environment, and in most cases polluted environments are characterized by resistance to synthetic and semi-synthetic antibiotics absent in nature (Manaia 2017). These determinants of resistance are very often present in environments where the related antibiotic is absent, or present in very low concentrations, suggesting different ecological or physiological factors play a role in the spread and persistence of the resistance.

Many ARGs are related with mobile genetic elements (e.g. plasmids), and gene acquisition elements (integrons) allowing their spread in the case of Horizontal Gene Transfer (HGT). The concomitant presence of a number of ecologically important genes in the same mobile genetic fragment can promote co-selection of the ARG even in the absence of a direct stressor (the molecule of the antibiotic), but simply because the ARG is directly linked to other genes within the same DNA fragment, and thus co-transferred by HGT.

A direct co-selection of ARGs in microbial communities exposed to diverse contaminations, from biocides to heavy metals, has been demonstrated (Seiler and Berendonk 2012). This co-factor can become predominant in settings where the release of these anthropogenic pollutants is concentrated and constant in time.

For example, sediments of rivers and lakes historically exposed to a release of Cd, Hg, Zn, and Cu are rich in ARB and ARGs, resistant even to last generation cephalosporins absolutely absent in these settings (Eckert et al. 2018).

Co-selection mechanisms dramatically increase the complexity of the cycle of antibiotic resistance, going from humans to the environment and back, because co-factors (e.g. pharmaceuticals, heavy metals) can support the selection and the persistence of ARGs even in the absence of a direct stress caused by antibiotic molecules.

Open waters (and even more their sediments) exposed to or historically contaminated by heavy metals can be thus considered environments where ARGs are selected within the resident microbiome, but other environments are also exposed to the same risk: civil and industrial wastewater treatments often abate nutrients by adding binding molecules of metals (e.g. Al or Fe) during the process (Di Cesare 2016a, b). This procedure can lead to processes of co-selection in favour of ARGs within the sewage treated, and to a final release into the environment of a high number of resistant bacteria. Alternative methodologies for the abatement of nutrients are available today and should be considered for the treatment of civil wastewater especially when ARG- and ARB-rich hospital wastewaters are present in the sewage.

3.1.2 Metal Presence and Co-Selection of Antibiotic Resistance Genes

Some heavy metals (e.g. Ni, Zn, Cu, Fe) are essential to metabolic reactions and are required as trace elements by organisms; others (e.g. Hg, Cd and Ag) have no biological role and are toxic to organisms, even at very low concentrations. In the latter case, they are longstanding and broad spectrum antimicrobial agents (Lemire et al. 2013). For example, metals with antimicrobial effects include compounds of mercury, arsenic and antimony, and salts of gold, silver, copper, potassium, zinc, tellurium and magnesium (Pal et al. 2017).

The mode of antibiotic action of metals is not fully understood, but they affect microbial cells through their chemical properties (thiophilicity, ability to perform redox reactions, function as a soft Lewis acid). Because of these properties, metals are used as a veterinary medicine. In fact, iron, cobalt, manganese, copper and zinc, for example, are used as nutritional additives in animal feed in livestock farming and fish production in Europe (Commission Regulation 1831/2003/EC, 2003). For instance, zinc and copper are used in pig and poultry production, because of their bactericidal and fungicidal properties (Nemecek et al. 2011). In particular, zinc oxide (up to 3000 mg/kg feed) is used in controlling post-weaning diarrhoea. Copper carbonate and other salts (up to 100 mg Cu per animal) can be used to treat mineral deficiency in food-producing animals. This metal also has antimicrobial characteristics; it is used, for example, as a fungicide in plants and animals and in the control of foot-rot in cattle and sheep. As they are excreted, metals can be found in animal manure and then in digestate. If they are used as fertilizers, they can reach the soil.

In a study conducted in Portugal (Alvarenga et al. 2015) different organic wastes (sewage sludge, agro-industrial sludge, municipal slaughterhouse sludge and paper mill waste), including different composts (mixed municipal solid waste compost, agricultural waste compost, agricultural waste and sewage sludge compost, and pig slurry digestate) used as agricultural soil amendments, were found to have high metal concentrations. In this study, sewage sludge was in line with Council Directive 86/278/EEC as regards agricultural use but contained more than 757 mg/kg of Zn. The same problem was found for the digestate, in which Zn was more than 2000 mg/kg. The composts had high concentrations of different metals (max concentrations: Cd 3 mg/kg; Cr: 99.1 mg/kg; Cu: 179 mg/kg; Hg: 0.63 mg/kg; Ni: 360 mg/kg; Pb: 202 mg/kg; Zn: 473 mg/kg) depending on the type of production. In any case, Cu Zn has been identified as the main metal contaminants in agricultural areas. They can reach soil due to the application of sewage sludge, manure or digestate from aerobic or anaerobic digestion of manure or compost.

In the European Union, a pre-assessment of the environmental impact of Zn and Cu feed amendments demonstrated that they derived from aquaculture and agriculture. The models foreseen that the no effect concentrations of Zn and Cu will be reach in some soil and water ecosystems within the next 10–50 years (Monteiro et al. 2010).

Fortunately, the EU Commission adopted in 2017 a decision to withdraw the marketing authorisations for veterinary products containing zinc oxide administered orally to food producing species due to its negative effects on animals' performance and on the environment, and to ban the use of zinc oxide by 2022. In fact, it increases antimicrobial resistance, changes piglets' microbiota and accumulates in vital organs and environmental areas.

Heavy metals can be found not only in agricultural areas but also in aquaculture, industrial runoff, sewage waters and urban soils (McLaughlin and Smolders 2001; Imfeld et al. 2011; Berg et al. 2012). Moreover, metals can be naturally present from geological sources (Cullen and Reimer 1989; Flemming and Trevors 1989; Smedley and Kinniburgh 2002). Most serious heavy metal contaminations in soils include, other than Cd and Zn, also Pb, Cu and Hg (Han et al. 2002). Once in the environment, metals persist and consequently tend to accumulate in food chains.

The Cd input to the agricultural soil can be also caused by inorganic fertilizers and pesticides. In fact, there can be a high concentration of Cd in phosphoric fertilizers because the phosphate rocks used as an essential feedstock in industrially produced fertilizer contain from 3 to 150 mg Cd/kg rock; it can be transferred from rocks to fertilizer in amounts that depend on the fertilizer manufacturing process (in some cases all the Cd is transferred; Roberts 2014).

Bacteria have several general or specific modes to resist metals: methylation, demethylation oxidation and reduction processes (Raja et al. 2018); generalized antimicrobial efflux systems for the expulsion of metal (Blanco et al. 2016); specific extracellular sequestration (using extracellular polymers or siderophores that can trap, precipitate or bind metal ions, reducing their uptake or increasing efflux by membrane transporters); intracellular specific physical sequestration of metal, binding a protein or other ligands to avoid damage to the metal-sensitive cellular target; additionally, microorganisms can use cytoplasmic proteins, such as bacterioferritin and metallothioneins to bind, sequester or store metals (Prabhakaran et al. 2016). In general, specific metal resistance mechanisms usually have a metal ion-specific response regulator, which controls the expression of structural resistance genes. For example, metal ion-specific efflux proteins or protein complexes and/or enzymes alter metal ions into a form less toxic to bacterial cells (Silver and Phung 1996). Metals can also bind to, or precipitate on, bacterial cell surfaces. Metals in this case interact with cell wall, involving proteins or cell-associated polysaccharides, such as lipopolysaccharide.

In the environment metals can select the resistant genes with a similar mode to the selection of antibiotic resistance, and these genes can be present in mobile genetic elements. In particular, co-selection of antibiotic and metal resistance is associated with Cu, Co, As, Cd, Cr, Hg, Pb, Ni and Zn (Zhao et al. 2019). Indeed, the association of metal and antimicrobial resistance is relatively common, since both resistance genes are frequently located on the same mobile genetic elements (Li et al. 2017). Consequently, it can be expected that the selective pressure by heavy metals contributes to the indirect co-selection of antibiotic resistance, particularly in environments contaminated with the two contaminants. These genes can encode, in the case of cross resistance, for a generic detoxifying mechanism that reduce in a non-specifically mode the intracellular concentrations of both antibiotics and metals. In the case of co-resistance, the resistance to metals and antibiotic involve separate genes, which are integrated on the same genetic element (Sanderson et al. 2016). One example is the class 1 integron-integrase gene (*intI*1). Integrons are genetic elements that contain a site-specific recombination system capable to catch, express and exchange gene cassettes. Gene cassettes of integrons can be integrated to bacterial chromosomes or plasmids through site-specific recombination catalyzed by the integrase, a tyrosine recombinase. They are commonly used as a proxy for anthropogenic impact in the environment (Gillings et al. 2015; Di Cesare et al. 2016b). In fact, *intI*1 gene was found to be associated to resistance genes for heavy metals, antibiotics and disinfectants. Its abundance can rapidly change in response to environmental pressures, because it is present in several bacterial species (that have rapid generation times), and it is often located on mobile genetic elements that can readily transfer among bacteria.

As metals can occur in agricultural soils, aquaculture areas, industrial runoff and sewage waters, they have an ample opportunity to mix with antibiotics, facilitating the proliferation of antibiotic resistance via co-selection of antibiotic and metal resistance genes. In fact, it was found that antibiotic-resistant bacteria are enriched in metal contaminated areas, and genes conferring co-selection to heavy metals and antibiotics are often found together in many clinical isolates (Baker-Austin et al. 2006; Fard et al. 2011; Ji et al. 2012; Seiler and Berendonk 2012).

Another issue is that metals contribute to the maintenance of antibiotic resistance (Hu et al. 2017). A systematic genomic analysis performed by Pal et al. (2015) highlighted that although biocides and metals have limited opportunities to promote Horizontal Gene Transfer of antibiotic resistance, these chemicals can select for antibiotic-resistant bacteria through chromosomal Biocide and Metal Resistance Genes. It was found that Cu has a direct role in the antibiotic resistance spread in bacteria isolated from agricultural soil polluted by Cu (Berg et al. 2005). More seriously, ARGs were more closely related to metals than residual antibiotics (Hu et al. 2016). Some authors found that the antibiotic selection pressure is transient (Wu et al. 2015). By contrast, metals can exert a long-term selection pressure because they resist degradation. Hu et al. (2017) noted that the significant enrichment of ARGs in soil that received Ni for more than 5 years may be ascribed to the increasing pressure forced by increasing Ni concentrations. The bioavailability and chemical form of metals, especially Zn, Cu, and Hg, can produce persistent selective pressures for ARGs for a longer time than most antibiotics, which can be, in some cases, degraded (Kolpin et al. 2002; Hu et al. 2017).

In addition, the higher presence of heavy metals in soil amended with manure may accelerate the dissemination of AR by enabling co-selection with antibiotics, especially in the cases of zinc and copper (Marcato et al. 2009; Brandt et al. 2010).

In the nanoparticles era, we also have to consider the possible interaction between these molecules and antibiotics. Nano-antimicrobials based on metals have been demonstrated to be more effective than antibiotics against resistant bacteria (Dastjerdi and Montazer 2010) and against multidrug-resistant bacteria. For example, silver nanoparticles were found to be active against MRSA (Methicillin-Resistant Staphylococcus Aureus) and MRSE (Methicillin-resistant Staphylococcus epidermidis). On the other hand, the same nanoparticles can promote antibiotic resistance. For example, silver nanoparticles, widely used in industry, consumer products and medical appliances due to their efficient antimicrobial properties, enhance bacterial resistance to antibiotics by promoting stress tolerance through induction of intracellular ROS (Kaweeteerawat et al. 2017).

3.1.3 Role of Nutrients and Other Organic Contaminants

Some nutrients may influence the presence of ARGs. For example, commonly used nitrogen fertilisers could affect them in the soil, producing shifts in the relative abundance of microorganisms and the soil ARG content (Forsberg et al. 2014). In particular, in soils with a high concentration of nitrogen β-lactamases are depleted whereas membrane transporters are enriched. High N levels favour particular organisms (such as copiotrophs, which prefer nutritionally rich environments), causing shifts in bacterial abundances, which in turn probably affect resistome composition.

The abundance of ARGs is positively correlated to wastewater quality. For example, the propagation of *tet*M (a cytoplasmic ribosome protection protein that confers tetracycline resistance by binding to the ribosome and remove the drug from its binding site) may be influenced by COD (chemical oxygen demand, which is correlated to organic content) and chlorophyll a (*chl*a). The higher COD could produce energy for the ARG spread.

Regarding surface waters, it is well known that *Chl*a is an important water quality indicator, which can reflect the water pollution level caused by human activities. Higher *chl*a concentration means worse water quality and higher ARG presence (Zhang et al. 2017). COD and NH_4^+-N levels were found to be linked to the ribosomal protection protein TRG in both manure-treatment paths and wastewater samples (Wang et al. 2016). Moreover, total ARG copies were correlated with total nitrogen (TN), total phosphorus (TP) and total organic carbon (TOC) in an eco-agricultural system (Cheng et al. 2016). Furthermore, water quality (COD, TN, ammonia, nitrate, and phosphate) in livestock lagoons was found to be positively correlated with TRGs (McKinney et al. 2010). Notably, most target ARGs were positively correlated with TOC, TN, TP, and NH_4^+-N, indicating that co-attenuation of TOC and nutrients with these ARGs occurred. TOC and TN were also positively correlated with ARGs in urban rivers (Zhou et al. 2017). Finally, the abundance of ARGs can diminish as water quality improves.

The organic matter of sediments is a relevant factor influencing ARG distribution in lakes (Devarajan et al. 2015; Wu et al. 2016).

As regards environmental parameters, Tao et al. (2014) found that dissolved oxygen could be one of the most important factors affecting the profiles of ARGs. The pH and temperature are also environmental factors that influence the presence of ARGs (Ma et al. 2011; Tang et al. 2015). Conductivity, salinity, and dissolved oxygen concentration in rivers are positively correlated with the abundance of resistance genes (Makowska et al. 2016). In particular, the *tet*A and *sul*2 gene abundance in bulk water was positively correlated with dissolved oxygen and negatively correlated with chlorophyll a (Di Cesare et al. 2015).

Anthropogenic activities may hasten the antibiotic resistance emergence and accumulation in different areas (Xiang et al. 2018). Among different organic contaminants, pentachlorophenol, an organochlorine compound used as a pesticide (herbicide, insecticide, fungicide, algaecide, used also in anti-fouling paints) and as a disinfectant was found to be associated with ARGs (Sharma and Thakur 2009). Although its use has declined due to its high toxicity and persistence, it has many applications, for example in agricultural seeds (for non-food uses), leather, wood preservation, paper mills etc. It was found the homology in integron gene cassettes and multi-drug resistant genes indicates the sharing of a common portfolio (degradation of pentachlorophenol and to confer antimicrobial resistance) by the bacterial plasmid. Pulp and paper effluents, which contain at the same time metals, lignin and chlorophenols, have been found to contain bacteria with a high tolerance to antibiotics (Chandra and Sankhwar 2011).

Polycyclic aromatic hydrocarbons (PAHs), which are of anthropogenic source from coal pyrolysis, liquid fossil fuels and biomass combustion, could serve as a selective stress to enrich ARGs. In fact, PAHs were found to have the capability to enrich ARGs in soil (Chen et al. 2017; Gorovtsov et al. 2018) and in coastal waters (Wang et al. 2017). ARGs in PAH-contaminated soils were found about 15 times more abundant than those in less contaminated ones. In this case, the ARGs were mainly of the efflux pump-encoding types associated with aromatic antibiotics (e.g., fluoroquinolones). Most of the ARGs enriched in the PAH-contaminated soils were not carried by plasmids, indicating the low possibility of their being transferred among bacteria.

Triclosan and Triclocarban, commonly used antimicrobial agents that enter wastewater treatment plants and the environment, can impact on antibiotic resistance in the environment (Carey and McNamara 2015; Carey et al. 2016). Triclosan is found in a wide range of products, including personal care products (soap, deodorant, toothpaste, shower gel etc.). Triclocarban is a polychlorinated, binuclear, aromatic antimicrobial agent commonly used in bar soaps, detergents, cosmetics and other personal care products to prevent them from cultivating bacteria and spoiling. The presence of these chemicals in the environment is often associated with a multidrug resistance. Moreover, triclosan induces horizontal gene transfer of resistance plasmids at the same concentration range (Jutkina et al. 2018)

Chlorhexidine, another disinfectant and antiseptic compound used both for skin disinfection before surgery and to sterilize surgical instruments, was found in a lab experiment to increase significantly the frequency with which antibiotic resistance got transferred (Jutkina et al. 2018).

For ensuring the safety and effectiveness of over-the-counter health care antiseptics (means products that may be sold without a doctor's prescription), the US Food and Drug Administration (FDA) has recently declare that triclosan (and other 23 active ingredients) needs a pre-market review because there was a lack of sufficient safety and efficacy data. It means that additional information on these products are needed to be market (FDA 2017) and at the moment these products are banned.

Polychlorinated biphenyls (PCBs), a class of organochlorine persistent pollutants structurally very similar to Triclosan, can cause co-selection of antibiotic-resistant bacteria in the same way as was found for metals and PAHs. PCBs are thus likely to cause the development of resistance to antibiotics. Although there are currently no data on the enrichment of ARGs in PCB-degrading bacteria, the ability to degrade PCBs was found to be strongly associated with HGT mechanisms (Gorovtsov et al. 2018).

3.1.4 Disinfection of Waters and Wastewaters

Disinfection is a crucial step in human water use and reuse as it should ensure the full removal of microbial pathogens. Chemical and physical disinfection treatments are broadly applied to drinking water and wastewater all over the world, although a growing concern about the impact on human health of disinfection by-products has resulted in the progressive abandonment of some processes, and in some cases (e.g. in Northern Europe, or in Switzerland) in the complete removal of a disinfection step at the end of treatment and its substitution by nano- or ultra-filtration of the water and by in-line bio-essays. Nevertheless, a final disinfection stage is still mandatory in many countries for both drinking water and wastewater (e.g. USA, Italy).

The impact of disinfection on the fate of resistant bacteria is difficult to generalize, as different treatments have different performances results. In any case, a reduction in ARBs is always comparable to or lower than that of the remaining microbial community, calling for an intense aspecificity of the methods applied (Di Cesare et al. 2016a, b, c). The same is true for ARGs, where the potential reduction caused by bacterial cell destruction is generally balanced by a concomitant increment of extracellular DNA, including ARGs (Guo et al. 2018).

Chlorination (performed by chlorine gas or hypochlorite, either of calcium or sodium) is the method most commonly used for drinking water and wastewater disinfection worldwide. It is reasonably cheap and ensures a high level of abatement of bacterial numbers. Chlorine chemistry in water is a classical treatment well explained in several reviews (Snoeyink and Jenkins 1980; Nikolaou et al. 2007). On the other hand, its use promotes the formation of highly toxic by-products (e.g., trihalomethanes, haloacetic acids) in the water treated. This has been causing a constant reduction in its use and in many systems it has been substituted by other disinfectants. Towards determinants of antibiotic resistance, chlorination acts in an aspecific way, determining bacterial inactivation by (i) damaging cell surface components and (ii) directly impairing intracellular functions, with the disinfectant deactivating DNA through pyrimidine dimerization and consequently inhibiting cell growth (Cho et al. 2010). Chlorination can easily abate bacterial numbers by 90–99.9% but it has no specific impact on ARBs and ARGs: the first are targeted like any other microbe, while the latter are mostly transposed from intra- to extracellular DNA (Di Cesare et al. 2016b). The risk of re-growth and potential co-selection of resistances is generally reduced by the persistence of the toxic action of the disinfectant (up to 72 h, Di Cesare et al. 2016a).

Peracetic acid (PAA) disinfection, introduced as a more powerful variant of old oxygen peroxide disinfection, is nowadays used as a possible alternative to chlorination (Gehr et al. 2003; Antonelli et al. 2013; Formisano et al. 2016). PAA's disinfection mechanism is still not fully understood: in general, the inactivation of microorganisms is caused by the release of active oxygen, promoting the denaturation of proteins as for H_2O_2, but increasing cell wall permeability by disrupting sulphydryl and sulphur bonds (Gehr et al. 2003, Turolla et al. 2017). The performance of PAA in reducing determinants of antibiotic resistance is comparable to chlorination, and its effect is aspecific. The strong abatement of bacterial numbers and the short duration of the disinfectant can result in an unwanted selection in favour of ARBs and ARGs through co-selection mechanisms in the water treated (Di Cesare et al. 2016a, b, Fiorentino et al. 2019) because of reduced competition for nutrients and thus limited environmental filtering, especially in drinking water and in wastewater treated for agricultural reuse (Di Cesare et al. 2016c).

Recently oxidation (disinfection by ozone, coupling the disinfection effects of the molecular ozone and HO radicals (HO•) has become used significantly in water disinfection. Although expensive when compared to PAA and chlorine, ozone has demonstrated an impressive performance in abating contaminants of emerging concern, and therefore seems to be a disinfection technology fitting well within the parameters imposed by new legislation in the field of water treatment. Unfortunately, recent studies have also demonstrated the formation of dangerous oxidation by-products, including N-nitrosodimethylamine (NDMA) and bromate (Zimmermann et al. 2011; Sgroi et al. 2014), for ozone. The inactivation of bacterial cells by ozonation is obtained by damaging cell surfaces (Cho et al. 2010). Furthermore, a potential selection in favour of pathogens and of ARBs was detected in pilot and full scale systems (Alexander et al. 2016), raising a general concern about the use of oxone for the production of water for reuse. Further treatments (e.g. filtration) could reduce the risk of ARB regrowth and co-selection of ARGs, and for this reason microfiltration is often coupled with ozone disinfection in new treatment plants.

UV-C radiation is widely used for drinking water production and civil wastewater disinfection (Liberti et al. 2002; Munir et al. 2011). Unlike chemical disinfections, UV-C induces direct photochemical damage to intracellular DNA with limited cell wall damage (Cho et al. 2010). UV-C disinfection is based on the sensitivity of microorganisms to UV radiation and is limited by the contact time and the turbidity of the water treated (low for drinking water, higher for wastewater). Although rather expensive, UV disinfection has the advantage of damaging DNA, thus reducing the potential co-selection and HGT of fully functional genes, including ARGs. To date, there is a lack of studies specifically targeting the efficiency in ARGs reduction by UV treatments, but a study comparing UV and chemical disinfections demonstrated a generally good performance of the methods in terms of water reuse (Di Cesare et al. 2016c).

Finally, advanced oxidation processes (AOPs) are taking more and more space in research and in pilot systems. This disinfection methodology is based on the formation of reactive oxygen species (ROS), mainly hydroxyl radicals (HO•), that can remove several contaminants in wastewater (Rizzo 2011) and inactivate microorganisms (Dunlop et al. 2010; Fiorentino et al. 2015). AOPs include homogeneous (e.g., UV/H_2O_2, photo-Fenton ($UV/Fe/H_2O_2$), O_3, O_3/H_2O_2 etc.) and heterogeneous (solid semiconductors + light source, e.g., UV/TiO_2) photocatalytic processes (Klamerth et al. 2010). Bacterial inactivation by ROS is produced by the progressive damaging of cell membranes by HO• (Dalrymple et al. 2010) and by the diffusion of superoxide radicals with intracellular activity (Kikuchi et al. 1997). Their impact on ARBs and ARGs is again aspecific, but the combined effect of HO• and of superoxide radicals seems to prevent regrowth and co-selection mechanisms (Fiorentino et al. 2018). Although extremely promising, AOPs are still too experimental and expensive for large-scale application.

To conclude, a number of technologies are nowadays applied for the abatement of bacteria (and specifically pathogens) in drinking water and treated wastewater, but none of them specifically targets resistance determinants, by directly preventing selection and co-selection of resistance genes. Moreover, the new technologies applied miss this point, being designed for the removal of a number of different pollutants. The use of membranes, coupled to chemical and physical disinfection could help in abating resistances, but with a significant increase in treatment installation and management costs.

3.2 Conclusions

The presence of biologically-active pharmaceuticals such as antibiotics in the environment is of growing concern, because some of these substances have shown direct effects on wildlife at or below the concentrations found in water and soil. If global environmental pollution by antibiotic continues, the spread and the stabilization of resistances in the environment is expected to continue as well.

There is a need to analyse concentrations of antibiotics together with resistance genes in order to understand how much an environment is contaminated, since the presence and/or the maintenance of these genes is also due to factors other than the presence of antibiotics or their metabolites.

There is an urgent need to understand the distribution of resistance determinants in bacterial communities, clarify resistance mechanisms, and to determine the environmental factors promoting their dissemination. The holistic parameters of the evolution of antibiotic resistance need to be clarified, to show the emergence and the dynamics of ARB and ARGs in both human and environmental compartments.

Direct selection pressure imposed by antibiotics, indirect pressures by co-selective agents, and horizontal gene transfer are fundamental drivers of the evolution and the spread of antibiotic resistance. More detailed studies on co-selection mechanisms are needed to fully understand this process, and its relative importance for the maintenance of ARGs in natural microbiomes. Current research is limited to case studies and to experimental evidence, in most cases statistically significant evidence of co-occurrence of resistances in heavily polluted environments. Mechanistic studies, nowadays feasible thanks to advances in sequencing and in data analysis, should make it possible not only to understand the co-selection processes but also to conduct a detailed risk assessment involving different environments and plan consequent mitigation strategies.

Most attention has been focused on the emergence of resistant organisms in human medicine and in agriculture. Much less attention has been given to antibiotic contamination of the other environments.

The presence of different pollutants in the environment (metals such as Cd, Co, Hg, Cr, Ni, Cu, As, Pb and Zn; PAHs; pentachlorophenol, disinfectant products etc.), can be co-factors enhancing the presence of ARGs. Currently, only a few factors that can enhance the presence of ARGs have been studied and comprehensive evidence for the ability of these factors to co-select antibiotic-resistance in the environment is still largely lacking. For example, only a few studies to date have linked persistent organic pollutants, including PAHs and PCBs, to the selection of ARGs in the environment. The monitoring of ARG distribution and its association with other contaminants as co-selective agents is therefore necessary.

What is most important is to understand the fate and effects of antibiotics in the environment at different concentrations, and co-selection, to identify the physical locations and the genetic systems where the most rapid changes occur, and to identify potential mitigation strategies. However, in many areas we still lack essential information about the origins, dissemination and maintenance of ARGs. To understand these dynamics, an integration of data sets spanning different biological, temporal and spatial scales and linking these with abiotic and biotic environmental factors is needed.

Effective environmental monitoring tools should ideally capture not only ARGs, but also mobile genetic elements and indicators of co-selective forces, such as metal resistance genes or the presence of other factors that may increase or maintain the presence of ARGs.

Improving sanitation, increasing access to clean water, and ensuring good governance, as well as increasing public health-care expenditure and better regulation of the private health sector are all necessary to reduce global antimicrobial resistance. Moreover, improved wastewater treatment processes, reduction of manufacturing emissions, consideration of environmental impacts in procurement and drug approval decisions, and better manure management could be the first steps to reduce the problem.

Finally, a standardization of resistance assessment is necessary to monitor antibiotic resistance at an international level, compare the resistance prevailing in different geographical regions, assess possible relationships between antibiotic resistance and antibiotic consumption, and acquire a temporal perspective on resistance dynamics. In particular, only a minority of environmental studies simultaneously quantify antibiotic residues together with other co-factors and associated ARGs to control for temporal and spatial variability. There is also a need to make uniform the metagenomic data presented in literature. In fact, it is common that there is any possibility to check/validate/use the metagenomic data of published articles either because they are totally missing information on the repository used (if any) to deposit them, or because the accession numbers provided are wrong, or because when correct they are simply related to metagenomes without any possibility to understand what is what. Considering that in the field of AMR studies on metagenomics of the 16S but even more studies on the metagenomics of the full resistome are essential, make metagenomic data public before publication would be in general requested and mandatory.

References

Alanis AJ (2005) Resistance to antibiotics: are we in the post-antibiotic era? Arch Med Res 36:697–705

Alexander J, Knopp G, Dötsch A, Wieland A, Schwartz T (2016) Ozone treatment of conditioned wastewater selects antibiotic resistance genes, opportunistic bacteria, and induce strong population shifts. Sci Tot Environ 559:103–112

Alvarenga P, Mourinha C, Farto M, Santos T, Palma P, Sengo J, Morais MC, Cunha-Queda C (2015) Sewage sludge, compost and other representative organic wastes as agricultural soil amendments: benefits versus limiting factors. Waste Manag 40:44–52

Aminov RI (2010) A brief history of the antibiotic era: lessons learned and challenges for the future. Front Microbiol 1:134

Andersson DI, Hughes D (2010) Antibiotics resistance and its cost: is it possible to reverse resistance? Nat Rev Microbiol 8(4):260–271

Antonelli M, Turolla A, Mezzanotte V, Nurizzo C (2013) Peracetic acid for secondary effluent disinfection: a comprehensive performance assessment. Water Sci Technol 68(12):2638–2644

Baker-Austin C, Wright MS, Stepanauskas R, McArthur JV (2006) Co-selection of antibiotic and metal resistance. Trends Microbiol 14:176–182

Ben Y, Fu C, Hu M, Liu L, Wong MH, Zheng C (2019) Human health risk assessment of antibiotic resistance associated with antibiotic residues in the environment: a review. Environ Res 169:483–493

Berendonk TU, Manaia CM, Merlin C, Fatta-Kassinos D, Cytryn E, Walsh F, Bürgmann H, Sørum H, Norström M, Pons MN, Kreuzinger N, Huovinen P, Stefani S, Schwartz T, Kisand V, Baquero F, Martinez JL (2015) Tackling antibiotic resistance: the environmental framework. Nat Rev Microbiol 13(5):310–317

Berg J, Tom-Petersen A, Nybroe O (2005) Copper amendment of agricultural soil selects for bacterial antibiotics resistance in the field. Lett Appl Microbiol 40(2):146–151

Berg J, Brandt KK, Al-Soud WA, Holm PE, Hansen LH, Sørensen SJ, Nybroe O (2012) Selection for cu-tolerant bacterial communities with altered composition, but unaltered richness, via long-term cu exposure. Appl Environ Microbiol 78:7438–7446

Blanco P, Hernando-Amado S, Reales-Calderon JA, Corona F, Lira F, Alcalde-Rico M, Bernardini A, Sanchez MB, Martinez JL (2016) Bacterial multidrug efflux pumps: much more than antibiotic resistance determinants. Microorganisms 4(1):pii E14

Brandt KK, Frandsen RJN, Holm PE, Nybroe O (2010) Development of pollution-induced community tolerance is linked to structural and functional resilience of a soil bacterial community following a five-year field exposure to copper. Soil Biol Biochem 42(5):748–757

Burridge LE, Weis JS, Cabello F, Pizarro JC, Bostick K (2010) Chemical use in salmon aquaculture: a review of current practices and possible environmental effects. Aquaculture 306(1–4):7–23

Carey DE, McNamara PJ (2015) The impact of triclosan on the spread of antibiotic resistance in the environment. Front Microbiol 5:780

Carey DE, Zitomer DH, Hristova KR, Kappell AD, McNamara PJ (2016) Triclocarban influences antibiotic resistance and alters anaerobic digester microbial community structure. Environ Sci Technol 50(1):126–134

Chandra R, Sankhwar M (2011) Influence of lignin, pentachlorophenol and heavy metal on antibiotic resistance of pathogenic bacteria isolated from pulp paper mill effluent contaminated river water. J Environ Biol 32(6):739–745

Chen B, He R, Yuan K, Chen E, Lin L, Chen X, Sha S, Zhong J, Lin L, Yang L, Yang Y, Wang X, Zou S, Luan T (2017) Polycyclic aromatic hydrocarbons (PAHs) enriching antibiotic resistance genes (ARGs) in the soils. Environ Pollut 220(Pt B):1005–1013

Cheng WX, Li JN, Wu Y, Xu LK, Su C, Qian YY, Zhu YG, Chen H (2016) Behavior of antibiotics and antibiotics resistance genes in eco-agricultural system: a case study. J Hazard Mater 304:18–25

Cho M, Kim J, Kim JY, Yoon J, Kim J-H (2010) Mechanisms of *Escherichia coli* inactivation by several disinfectants. Water Res 44:3410–3418

Cullen WR, Reimer KJ (1989) Arsenic speciation in the environment. Chem Rev 89:713–764

Curiao T, Marchi E, Viti C, Oggioni MR, Baquero F, Martinez JL, Coque TM (2015) Polymorphic variation in susceptibility and metabolism of triclosan-resistant mutants of Escherichia coli and *Klebsiella pneumoniae* clinical strains obtained after exposure to biocides and antibiotics. Antimicrobiol Agents Chem 59:3413–3423

Czekalski N, Gascón Díez E, Bürgmann H (2014) Wastewater as a point source of antibiotic-resistance genes in the sediment of a freshwater lake. ISME J 8(7):1381–1390

D'Costa VM, King CE, Kalan L, Morar M, Sung WW, Schwarz C, Froese D, Zazula G, Calmels F, Debruyne R, Golding GB, Poinar HN, Wright GD (2011) Antibiotic resistance is ancient. Nature 477(7365):457–461

Dalrymple OK, Stefanakos E, Trotz MA, Goswami DY (2010) A review of the mechanisms and modeling of photocatalytic disinfection. Appl Cat B 98:27–38

Dastjerdi R, Montazer M (2010) A review on the application of inorganic nano-structured materials in the modification of textiles: focus on antimicrobial properties. Colloids Surf B Biointerfaces 79:5–18

Devarajan N, Laffite A, Graham ND, Meijer M, Prabakar K, Mubedi JI, Elongo V, Mpiana PT, Ibelings BW, Wildi W, Poté J (2015) Accumulation of clinically relevant antibiotic-resistance genes, bacterial load, and metals in freshwater lake sediments in central europe. Environ Sci Technol 49(11):6528–6537

Di Cesare A, Eckert EM, Teruggi A, Fontaneto D, Bertoni R, Callieri C, Corno G (2015) Constitutive presence of antibiotic resistance genes within the bacterial community of a large subalpine lake. Mol Ecol 24(15):3888–3900

Di Cesare A, Fontaneto D, Doppelbauer J, Corno G (2016a) Fitness and recovery of bacterial communities and antibiotic resistance genes in urban wastewaters exposed to classical disinfection treatments. Environ Sci Technol 50(18):10153–10161

Di Cesare A, Eckert EM, D'Urso S, Bertoni R, Gillan DC, Wattiez R, Corno G (2016b) Co-occurrence of integrase 1, antibiotic and heavy metal resistance genes in municipal wastewater treatment plants. Water Res 94:208–214

Di Cesare A, Eckert EM, Corno G (2016c) Co-selection of antibiotic and heavy metal resistance in freshwater bacteria. J Limnol 75(s2):59–66

Di Cesare A, Eckert EM, Rogora M, Corno G (2017) Rainfall increases the abundance of antibiotic resistance genes within a riverine microbial community. Environ Pollut 226:473–478

Dunlop PSM, Sheeran CP, Byrne JA, McMahon MAS, Boyle MA, McGuigan KG (2010) Inactivation of clinically relevant pathogens by photocatalytic coatings. J Photochem Photobiol 216:303–310

Eckert EM, Di Cesare A, Coci M, Corno G (2018) Persistence of antibiotic resistance genes in large subalpine lakes: the role of anthropogenic pollution and ecological interactions. Hydrobiologia 824(1):93–108

Fard RM, Heuzenroeder MW, Barton MD (2011) Antimicrobial and heavy metal resistance in commensal enterococci isolated from pigs. Vet Microbiol 148:276–282

FDA, Food and Drug Administration (2017) FDA in brief: FDA issues final rule on safety and effectiveness for certain active ingredients in over-the-counter health care antiseptic hand washes and rubs in the medical setting. https://www.fda.gov/newsevents/newsroom/fdainbrief/ucm589474.htm

Fiorentino A, Ferro G, Alferez MC, Polo-Lopez MI, Fernandez-Ibanez P, Rizzo L (2015) Inactivation and regrowth of multidrug resistant bacteria in urban wastewater after disinfection by solar-driven and chlorination processes. J Photochem Photobiol 148:43–50

Fiorentino A, Cucciniello R, Di Cesare A, Fontaneto D, Prete P, Rizzo L, Corno G, Proto A (2018) Disinfection of urban wastewater by a new photo-Fenton like process using cu-iminodisuccinic acid complex as catalyst at neutral pH. Water Res 146:206–215

Fiorentino A, Di Cesare A, Eckert EM, Rizzo L, Fontaneto D, Yang Y, Corno G (2019) Impact of industrial wastewater on the dynamics of antibiotic resistance genes in a full-scale urban wastewater treatment plant. Sci Tot Environ 646:1204–1210

Flemming CA, Trevors JT (1989) Copper toxicity and chemistry in the environment: a review. Water Air Soil Pollut 44:143–158

Formisano F, Fiorentino A, Rizzo L, Carotenuto M, Pucci L, Giugni M, Lofrano G (2016) Inactivation of *Escherichia coli* and enterococci in urban wastewater by sunlight/PAA and sunlight/H_2O_2 processes. Process Saf Environ Prot 104(Part A):178–184

Forsberg KJ, Patel S, Gibson MK, Lauber CL, Knight R, Fierer N, Dantas G (2014) Bacterial phylogeny structures soil resistomes across habitats. Nature 509:612–616

Gehr R, Wagner M, Veerasubramanian P, Payment P (2003) Disinfection efficiency of peracetic acid, UV and ozone after enhanced primary treatment of municipal wastewater. Water Res 37:4573–4586

Gillings MR, Gaze WH, Pruden A, Smalla K, Tiedje JM, Zhu YG (2015) Using the class 1 integron-integrase gene as a proxy for anthropogenic pollution. ISME J 9:1269–1279

Gorovtsov AV, Sazykin IS, Sazykina MA (2018) The influence of heavy metals, polyaromatic hydrocarbons, and polychlorinated biphenyls pollution on the development of antibiotic resistance in soils. Environ Sci Pollut Res Int 25(10):9283–9292

Grenni P, Ancona V, Barra Caracciolo A (2018) Ecological effects of antibiotics on natural ecosystems: a review. Microchem J 136C:25–39
Guo X, Yang Y, Lu D, Niu Z, Feng J, Chen Y, Tou F, Garner E, Xu J, Liu M, Hochella MF (2018) Biofilms as a sink for antibiotic resistance genes (ARGs) in the Yangtze Estuary. Water Res 129:277–286
Han F, Kingery W, Selim H (2001) Accumulation, redistribution, transport and bioavailability of heavy metals in waste amended soils. In: Iskander I, Kirkham M (eds) Trace elements in soil: bioavailability, fluxes and transfer. CRC, Boca Raton, pp 141–168
Han FX, Banin A, Su Y, Monts DL, Plodinec JM, Kingery WL et al (2002) Industrial age anthropogenic inputs of heavy metals into the pedosphere. Naturwissenschaften 89:497–504
Hu HW, Wang JT, Li J, Shi XZ, Ma YB, Chen D, He JZ (2016) Field-based evidence for copper contamination induced changes of antibiotics resistance in agricultural soils. Environ Microbiol 18(11):3896–3909
Hu HW, Wang JT, Li J, Shi XZ, Ma YB, Chen DL, He JZ (2017) Long-term nickel contamination increases the occurrence of antibiotics resistance genes in agricultural soils. Environ Sci Technol 51(2):790–800
Imfeld G, Bringel F, Vuilleumier S (2011) Bacterial tolerance in contaminated soils: potential of the PICT approach in microbial ecology. In: Amiard-Triquet C, Rainbow PS, Roméo M (eds) Tolerance to environmental contaminants. CRC Press, Boca Raton, Chapter 14
Ji X, Shen Q, Liu F, Ma J, Xu G, Wang Y, Wu M (2012) Antibiotic resistance gene abundances associated with antibiotics and heavy metals in animal manures and agricultural soils adjacent to feedlots in Shanghai; China. J Hazard Mater 235–236:178–185
Johnson TA, Stedtfeld RD, Wang Q, Cole JR, Hashsham SA, Looft T et al (2016) Clusters of antibiotic resistance genes enriched together stay together in swine agriculture. MBio 7:e02214–e02215
Ju F, Beck K, Yin X, Maccagnan A, McArdell CS, Singer HP, Johnson DR, Zhang T, Bürgmann H (2018) Wastewater treatment plant resistomes are shaped by bacterial composition, genetic exchange, and upregulated expression in the effluent microbiomes. ISME J 13:346–360. https://doi.org/10.1038/s41396-018-0277-8
Jutkina J, Marathe NP, Flach CF, Larsson DGJ (2018) Antibiotics and common antibacterial biocides stimulate horizontal transfer of resistance at low concentrations. Sci Total Environ 616(617):172–178
Kaweeteerawat C, Na Ubol P, Sangmuang S, Aueviriyavit S, Maniratanachote R (2017) Mechanisms of antibiotic resistance in bacteria mediated by silver nanoparticles. J Toxicol Environ Health A 80(23–24):1276–1289
Kikuchi Y, Sunada K, Iyada T, Hashimoto K, Fujishima A (1997) Photocatalytic bactericidal effect of TiO2 thin films: dynamic view of the active oxygen species responsible for the effect. J Photochem Photobiol A 160:51–56
Klamerth N, Rizzo L, Malato S, Maldonado MI, Aguera A, Fernandez-Alba AR (2010) Degradation of fifteen emerging contaminants at μg L^{-1} initial concentrations by mild solar photo-Fenton in MWTP effluents. Water Res 44:545–554
Kolpin DW, Furlong ET, Meyer MT, Thurman EM, Zaugg SD, Barber LB, Buxton HT (2002) Pharmaceuticals, hormones, and other organic wastewater contaminants in US streams, 1999-2000: a national reconnaissance. Environ Sci Technol 36(6):1202–1211
Li LG, Xia Y, Zhang T (2017) Co-occurrence of antibiotic and metal resistance genes revealed in complete genome collection. ISME J 11(3):651–662
Liberti L, Notarnicola M, Petruzzelli D (2002) Advanced treatment for municipal wastewater reuse in agriculture. UV disinfection: parasite removal and by-product formation. Desalination 152:315–324
Lemire JA, Harrison JJ, Turner RJ (2013) Antimicrobial activity of metals: mechanisms, molecular targets and applications. Nat Rev Microbiol 11:371–384

Ma Y, Wilson CA, Novak JT, Riffat R, Aynur S, Murthy S, Pruden A (2011) Effect of various sludge digestion conditions on sulfonamide, macrolide, and tetracycline resistance genes and class I integrons. Environ Sci Technol 45(18):7855–7861

Manaia CM (2017) Assessing the risk of antibiotic resistance transmission from the environment to humans: non-direct proportionality between abundance and risk. Trends Microbiol 25(3):173–181

Makowska N, Koczura R, Mokracka J (2016) Class 1 integrase, sulfonamide and tetracycline resistance genes in wastewater treatment plant and surface water. Chemosphere 144:1665–1673

Marcato CE, Pinelli E, Cecchi M, Winterton P, Guiresse M (2009) Bioavailability of Cu and Zn in raw and anaerobically digested pig slurry. Ecotoxicol Environ Saf 72(5):1538–1544

Martinez JL (2014) General principles of antibiotic resistance in bacteria. Drug Discov Today Technol 11:33–39

McKinney CW, Loftin KA, Meyer MT, Davis JG, Pruden A (2010) Tet and sul antibiotic resistance genes in livestock lagoons of various operation type, configuration, and antibiotic occurrence. Environ Sci Technol 44(16):6102–6109

McLaughlin MJ, Smolders MJ (2001) Background zinc concentrations in soil affect the zinc sensitivity of soil microbial processes—a rationale for a metalloregion approach to risk assessments. Environ Toxicol Chem Nov 20(11):2639–2643

Melnyk A, Wong A, Kassen R (2015) The fitness costs of antibiotic resistance mutations. Evol Appl 8:273–283

Monteiro SC, Boxall ABA (2010) Occurrence and fate of human pharmaceuticals in the environment. Rev Environ Contam Toxicol 202:153–154

Munir M, Wong K, Xagorarak I (2011) Release of antibiotic resistant bacteria and genes in the effluent and biosolids of five wastewater utilities in Michigan. Water Res 45:681–693

Murray AK, Zhang L, Yin X, Zhang T, Buckling A, Snape J, Gaze WH (2018) Novel insights into selection for antibiotic resistance in complex microbial communities. MBio 9(4):e00969–e00918

Nemecek T, Dubois D, Huguenin-Elie O, Gaillard G (2011) Life cycle assessment of Swiss farming systems: I. Integr Org Farm Agric Syst 104:217–232

Nikolaou A, Rizzo L, Selcuk H (2007) Control of disinfection by-products in drinking water systems. Nova Science Publishers, Hauppauge

Pal C, Bengtsson-Palme J, Kristiansson E, Larsson DG (2015) Co-occurrence of resistance genes to antibiotics, biocides and metals reveals novel insights into their co-selection potential. BMC Genomics16:964

Pal C, Bengtsson-Palme J, Rensing C, Kristiansson E, Larsson DGJ (2014) BacMet: antibacterial biocide and metal resistance genes database. Nucleic Acids Res 42:D737–D743

Pal C, Asiani K, Arya S, Rensing C, Stekel DJ, Larsson DGJ, Hobman JL (2017) Metal resistance and its association with antibiotic resistance. Adv Microb Physiol 70:261–313

Prabhakaran P, Ashraf MA, Aqma WS (2016) Microbial stress response to heavy metals in the environment. RSC Adv 6:109862–1098772

Perry JA, Wright GD (2013) The antibiotic resistance "mobilome": searching for the link between environment and clinic. Front Microbiol 4:138

Rizzo L (2011) Bioassays as a tool for evaluating advanced oxidation processes in water and wastewater treatment. Water Res 45:4311–4340

Roberts TL (2014) Cadmium and phosphorus fertilizers: the issues and the science. 2nd international symposium on innovation and Technology in the Phosphate Industry. Procedia Eng 83:52–59

Salyers AA, Amábile-Cuevas CF (1997) Why are antibiotic resistance genes so resistant to elimination? Antimicrob Agents Chemother 41:2321–2325

Sanderson H, Fricker C, Brown RS, Majury A, Liss SN (2016) Antibiotics resistance genes as an emerging environmental contaminant. Environ Rev 24:205–218

Raja SE, Praveen KR, Baskar G (2018) Microbial transformation of heavy metals. In: Varjani S, Gnansounou E, Gurunathan B, Pant D, Zakaria Z (eds) Waste bioremediation. Energy, environment, and sustainability. Springer, Singapore, pp 249–263

Seiler C, Berendonk TU (2012) Heavy metal driven co-selection of antibiotic resistance in soil and water bodies impacted by agriculture and aquaculture. Front Microbiol 3:399

Sgroi M, Roccaro P, Oelker GL, Snyder SA (2014) N-nitrosodimethylamine formation upon ozonation and identification of precursors source in a municipal wastewater treatment plant. Environ Sci Technol 48:10308–10315

Sharma A, Thakur IS (2009) Identification and characterization of integron mediated antibiotic resistance in pentachlorophenol degrading bacterium isolated from the chemostat. J Environ Sci 21(6):858–864

Silver S, Phung LT (1996) Bacterial heavy metal resistance: new surprises. Annu Rev Microbiol 50:753–789

Smedley PL, Kinniburgh DG (2002) A review of the source, behaviour and distribution of arsenic in natural waters. Appl Geochem 17:517–568

Snoeyink VL, Jenkins D (1980) Water chemistry. Wiley, New York

Tang X, Lou C, Wang S, Lu Y, Liu M, Hashmi MZ, Liang X, Li Z, Liao Y, Qin W, Fan F, Xu J, Brookes PC (2015) Effects of long-term manure applications on the occurrence of antibiotics and antibiotics resistance genes (ARGs) in paddy soils: evidence from four field experiments in south of China. Soil Biol Biochem 90:179–187

Tao CW, Hsu BM, Ji WT, Hsu TK, Kao PM, Hsu CP, Shen SM, Shen TY, Wan TJ, Huang YL (2014) Evaluation of five antibiotics resistance genes in wastewater treatment systems of swine farms by real-time PCR. Sci Total Environ 496:116–121

Turolla A, Sabatino R, Fontaneto D, Eckert EM, Colinas N, Corno G, Citterio B, Biavasco F, Antonelli M, Mauro A, Mangiaterra G, Di Cesare A (2017) Defence strategies and antibiotic resistance gene abundance in enterococci under stress by exposure to low doses of peracetic acid. Chemosphere 185:480–488

von Wintersdorff CJH, Penders J, vanNiekerk JM, Mills ND, Majumder S, van Alphen LB, Savelkoul PHM, Wolffs PFG (2016) Dissemination of antimicrobial resistance in microbial ecosystems through horizontal gene transfer. Front Microbiol 7:173

Zaman SB, Hussain MA, Nye R, Mehta V, Mamun KT, Hossain N (2017) A review on antibiotic resistance: alarm bells are ringing. Cureus 9(6):e1403

Zhang K, Niu ZG, Lv Z, Zhang Y (2017) Occurrence and distribution of antibiotic resistance genes in water supply reservoirs in Jingjinji area, China. Ecotoxicology 26(9):1284–1292

Wales AD, Davies RH (2015) Co-selection of resistance to antibiotics, biocides and heavy metals, and its relevance to foodborne pathogens. Antibiotics (Basel) 4:567–604

Wang J, Ben W, Yang M, Zhang Y, Qiang Z (2016) Dissemination of veterinary antibiotics and corresponding resistance genes from a concentrated swine feedlot along the waste treatment paths. Environ Int 92-93:317–323

Wang J, Wang J, Zhao Z, Chen J, Lu H, Liu G, Zhou J, Guan X (2017) PAHs accelerate the propagation of antibiotic resistance genes in coastal water microbial community. Environ Pollut 231(Pt 1):1145–1152

Wu D, Huang ZT, Yang K, Graham D, Xie B (2015) Relationships between antibiotics and antibiotics resistance gene levels in municipal solid waste leachates in Shanghai, China. Environ Sci Technol 49(7):4122–4128

Wu D, Lu R, Chen Y, Qiu J, Deng C, Tan Q (2016) Study of cross-resistance mediated by antibiotics, chlorhexidine and Rhizoma coptidis in *Staphylococcus aureus*. J Glob Antimicrob Resist 7:61–66

Xiang Q, Chen QL, Zhu D, An XL, Yang XR, Su JQ, Qiao M, Zhu YG (2018) Spatial and temporal distribution of antibiotic resistomes in a peri-urban area is associated significantly with anthropogenic activities. Environ Pollut 235:525–533

Yamaji R, Friedman CR, Rubin J, Suh J, Thys E, McDermott P, Hung-Fan M, Riley LW (2018) A population-based surveillance study of shared genotypes of *Escherichia coli* isolates from retail meat and suspected cases of urinary tract infections. mSphere 3(4):pii: e00179-18

Zhao Y, Cocerva T, Cox S, Tardif S, Su JQ, Zhu YG, Brandt KK (2019) Evidence for co-selection of antibiotic resistance genes and mobile genetic elements in metal polluted urban soils. Sci Total Environ 656:512–520

Zhou ZC, Zheng J, Wei YY, Chen T, Dahlgren RA, Shang X et al (2017) Antibiotic resistance genes in an urban river as impacted by bacterial community and physicochemical parameters. Environ Sci Pollut Res Int 24:23753–22376

Zimmermann SG, Wittenwiler M, Hollender J, Krauss M, Ort C, Siegrist H, von Gunten U (2011) Kinetic assessment and modeling of an ozonation step for full-scale municipal wastewater treatment: micropollutant oxidation, by-product formation and disinfection. Water Res 45:605–617

Microbial Resistance to Antibiotics

4

Martha Premlatha

Abstract

Development of Antibiotic Resistance in bacteria has become a major public health concern. Bacteria from Clinical and Non Clinical settings have become resistant to several antibiotics. Self-medication, Excessive use of antibiotics and Incomplete antibiotic treatment is the Primary cause of resistance to antibiotics. Studies have shown that the Prophylactic use of "non priority" antimicrobials in Animal husbandry has lead to transmission of resistant bacteria Via Food chain and Environment affecting the human health.

Development of Intrinsic antibiotic resistance to bacteria is mediated by various mechanisms like Minimization of Intracellular concentration of antibiotic due to poor penetration into bacterium or Antibiotic Efflux, Modification of antibiotic target by gene mutation and Inactivation of antibiotic by hydrolysis. Bacteria can also acquire resistance to antibiotics via mutation in Chromosomal genes and by horizontal transfer.

It is estimated that 700,000 deaths occur annually due to AMR Worldwide affecting every country. Keeping in view of this current status of AMR, there is need for the development of Genetically engineered phages to deliver antimicrobial agents to bacteria; Synthesis of Metal Nanoparticles which affect the bacterial cell and their metabolic pathways; Production of Antimicrobial Peptides as they are Host Defense Effector molecules in the living organism and Synthesis of New antibiotics and use of Prebiotics and Quroum sensing Inhibitors as an Alternative Approach to Combat Antibiotic resistance or to treat Pathogenic bacterial Infection. Preventive measures have been under taken at National level and International level.

M. Premlatha (✉)
Amity Institute of Virology, AUUP, Noida, India
e-mail: premlathanewdelhi@gmail.com

S. M. Mandal, D. Paul (eds.), *Bacterial Adaptation to Co-resistance*,
https://doi.org/10.1007/978-981-13-8503-2_4

Keywords
Antimicrobials · Alternative approaches to antibiotic therapy · Antibacterial Resistance

4.1 Introduction

Microbial Resistance is the ability of the microbe to withstand against the antibiotics or biocides that kill or control them. Many antimicrobials have been developed to destroy the disease producing microbes. The most commonly used antimicrobials are antibiotics which target bacteria. Antibiotics are manufactured at an estimated scale of about 100,000 tons annually worldwide. Emergence of Resistance among the important bacterial pathogens like E. coli, K. pneumoniae, S. aureus, S. pneumoniae, Salmonella spp., Shigella spp., C. difficile, M. tuberculosis has become a major public health concern.

Antibiotic resistance is a consequence of evolution via Natural Selection. As a result, the medicines become ineffective and infections persist in the body, increasing the risk of Infection. Without effective antimicrobials for treatment of infections, medical procedures such as organ transplantation, Cancer chemotherapy, Diabetes Management and Surgery are at high Risk. Antimicrobial resistance increases the cost of Healthcare with prolonged duration in the Hospital.

Since the 1940s, these drugs have had an impact on the reduction of infectious diseases. However, these drugs have been widely used for so long that the infectious organisms to which the antibiotics are designed to kill have adapted to them, making the drugs ineffective (Table 4.1). Each year in the United States, at least two million people become infected with bacteria that are resistant to antibiotics and at least 23,000 people die each year as a direct result of these infections. In India, about 57,000 neonatal sepsis deaths occur annually due to antibiotic resistance. According to recent studies, Antibiotic resistance is estimated to cause around 300 million premature deaths by 2050, with a loss of up to $100 trillion (£64 trillion) to the global economy.

4.2 Dissemination of Antibiotic Resistance in Clinical and Non Clinical Settings

Antibiotic resistant bacteria are present in Agricultural Soil which may spread into the food chain. Antibiotic use in foo animals can lead to resistant infections in humans. These Bacteria cause some of the most common occurring infections in community, in hospitals or can be transmitted through Food chain (Linton 1997 and Doyle 2006) which is depicted in (Fig. 4.1). 'Resistome' is a broad term that describes the presence of all the antibiotic resistance genes found in free living organisms in the Environment or Commensals. Epidemiological studies have demonstrated that there is direct relationship between the Antibiotic Consumption and the emergence or dessimination of the resistant bacterial strains.

Table 4.1 The Timeline of Antibiotic Development and the Evolution of Resistance

1928	Discovery of Penicillin by Alexander Fleming from *Penicillium notatum*
1942	Mass Production of Penicillin and its clinical use to treat Allied troops in Europe during World War II
1943	Isolation of Streptomycin from *S. griseus* by Waksman and its use in the treatment of Tuberculosis.
1945	Production of Tetracycline from actinomycetes soil bacteria and used as broad spectrum antibiotic
1947	Infections caused by Penicillin resistant *S. aureus*
1949	Waksman isolated a second Amino glycoside, neomycin from *Streptomyces fradiae* and approved for human use.
1952	Erythromycin was Isolated from bacteria *Saccharopolyspora erythraea* and launched commercially
1955	Erythromycin resistant to staphylococci was observed
1956	Vancomycin the first glycopeptide is approved for human use
1957	Rifamycin group of antibiotics was discovered. Rifampicin was isolated from *Amycolaptopsis rifamycinica*
1961	Methicillin Resistant *S. aureus* (MRSA) was first observed
1962	Discovery of Quinolones and Streptogramins
1967	Resistance to Penicillin observed in *N. gonorrhoeae* and *S. pneumoniae*
1970	First generation Cephalosporin e.g. Cefalexin was produced
1977	Penicillin globally recognized as ineffective in the treatment of *S. pneumoniae* infections
1982	Resistance of MRSA to cephalosporins was observed
1983	Resistance to penicillin observed in Enterococcus faecium
1986	Norfloxacin is developed as first fluroquinone from Nalidixic acid
1987	Intrinsic Resistance to Vancomycin observed in enterococci
1989	Development of Linezolid as a novel oxazolidinone
1990	MDR *P. aeruginosa* and *M. tuberculosis* observed in hospital infection globally, 50% of MRSA strains resistant to fluoroquinolones
1992	More than 95% of all MRSA strains worldwide are resistant to Penicillin, Ampicillin
1993	Resistance to Carbapenems observed in *K. pneumoniae*
1997	Resistance to Vancomycin observed in *S. aureus*
1999	Resistance to Linezolid class of antibiotics to enterococci
2000	Linezolid was approved for human use
2000	XDR extensively drug resistant tuberculosis
2001	Linezolid resistance in *S. aureus* was reported
2003	Daptomycin was approved for human use
2004	Pan-drug resistant *Acinetobacter* and *Pseudomonas*
2007	Chromosomal resistance rendered a class of drugs fluoroquinolones, ineffective in *N. gonorrhoeae* infections
2008	NDM-I was first detected in a *Klebsiella pneumoniae* isolate of a Swedish patient of Indian origin in 2008 which is resistant to Carbapenems through HGT
2010	Ceftaroline was produced
2011	Ceftaroline resistant was observed in staphylococcal infections
2012	WHO reported 170,000 people died from resistant tuberculosis infection
2013	Mutation in the ribosomal protein S5 is responsible for the resistance of *Neisseria gonorrhoeae* strains. Decreased susceptibility to spectinomycin, cefixime and ceftriaxone was observed
2014	WHO has identified super bug out breaks like *Klebsiella pneumonia* and gonorrhea strains all over the world
2017	Patient dies from a *Klebsiella pneumoniae* strain resistant to every available antibiotic in the US
2018	Global action plan on Antibiotic resistance

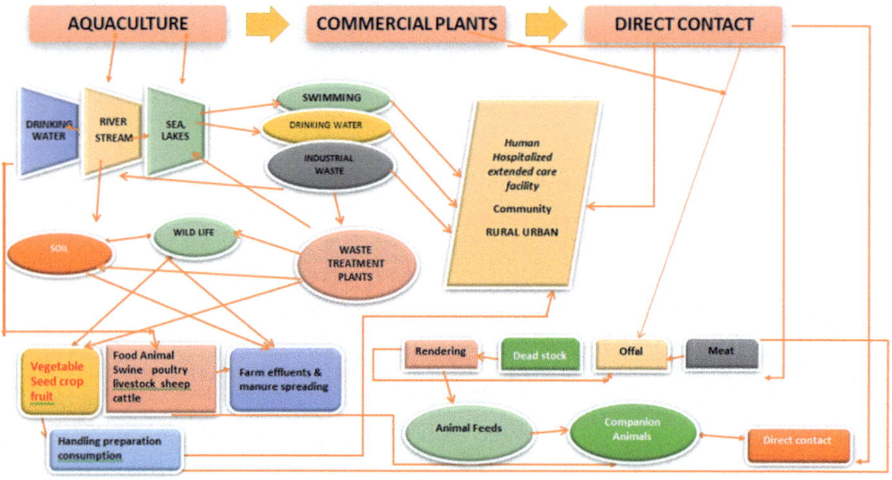

Fig. 4.1 Dissemination of antibiotic resistance in clinical and non clinical settings. (Source: Modified from Microbiol Mol. Biol. Rev. 2010 Sept;74(3):417–433.doi.10.1128/MMBR)

4.2.1 Definition of Antibiotic Resistance

The Concentration of drug at the site of Infection must inhibit the organism and also should not be toxic to the human cells. AR is defined as microorganisms that are not inhibited by usually achievable systematic concentration of antimicrobial agent with normal dosage schedule and or fall in the minimum inhibitory Concentration Range.

$$\text{Antibiotic Resistance (DR)} = \text{MIC/MCC} > \text{Toxic Plasma Concentration}$$

4.3 Causes of Antibiotic Resistance

4.3.1 Overuse and Inappropriate Use of Antibiotics for Viral Infections

In many countries, antibiotics are available over the counter without a prescription. This lack of regulation results in easy access to more number of cheap drugs, which promotes overuse. Procurement of these drugs on online has also made them easily accessible. Common cold, Sore throats, Acute Bronchitis, Ear infections are mostly Viral which can't be cured by antibiotics.

4.3.2 Indiscriminate use of Antibiotics

Incorrectly prescribed antibiotics also contribute to the promotion of resistant bacteria. One U.S. study by Bartlett et al. (2013) reported that a pathogen was defined in only 7.6% of 17,435 patients hospitalized with community-acquired pneumonia (CAP). In contrast, investigators at the Karolinska Institute in Sweden were able to identify the probable pathogen in 89% of patients with CAP using polymerase chain reaction [PCR] and semi-quantitative PCR, where the diagnostic tests also have a role in the use of antibiotics.

Sub-inhibitory and sub-therapeutic antibiotic concentrations can promote the development of antibiotic resistance by alterations in gene expression. Changes in antibiotic-induced gene expression can increase virulence, while increased mutagenesis and HGT promote antibiotic resistance It was observed that Low levels of antibiotics have been shown to contribute to strain diversification e.g. *Pseudomonas aeruginosa*. Sub-inhibitory concentrations of piperacillin and/or tazobactam have also been shown to induce broad proteomic alterations in *Bacteroides fragilis* (Viswanathan 2014).

Global data base for the period from 2000 to 2017 on the frequency of Non-prescription sale and supply of antibiotics in community Pharmacies has been reported by Pourmand et al. (2017). South America has the highest incidence of Non-Prescription supply of Antibiotics.

4.3.3 Farming

Humans ingest the antibiotics used in livestock through food. The transfer of resistant bacteria to humans by farm animals was observed by Bartlett et al. (2013) where high rates of antibiotic resistance were found in the intestinal flora of both farm animals and farmers.

The agricultural use of antibiotics also affects the environmental microbiome. Up to 90% of the antibiotics given to livestock are excreted in urine and stool, which are widely dispersed through fertilizer, groundwater, and surface runoff.

4.3.4 Availability of Few New Antibiotics

Antibiotic development is no longer profitable for the pharmaceutical industry. Because antibiotics are used for relatively short periods and are often curative.

Many new antibiotics fail to go beyond a few countries and are therefore not available in other countries where it is necessary.

4.3.5 Regulatory Barriers

Reduction of new antibiotic approvals and difficulties in pursuing regulatory approval which include: bureaucracy, absence of clarity, differences in clinical trial requirements among countries, changes in regulatory and licensing rules, and ineffective channels of communication.

4.3.6 Climatic Change and Population Density

Epidemiologists from Boston Children's Hospital and the University of Toronto found that higher local temperatures and population densities correlated to a greater level of antibiotic resistance among a number of common bacterial strains. MacFadden et al. (2018) have conducted a study during the period from 2013 to 2015 by assembling a large data base of U.S antibiotic resistance information related to *E. coli*, *K. pneumoniae* and *S. aureus* from various sources including hospital, laboratory and disease surveillance. The data that was obtained from samples isolated from the people with resistant infection was compared to Latitude Coordinates and as well as to mean and median local temperatures. Local average minimum temperature increases of 10 degrees Celsius were linked to surges of 4.2, 2.2 and 3.6% in resistant strains of *E. coli*, *K. pneumoniae* and *S. aureus* respectively. It was also found that an increase of 10,000 people per square mile was related to 3 and 6% increases in resistance to *E. coli* and *K. pneumoniae* respectively indicating that population density also plays an important role.

4.4 Intrinsic Resistance Mechanisms for Survival of Bacteria

(a) Drugs such as tetracycline or erythromycin are pumped back out of bacterial cells through efflux pump proteins to keep intracellular drug concentrations below therapeutic level.
(b) The antibiotic is destroyed by chemical modification by an enzyme that is released by the resistant bacteria. This is depicted in the picture (b) Secretion of the b-lactamase into the periplasmic space to hydrolyse penicillin molecules before they reach their PBP targets in the cytoplasmic membrane of this Gram-negative bacterium.
(c) The aminoglycoside antibiotic kanamycin can be enzymatically modified at three sites by different mechanism of enzymatic processing – N-acetylation, O-phosphorylation or O-adenylylation – to block recognition by its target on the ribosome.

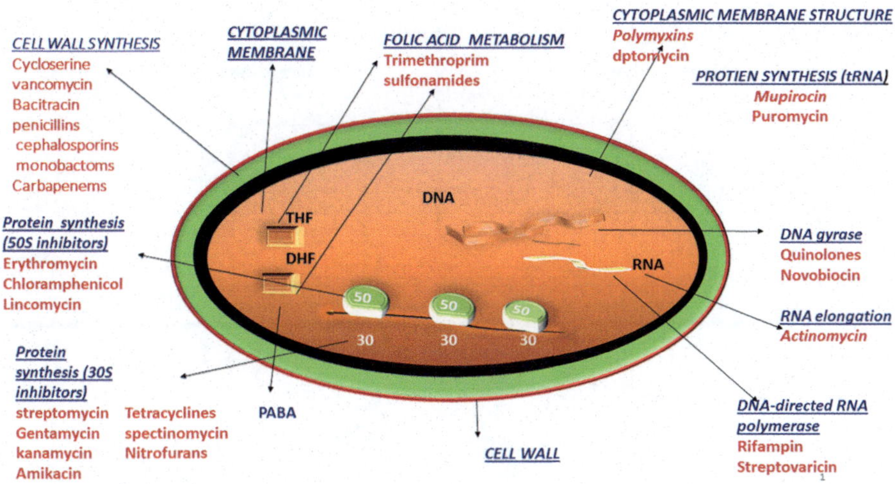

Fig. 4.2 Antibacterial agents and their mode of action

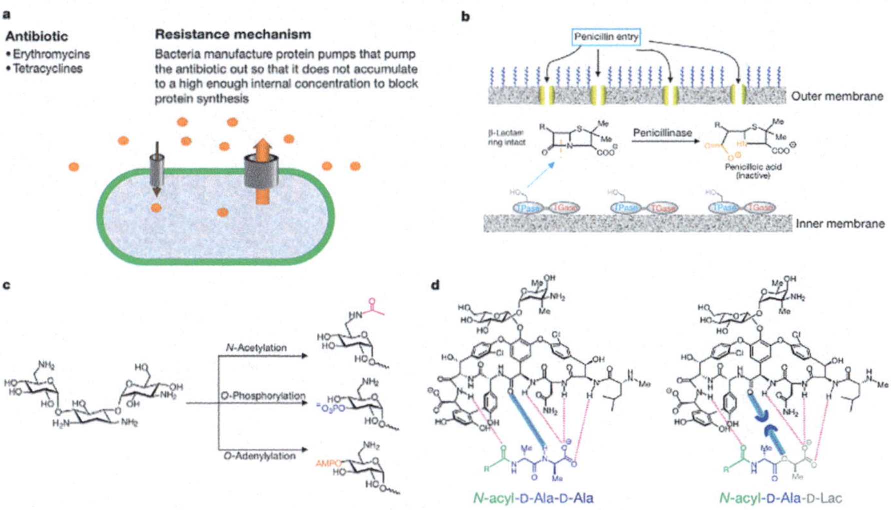

Fig. 4.3 Intrinsic resistance mechanisms for survival of bacteria

(d) Reprogramming of the target structure in the bacterium to have a low affinity for antibiotic recognition. Here, the switch from the amide linkage in the D-Ala-D-Ala peptidoglycan termini to the ester linkage in the D-Ala-D-Lac termini is accompanied by a 1000-fold drop in drug-binding affinity. All these mechanisms are shown in the Fig. 4.3.

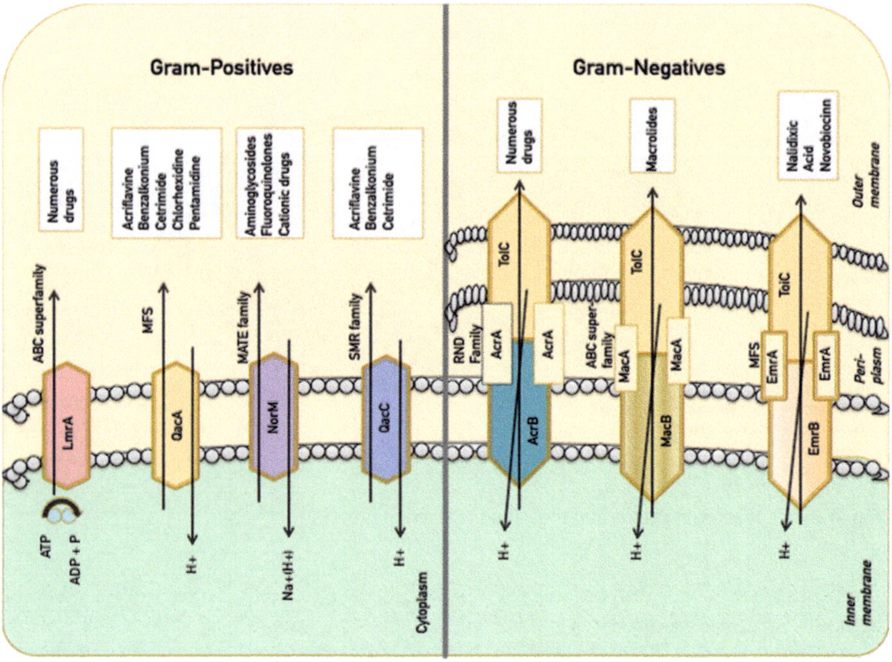

Fig. 4.4 Efflux pumps in bacteria that contribute to Multidrug Resistance. (Source: Modified from Microbiiol Spectr 2016. Apr 4(2):10.1128/microbiol.spec.VMBF.0016)

Five major families of efflux pumps in gram positive and gram negative bacteria are depicted in Fig. 4.4

1. ABC super family ATP binding cassette
2. MFS Major Facilitator Superfamily
3. MATE Multidrug and toxic compound extrusion family
4. SMR Small multidrug resistance family
5. RND Resistance nodulation division family

4.5 Acquired Resistance Mechanism

Numerous antimicrobial resistance phenotypes result from acquisition of external genes that may provide resistance to an entire class of antimicrobials. These genes are frequently associated with large transferable extra chromosomal DNA element called Plasmids on which the other mobile DNA elements such as transposon and integron (Fig. 4.5). Transposons are the mobile genetic elements that can carry resistance genes and possess transposase activity providing the recombination of Resistance genes with plasmid or the chromosome. Integrons consist of integrase

Fig. 4.5 Expression of gene cassettes for antibiotic resistance determinants in Integron. (Source: Modified from Infect. Ecol Epidemiol 2015:510:3402/ieev528564)

enzyme encoded by *intI* gene, a recombination site recognised by Integrase and a Promoter which are necessary for the expression of Gene Casettes present in the Integron. These arrangements help in the acquisition of Antibiotic resistance genes in the bacterial Genome especially in plasmids. Through process of Conjugation the resistant genes present in the plasmid are spread through Transposon or Integron to other strain or species. Integrons are ubiquitous in nature. The number of Gene casettes in 50m^2 soil has been estimated to be >2000 (Michael et al. 2004). Gene encoding the Integrase intI is induced by SOS response. Antibiotics such as Trimethoprim, quinolones, Beta-Lactams are known to induce SOS response. The basic structure of Integron is depicted in Fig. 4.5. The gene intI encodes a site specific integrase which can excise and integrate gene casettes at site specific integration site attI. For example, this Integron contains 3 Gene Casettes denoted as GC1, GC2, GC3. Expression of the gene Casettes is induced by the promoter Pc. This Integron contains 2 Conserved regions at the 3′ end, for quaternary ammonium compound resistance gene qacEΔI and sulphonamide resistance gene SulI.

Emergence of Resistance mechanism of several bacterial pathogens are discussed below:

4.6 Methicillin Resistance Staphylococcus aureus (MRSA)

People with MRSA are estimated to be 64% are likely to die than people with a nonresistant form of infection. MRSA has increased significantly in Hospitals worldwide.

MRSA strains, especially has a diversity in the structure of the genetic element that confers resistance; the staphylococcal cassettes chromosome mec (SCC mec) which carries the methicillin resistance gene, mecA which encodes the 76 KDa Pencillin binding proeins PBP2a(PBP2). The mecA has been originated from S.sciuri. The two genes ccrA and ccrB present on mec element from one isolate have shown to code for recombinase proteins that are capable of excising and integrating the mec element into the chromosome (Katayama et al. 2000). These genetic elements frequently carry resistance genes for other antimicrobials such as aminoglycosides (gentamicin, tobramycin) erythromycin and tetracycline.

Hospital acquired MRSA are likely to carry antimicrobial resistance to additional drugs including fluoroquinolones.

S. aureus is a colonizer of the skin, and body entry portals (ears, eyes, nasal passage). Any cut in the skin or colonization of the individual with compromised immune system can provide an entry for these bacteria to cause infection.

The disease is progressed by production of toxins like, α toxin, β toxin, epsilon toxin, exfoliate toxins, staphylococcal enterotoxin, toxic shock syndrome toxin, hyaluronidase, lipases, coagulases, staphylokinases and certain surface proteins like clumping factor collagen binding, fibronectin binding protein, capsular polysaccharide adhesin and protein A promotes in colonization that causes tissue invasion and destruction.

Strains of MRSA isolated in hospitals have now spread into the community. Methicillin resistance in *S. aureus* is mediated by the mecA gene, which encodes for a novel penicillin binding protein PBP-2A. In MRSA, exposure to methicillin inactivates the 4 high binding affinity PBPs normally present. PBP-2A, which displays a low affinity for methicillin, takes over the function of these PBPs, permitting the cell to grow.

Regulation of the Methicillin-resistant phenotype and production of PBP = are influenced by the action of other genes. Two genes located upstream from mecA-mecR1 and mec1-control expression PBP-2A have displayed efficacy against MRSA in vivo.

4.7 Vancomycin Resistant Enterococci (VRE)

Enterococci cause a wide range of illnesses among hospitalized patients, other health care settings including blood stream, surgical site and urinary tract infections. The most frequent Enterococcal infections are urinary tract infections. Enterococci are intrinsically resistant to β-lactam agents and aminoglycosides and were the first bacteria to acquire vancomycin resistance. The greatest threat posed by vancomycin resistant enterococci is the potential to transfer their resistance genes to more pathogenic gram positive bacteria causing a significant health problem.

It was observed by Arthur et al. (1996) that a vancomycin-resistant strain reveals the presence of several different phenotype of glycopeptides resistance. E.g. vancomycin and or teicoplanin-inducible strain. The gene encoding vancomycin resistance is often found on a transposon that is relatively easily transferred to other enterococcal species by conjugation. Vancomycin-resistant enterococci with the van B resistance phenotype have variable levels of vancomycin resistance and are susceptible to teicoplanin. The van B resistance phenotype is also inducible by vancomycin, but not by teicoplanin and exposure to vancomycin also produces teicoplanin resistance.

The genes encoding the van B resistance phenotype are more commonly chromosomal but can also be transferred by conjugation. Quintilani et al. (1993). The van C resistance phenotype, which consists of relatively low levels of vancomycin resistance without teicoplanin resistance, is caused by chromosomally encoded genes that are found in all strains of Enterococcus casseliflavus, Enterococcus gallinarum and Enterococcus flavescens which are not transported.

The fourth phenotype is Van D which is similar to Van B (Vancomycin 64 ug/ml; teicoplanin MIC 4 ug/ml).

Both Van A and Van B resistance phenotype are made by the products of a group of genes that encode a two component regulatory system, which causes inducible resistance.

Another set of genes encodes a group of enzymes that enable the enterococci to synthesize cell wall precursors ending in D-alanine lactate, instead of D-alanine-D-alanine which is the usual vancomycin binding site.

4.8 Drug Resistant Streptococcus Pneumonia (DRSP)

S. pneumonia is a major cause of bacterial pneumonia and meningitis, as well as blood stream, ear & sinus infections. Resistant S. pneumoniae infections complicate the treatment resulting in 1.2 million illnesses and 7000 deaths/year according to Centers for Disease Control and Prevention US 2013.

DRSP is most commonly resistant is to penicillin as well as to one or more antibiotics (Erythromycin, Trimethoprim/ Sulfamethoxozole, vancomycin, tetracycline, chlorampenical Ofloxacin) in 30% of cases according to Kim et al. (2016).

There are more than 90 strains (serotypes) of pneumococcus bacteria. The serotypes 6A, 6B, 9 V, 14, 19A, 19F and 23F accounted for most DRSP before the introduction of the 7 valent pneumococcal conjugate vaccine PCV7 Prevnar in U.S. (2000) PCV 13 also gave protection to 13 strains. Use of PCV 7 and PCV 13 has not only prevented pneumococcal disease but also reduced antibiotic resistance by blocking the transmission of resistant *S. pneumoniae* strains.

Alterations in the target enzymes for β-lactam antibiotics, the penicillin binding protein have been recognized as a major resistance mechanism in *Streptococcus pneumoniae*. Mutations in PBPs that confer a reduced affinity to β-lactams have been identified in laboratory mutants and clinical isolates. Clinical isolates display a mosaic structure of the affected PBP genes resulting in interspecies gene transfer and recombination events.

4.9 Antibacterial Resistance *in E.coli*

Although E. coli is part of the normal flora in the Intestine of humans and animals it is the commonest cause of community and hospital acquired urinary tract and blood stream infections in all age groups. It is also one of the leading causative agents of foodborne infections worldwide.

Resistance in E. coli readily develops either through mutations, E.g. fluoroquinolone resistance, or by acquisition of mobile genetic elements, E.g. broad spectrum penicillins (e.g. ampicillin or amoxicillin) and resistance to third-generation cephalosporins. Resistance to third-generation cephalosporins is mainly conferred by enzymes known as Extended spectrum beta-lactamases (ESBLs); these enzymes destroy many beta-lactam antibacterial drugs. ESBLs are transmissible between bacteria and even between bacterial species. Because E. coli strains that have ESBL are generally also resistant to several other antibacterial drugs Gram negative enterobacteriaceae with resistance to carbapenem by NewDelhi Metallo beta Lactamases(NDM-1) was studied by different researchers from India, Pakistan & U.K as it caused a severe threat in 2010 (Kumarswamy 2010).

4.10 Antibacterial Resistance to Salmonella

Salmonella enterica which causes Non typhoidal Food borne Infection and enteric typhoid fever in humans is associated with more than 1,200,000 illnesses annually and out of these 100,000 infection are due to antibiotic resistant salmonella. 36,000 and 33,000 illnessses/year are resistant to Ceftriaxone and Ciprofloxacin respectively. The horizontal transmission of resistance has an important role in the dissemination of antibiotic resistance in S. enterica. The emergence of S.typhimurium defintive type DT104 as a Multi drug resistant pathogen causing food borne out breaks has Chromosomally encoded resistance to more than 5 Antibiotics i.e., Pencillin, Chloramphenicol, Florfenicol, Streptomycin, Sulfonamide and tetracycline. Typhimurium DT104 was found to be associated with mobile non conjugative plasmids. Fluroquinolone resistance determining region (QRDs) of the genes that code for Gyrase gyrA and gyrB and topoIsomeraseIV which are the targets for Fluroquinolones in bacterial cells resulting in gene mutation leading to resistance to Fluroquinolones. Emergence of plasmid mediated ESBLs namely CTX-M is commonly found in Salmonclla spp. and associated with Ceftoxamine hydrolysis. These genes are transferred horizontally via Conjugation plasmids &Transposons resulting in CTX-M ESBLs.

4.11 Alternative Approaches to Combat Antibiotic Resistance

4.11.1 Bacteriophages

Phages play an important role in combating infections caused by variety of pathogens in humans. Development of phage therapy is an alternative approach to improve the treatment of bacterial infections. Bacteriophages were discovered by English Microbiologist Twort in 1915 but "the bacteriophage phenomenon" era began after the pioneer work done by a French-Canadian Microbiologist Felix d'Herelle in 1917. They are very specific and found abundance in Nature.

He has proved that phage titers increased in disease progression and peaked during recovery in humans. The safety of the phage suspension was tested on the patients suffering from bacillary dysentery and cholera. Later, phage therapy was applied in wound treatment. Lytic bacteriophages were able to kill antibiotic– resistant bacteria at the end of phage infection cycle and use two component lysis systems to release progeny virions by the destruction of the bacteria cell wall.

Certain lytic bacteriophages use single proteins, amurins to inhibit the synthesis of peptidoglycan. The phages utilize two types of proteins to kill the host cell.

(a) Holins
(b) Endolysins

Holins are involved in the host cell lysis triggering process and their role is to perforate the host cytoplasmic membrane and thus to cooperate with endolysin by giving them an access to bacterial peptidoglycan. Therefore, holins determine the time of bacterial lysis. They act at precise time point and control bacterial murein accessibility for phage endolysins and synchronize the activity of the holin-lysin system with the late phase events of the phage replication cycle (Dewey et al. 2010 and Shi et al. 2012).

Every holin has one hydrophobic transmembrane domain (TMD) and a C terminal hydrophilic domain that carries a high electric charge.

There are 3 classes of holins

Class I: Protein with more than 95 amino acid residues in length and three TMDs.
E.g. Staphylococcus aureus bacteriophage p68 hol 15 protein and E. coli phage λS105 protein.
Class II: 65–95 amino acid residues in length and two TMDs.
E.g. Lambdoid phage 21 S protein, Clostridium perfringens bacteriophage 3626 hol 3626.
Class III: Only one TMD, E.g. Phage Φ CP26F holin S105 the product of the phage λS gene expression, localizes the plasma membrane at a proper time point, and forms lethal lesion in lipid bilayer.

Phage endolysin are responsible for cell wall degradation.

Bacteriophages hydrolyse the peptidoglycan of the infected bacteria. Endolysins perform activities of endopeptidase, amidase, glycosidase or lytic transglycosylase to kill bacterial cell by murein destruction.

At the end of the phage replication cycle, endolysin promote the release of progeny virions. Endolysins directed against Gram negative bacteria have different structures with that of gram positive ones.

4.12 Interaction Between Bacteriophages and Host Immune System During Phage Therapy

An oral administration of phages during phage therapy of bacterial infection caused by staphylococcus, Klebsiella, Escherichia, proteus and pseudomonas induces the production of antibodies. No immunological complications were observed with the intake of large amounts of phages (Sarkar et al. 2012). Intravenous administration of bacterial pathogens strongly stimulate both innate and adaptive immunity (Merril et al. 2006).

Dabrowska et al. (2005) observed that phages can penetrate into the circulation irrespective of the route of administration. If there are no host bacteria for specific phages, they are rapidly removed from the blood and internal organs by phagocytic cells. Moreover, bacterial predators are internalized and eliminated by cells of the reticuloendothelial system of liver and spleen. It was observed that Kupffer cells can phagocytose phages four times faster than the spleen macrophages which suggests that arrested bacterial pathogens in spleen may stimulate lymphocytes to produce antibodies.

It was observed by Weber Dabrowska et al. (2002) that patients subjected to phage therapy were characterized by the decrease in number of mature neutrophils and the increased number of neutrophil precursors in the peripheral blood. These results indicated that phage preparations could activate innate immune response, in the clearance of bacterial Infection.

4.12.1 Delivery of Phages

Singla et al. (2016) evaluated the liposome entrapped Klebsiella pneumonia phage particles in different organs of Balb/C Mice. It was observed that liposome – entrapped phages were more stable in blood and organs than the free phages. Because of the longer bioretention rate, these phages can be used in the treatment of K. pneumonia infections in lung and kidney infections.

Oral Administration of phages helps in elimination of diarrheic pathogens like Salmonella spp. Clostridium difficile and *E. coli*. These phages were used to control the spread of *Campylobacter jejuni* and *E. coli* infections in chickens. A 15 year old patient with Cystic Fibrosis with a disseminated Mycobacterium absceus infection was treated with effective lytic Phage derivatives. It was observed that Intravenous Phage treatment was tolerated with clinical improvement Rebekah et al. (2019).

4.13 Genetically Engineered Phages to Deliver Antimicrobial Agents to Bacteria

4.13.1 Lytic Phage Therapy

Animal studies by Smith and Huggins (1983) demonstrated that a single intramuscular dose of an anti-k 1 phage preparation (3×10^8 PFU) was more effective than multiple dose of antibiotics in protecting mice from an *E. coli* 018: K 1: H7 infection. Later, Studies by Smith et al. (1987) showed that a single oral dose of *E. coli*. Specific phage could successfully reduce the number of bacteria in the alimentary tract of calves, piglets and lambs infected with Enteropathogenic *E. coli* strain. Use of genetically engineered Phages as an alternative therapy to treat bacterial infections was studied by Caroline et al. (2003).

Phage with lytic activity against clinical isolates of *Enterococcus faecium* was shown to be effective in rescuing mice from lethal vancomycin resistant *E. faecium* infection.

The engineering of the filamentous phage M13 to carry an integrin binding peptide and a fragment of the polymorphic membrane protein D from the sexually transmitted pathogen Chlamydia trachomatis (ct) to eliminate Chlamydia trachomatis infection. Keeping in view of this study, these modified phages were able to significantly reduce the infection in both primary endocervical and Hela cells.

4.14 Non-lytic Phage to Deliver DNA Encoding Bactericidal Proteins

This Approach Utilises non lytic phages to deliver DNA encoding bactericidal Proteins (Norris et al. 2000). In bacteria, programmed cell death is mediated through "addiction modules" consisting of two components, a stable toxin and an unstable antidote that antagonizes the toxic effect. e.g. pem 1-pem k genes of plasmid R100, the phd-doc genes of phage pl and the ccdA-ccdB genes of plasmid F.

Holcik and Iyer (1997) observed that addiction modules help in the stability of the extrachromosomal elements by selectively killing plasmid-free cells, resulting in the multiplication of plasmid harboring cells in the population.

4.15 Genetically Modified Viruses

Several companies have used bacteriophages by using the CRISPR gene-editing system to kill specific antibiotic resistant bacteria (Reardon 2017).

Locus and several other companies developed phages that make the bacterial immune system known as CRISPR against itself. In Locus' phages target bacteria which are resistant to antibiotics, the CRISPR system includes DNA with instructions for modified guide RNAs that home in on part of antibiotic-resistance gene. Once the phage infects a bacterium, the guide RNA latches on the resistance gene and the production of the enzyme cas3 in the bacterium normally kill phages, to destroy that genetic sequence, instead cas3 destructs all the DNA killing the bacterium.

Eligo Bioscience in Paris applied similar approach. It has removed all the genetic instructions that allow phages to replicate and the insert DNA which encodes guide RNAs and the bacterial enzyme Cas9. Cas9 cuts the bacterium's DNA at a designated spot, which triggers the bacterium to self-destroy. This system can be applicable in future to target human gut pathogens.

4.16 New Antibiotic Effective Against Drug-Resistant Bacteria

Maffoli et al. (2017) have discovered a new antibiotic which is effective against drug resistant bacteria: pseudouridimycin. The new antibiotic was produced by a microbe found in a soil sample collected in Italy. This new antibiotic kills a broad spectrum of drug-sensitive and drug-resistant bacteria in a test tube and cured bacterial infections in mice.

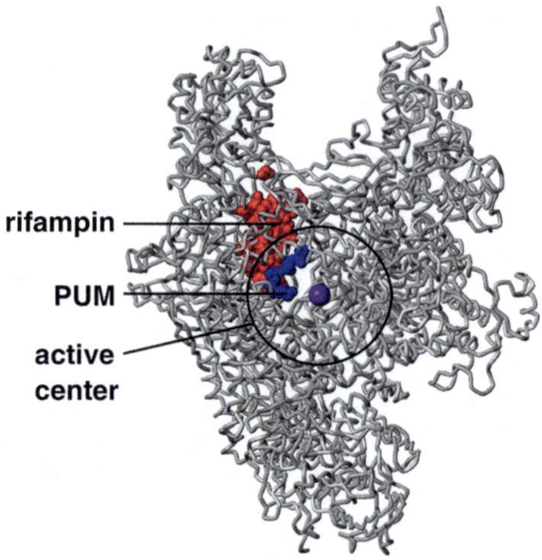

Fig. 4.6 Structure of bacterial RNA polymerase, showing the binding sites for the new antibiotic pseudouridimycin (PUM) and the current antibacterial drug rifampin (Rif). (Source: Rutgers University. https://www.sciencedaily.com/releases/2017/06/170615142842.htm)

Pseudouridimycin inhibits bacterial RNA polymerase, the enzyme responsible for bacterial RNA synthesis. The binding site and mechanism of action of pseudouridimycin is different from that of rifampin, pseudouridimycin exhibits no cross-resistance with rifampin, functions additively when co-administered by rifampin. Its spontaneous resistance rate is just one-tenth of the spontaneous resistance rate of rifampin.

Pseudouridimycin functions as nucleoside-analog inhibitor of bacterial RNA polymerase, in the synthesis of RNA. This new antibiotic binds tightly to the NTP binding site on bacterial RNA polymerase (Fig. 4.6). Pseudouridimycin selectively inhibits bacterial RNA polymerase but not human RNA polymerases.

Alterations of the NTP binding site that disrupts the binding of the new antibiotic also alters the RNA polymerase activity resulting in the dead bacteria instead of resistant bacteria. "Nucleoside-analog inhibitors that selectively inhibit viral nucleotide polymerases in Viral infections may have the same effect in the treatment of bacterial diseases by inhibiting the bacterial RNA Polymerases.

Simon et al. 2018 observed that patients having a disease with Gram negative bacteria could be treated with siderophore-based drug cefiderocol. Cefiderocol binds to Iron and is transported through the outer membrane by the bacterium own Iron-Transport system These Iron channels will make the drug to bypass the bacteria's porin channels and again repeat entry even if the bacteria has evolved efflux pumps. This drug was found to be effective in Phase 2 trials.

4.17 Metal Nanoparticles

It is well known that some metals have antimicrobial properties; therefore, exploring metal nanoparticles as a new antimicrobial therapy could eliminate antibiotic-resistant bacteria. There are several ways by which metal nanoparticles can affect bacterial survival. Silver-containing antimicrobials can exert a physical stress on bacterial cells. Other evidence suggests that gallium can be effective in interfering with bacterial metabolic pathways by interrupting bacterial metal ion uptake (Minandri et al. 2014), which in turn affects biofilm-forming *P. aeruginosa in vitro*. Ahamad et al. (2007) synthesised the metallic nanoparticles of silver by reducing the aqueous Ag+ ion in the Culture supernatants of Klebsiella Pneumoniae. These Nanoparticles were evaluated for their antimicrobial activities with antibiotics such as Pencillin G, Amoxicillin, erythromycin, clindamycin and vancomycin against S. aureus & E. coli. The higest antimicrobial effect was observed for vancomycin, amoxycillin and Pencillin G against S. aureus.

4.18 Antimicrobial Peptides

Hwang et al. (2017) have shown that laboratory-engineered *E. coli* is able to secrete antimicrobial peptides in response to quorum-sensing molecules released by *P. aeruginosa*. These antimicrobial peptides were able to degrade biofilms formed by *P. aeruginosa*.

Using antimicrobial peptides on their own is another approach. For instance, pexiganan, a natural peptide was shown to be effective in killing both Gram-positive and Gram-negative bacteria (Flamm et al. 2015). Endogenous Cationic Antimicrobial Peptides (CAMPs) such as Defensins, Cathelicidins and Kinocidins which are produced by all MicroOrganisms are most effective components of Antimicrobial host defence which Contribute to Innate immunity. Pheromonicin Agr D1 is an engineered peptide that consists of the Staph aureus Pheromone gene AgrD1 which is fused to Colicin Ia and was found to be active against both Methicillin Sensitive S.aureus and Methicillin Resistant S.aureus (Qiu et al. 2003). The peptide dendrimers with Amphiphillic Properties and Postive Charges show high antimicrobial effect against E.coli, S.aureus, Candida albicans because of the interaction with bacterial cell membrane (Janiszewska and Urbanczyk 2007).

4.19 Probiotics

Probiotics such as B.Subtilis, Lactobacillus strains, Saccharomyces yeast have antimicrobial activities against pathogenic bacteria like Salmonella. Some strains of B.subtilis were found to be effective against S.enteritidis attachment and invasion of intestinal epithelial cells (Thirabunyanon and Thongwittaya 2012). It was reported by Millette et al. (2008) that bacteriocin producing Pediococcus acidilactici (Pediocin PA-1 Producer) and L.lactis (nisinZ Producer) could reduce colonization of Vancomycin resistant enterococci in the intestine of mice.

4.20 Prebiotics

Prebiotics are non digestable carbohydrate substrates that selectively promote the growth of Probiotics. Eg. These fermentable carbohydrates produce short chain fatty acids in the intestine and help in the growth of intestinal probiotic bacteria like Bifidbacterium &Lactobacillus.

4.21 Quorum Sensing Inhibitors

Quorum sensing is a bacterial communication process that depends on the bacterial population density. It involves small diffusible signalling molecules which activate the expression of genes that control a variety of functions like bioluminiscence, virulence, biofilm formation, sporulation. Quroum sensing is responsible for virulence in clinically important bacteria. Inhibition of Q.S appears to be a promising strategy to control these pathogens.

4.22 Preventive Measures to Combat Antimicrobial Resistance

In May 2015, the World Health Assembly endorsed a global action plan to tackle Antimicrobial Resistance and set out five strategic objectives:

- Improve awareness and understanding of antimicrobial resistance;
- Strengthen knowledge through surveillance and research;
- Reduce the incidence of infection;
- Optimize the use of antimicrobial agents; and
- Develop the economic case for sustainable investment that takes account of the needs of all countries, and increase investment in new medicines, diagnostic tools, vaccines and other interventions.

Preventing infections in the first place is another strategy in tackling AMR. Access to clean water supply and effective healthcare systems will substantially reduce the burden of AMR by limiting the spread of infections and decreasing the overall number of infected individuals. Furthermore, improving the hygiene and sanitation in hospitals can lower the number of cases associated with multiple-drug resistant bacteria.

Monitoring Programs at International and National Levels

(i) The Global Antimicrobial Resistance surveillance system (GLASS)
(ii) Global Antibiotic Research and Development Partnership (GARDP)
(iii) Interagency Coordination Group of Antimicrobial Resistance (IACG)
(iv) European Antimicrobial Resistance Surveillance Network (EARS Net)
(v) Resistance MAP: It is a Website by the Center for Disease Dynamics, Economic and policy which provides data on antimicrobial resistance at global level

National Level: National Centre for Microbial Resource (NCMR) and National Centre for Cell Science (NCCS), Pune Established a biorepository of Drug resistant microbes for Collection, Storage, Maintenance, Preservation and Characterisation in India.

Red Line Campaign: In India, Union Ministry of Health affairs has made mandatory to display a 5 mm thick red band on package of Medicines prescribed by the Doctor only.

References

Ahamad RS, Fakhimi A, Hamid RS et al (2007) Synthesis and effect of silver nano particles on the anti bacterial activity of different antibiotics against S. aureus and E. coli. Nanomed Nanotechnol Biol Med 3:168–171. https://doi.org/10.1016/j.nano.2007.02.001

Arthur M, Reynolds PE, Depardien F et al (1996) Mechanisms of glycopeptides resistance in enterococci. J Infect Dis 32:11–16

Bartlett JG, Gilbert DN, Spellberg B (2013) Sevenways to preserve the miracle of antibiotics. Clin Infect Dis 56(10):1445–1450

Caroline W, Kasman LM, David AS, Phillip AW, Dolan JW, Schmidt MG, Norris JS (2003) Antimicrob Agents Chemother 47(4):1301–1307. https://doi.org/10.1128/AAC.47.4.1301-1307.2003

Dabrowska K, Switala-jelenk, Opolski A et al (2005) Bacteriophage penetration in vertebrates. J Appl Microbiol 98:7–13. https://doi.org/10.1073/pnas.0914030107

Dewey JS, Savva CG, White RL, Vitha S, Holzenburg A, Young R (2010) Micron scale holes terminate the phage infection cycle. Proc Natl Acad Sci USA 107:2219–2223. https://doi.org/10.1073/obas.091430107

Doyle MP (2006) Antimicrobial resistance: implications for the food system. Compr Rev Food Sci Food Saf 5:71–137

Flamm RK, Rhomberg PR, Simpson KM et al (2015) In vitro spectrum of pexiganan activity when tested against pathogens from diabetic foot infections and with selected resistance mechanisms. Antimicrob Agents Chemother 59(3):1751–1754

Holcik M, Iyer VN (1997) Conditionally Lethal genes associated with bacterial plasmids. Microbiology 143:3403–3416

Hwang IY, Koh E, Wong A et al (2017) Engineered probiotic Escherichia coli can eliminate and prevent Pseudomonas aeurginosa gut infection in animal models. Nat Commun 8:15028

Janiszewska J, Urbanczyk LZ (2007) Amphillic dendrimeric peptides as model non sequential Pharmacophores with Antimicrobial properties. J Mol Microbiol Biotechnol 13:220–225

Katayama Y, Ito T, Hiramatsu K (2000) A new class of genetic element, Staphylococcus casette chromosome mec, enccodes methicillin resistance in Staphylococcus aureus. Antimicrob Agents Chemother 44(6):1549–1555. https://doi.10.1128/aac44.6.1549-1555.2000

Kumaraswamy KK, Toleman MA, Walsh TR et al (2010) Emergence of a new antibiotic resistance mechanism in India, Pakistan and the U.K. a molecular, biological and epidemiological study. Lancet Infect Dis 10(9):597–602

Kim L, Mc Gee L, Tom Czyk S, Beall B (2016) Biological and epidemiological features of antibiotic-resistance streptococcus pneumonia in pre-and post conjugate vaccine era. Clin Microbiol Rev 29(3):525–552

Linton AH (1997) Antibiotic resistance: the present situation reviewed. Vet Rec 100:354–360

MacFadden DR, McGough SF, Fisman D et al (2018) Antibiotic resistance increases with local temperature. Nat Clim Chang 8(6):510–514. https://doi.org/10.1038/541558-018-0161-6

Maffoli SI, Yuzhang DD et al (2017) Antibacterial nucleoside Analog inhibitor of bacterial polymerase. Cell 169(7):1240. http://dx.doi.org/10.1016/j.cell.2017.05.042

Merril CR, Scholl D, Adhya S (2006) Phage therapy. In: Calender R (ed) The bacteriophages. Oxford University Press, New York, pp 725–741

Michael CA, Gillings MR, Holmes AJ et al (2004) Mobile gene casettes:a fundamental resource for bacterial evolution. Am Nat 164:1–12

Millette M, Cornut G, Dupont C et al (2008) Capacity of Human nisin- and pediocin producing lactic acid bacteria to reduce intestinal Colonization by Vancomycin-resistant enterococci. Appl Environ Microbiol 74:1997–2003

Minandri F, Bonchi C, Frangipani E et al (2014) Promises and failures of gallium as an antibacterial agent. Future Microbiol 9(3):379–397

Norris JS, West water C, Schofield D (2000) Prokaroytic Gene therapy to combat Multi drug resistant bacterial infection. Gene Ther 7:723–725

Pourmand A, Mazer-Amirshahi M, Jasani G, May L (2017) Emerging trends in antibiotic resistance: implications for emergency medicine. Am J Emerg Med 35(8):1172–1176. https://doi.org/10.1016/j.ajem2017.03.010.Epub

Qiu XQ, Wang H, Lu XF et al (2003) An Engineered multidomain bactericidal peptide as a Model for targetted antibiotics against specific bacteria. Nat Biotechnol 21:1480–1485

Quintilani R, Evers S, Courvalin P (1993) The vanβ gene confers various levels of self-transferable resistance to vancomycin in enterococci. J Infect Dis 167:1220–1223

Rebekah M, Dedrick CA, Gurerrero B et al (2019) Engineered bacteriophages for treatment of a patient with disseminated drug resistant Mycobacterium abscessus. Nat Med 25(5):730. https://doi.org/10.1038/S41591-019-0437-Z

Reviews on Antimicrobial Resistance. Tackling drug-resistant infections globally: final report and recommendations. 2016. Available from: https://amr-review.org/home.html

Reardon S (2017) Modified viruses deliver death to antibiotic resistant bacteria. Nature 546:586–587. https://doi.org/10.1038/nature2017.22173

Sarker SA, McCallin S, Barnetto C et al (2012) Oral T-4 like phage CockTail application to healthy adult volunteers from Bangladesh. Virol 434:222–232. https://doi.org/10.1016/j.virol.2012.09.002

Shi Y, YanY JW, Du B, Meng Y, Wang H, Sun J (2012) Characterization and determination of holing protein of streptococcus suis bacteriophage SMP in heterologous host. Virol J 9:70. https://doi.org/10.1186/1743-4229-70

Simon P, David VV, Roger E, et al (2018) Cefiderocol versus imipenem-cliastatin for the treatment of complicated urinary tract infections caused by Gram negative uropathogens; a phase2 randomised, doubleblind, noninferiority trial. Lancet Infect Dis. https://doi.org/10.1016/s1473-3099(18)30554-1

Singla S, Harjari K, Razak WS, Katare OP, Chhibber S (2016) Phospholipid vegides encapsulated bacteriophage: a novel approach to enhance phage biodistribution. J Virol Method 236:68–76. https://doi.org/10.1186/1743-422x-9-70

Smith HW, Huggins MB (1983) Effectiveness of Phages in treating experimental *E. coli* diarrhoea in calves, piglets and lambs. J Gen Microbiol 129(8):2659–2675

Smith HW, Huggins MB, Shaw KM (1987) The control of Experimental *E. coli* diarrhoea in calves by means of bacteriophages. J Gen Microbiol 133(5):1111–1126

Thirabunyanon M, Thongwittaya N (2012) Protection activity of a novel probiotic strain of Bacillus subtilis against Salmonella Enteritidis infection. Res Vet Sci 93:74–81. https://doi.org/10.1016/j.rvsc.2011.08.008

Viswanathan VK (2014) Off label abuse of antibiotics by bacteria. Gut Microbes 5(1):3–4. https://doi.org/10.4161/gmic.2807

Weber Dabrowska B, Zimecki M, Mulczyk M et al (2002) Effect of phage therapy on the turnover and function of Pheripheral neutrophils. FEMS Immunol Med Microbiol 34:135–138. https://doi.org/10.1111/j.1574-695X.2002.tb00614.x

Do Non-medical Uses of Antibiotics Develop Cross-Resistance in Clinical Pathogens?

5

Lalitha Cheepurupalli, Sudarshan Singh Rathore, Thiagarajan Raman, Niranjana Sri Sundaramoorthy, and Jayapradha Ramakrishnan

Abstract

Abuse of antibiotics and drastic decline in the development of new antibiotics has been attributed to the rapid evolution of antibiotic resistant pathogens. The applications of antibiotics for agriculture practices, growth promotion of animal, poultry and aquaculture has raised concerns in recent years for their role in the development of antimicrobial resistance in clinically relevant pathogens. In this chapter we have discussed the non-medical uses of antibiotics and its impact on antimicrobial resistance in clinical settings. Extensive research such as evidences for transfer of resistant genes and plasmids to clinically relevant pathogens is required to correlate non-medical use of antibiotics and evocation of antimicrobial resistance mechanisms.

Keywords

Antimicrobial resistance · Agriculture · Biocides

5.1 Introduction

Antibiotics are the chemical substances from natural sources and synthetic or semi-synthetic which interfere with growth of microbes and are used to treat infections in humans and animals (O'Neill 2015). Antimicrobial resistance (AMR) is one of the

L. Cheepurupalli (✉) · S. S. Rathore · N. S. Sundaramoorthy · J. Ramakrishnan (✉)
Centre for Research in Infectious Diseases (CRID), School of Chemical and Biotechnology (SCBT), SASTRA University, Thanjavur, Tamil Nadu, India
e-mail: ch.lalitha5@gmail.com; kavijayashal@gmail.com; antibioticbiology@gmail.com

T. Raman
Department of Advanced Zoology and Biotechnology, Ramakrishna Mission Vivekananda College, Chennai, Tamil Nadu, India

S. M. Mandal, D. Paul (eds.), *Bacterial Adaptation to Co-resistance*,
https://doi.org/10.1007/978-981-13-8503-2_5

most critical problem with substantially increased morbidity worldwide (Smith and Coast 2002). The evolution of various drug resistance mechanisms and spread of resistant bacteria is recognized as a growing threat worldwide. Pathogens harbouring extended spectrum beta-lactamase like carbapenem resistant Gram negative bacteria, methicillin-resistant *Staphylococcus aureus* (MRSA), *Vancomycin resistant Enterococci* (VRE), *Streptococcus pneumoniae* are major emerging threats to public health (Levy 2002b). The Centre for Disease Control (CDC) estimates that more than half of mortality due to bacterial infection are due to the infections of resistant bacteria (Wilson 2004). The drug resistant bacteria and genes can be easily transferred to animals to humans through food chain (da Costa et al. 2013). In Swedish Veterinary Antimicrobial Resistance Monitoring Report (SVARM), the observance of cross-resistance among antibiotics was confirmed (SANCO 2002). The transfer of resistance may occur through exchange of part of the bacterial DNA that carries resistant genes. AMR can be inherent or acquired. Inherent resistance is when the bacteria is naturally lack the susceptible to a drug because of the poor penetration of the agent to bacterial cell or lack of affinity between antibacterial agent and bacteria. Acquired resistance is when the susceptibility of the bacteria to the drug later becomes resistant by transferring of resistance from other antibiotic resistant strains (Romero et al. 2012). The mechanisms by which transfer happens can be by conjugation (plasmid transfer through pilus), transformation (assimilation of DNA from external environment) and transduction (genetic material acquired from bacteriophage). The discrete mobile genetic units called cassettes are present and a family of receptor elements called integrons, that provide the site into which gene cassettes and integrase, the enzyme responsible for gene movement are integrated. This can allow horizontal mobility of AMR, thus adapting to the environmental changes (Sekizaki et al. 2001). Some microorganisms acquire genes like beta-lactamase, which destroys beta-lactams and few other acquire efflux pumps that expel the antibacterial agent preventing from penetration in to the cell (Lawson et al. 2002; Kumarasamy et al. 2010; Sun et al. 2014). The horizontal mobility of AMR occur in an animal's gastrointestinal tract, the main reservoir of bacterial multiplication and exchange of genetic information. Antibiotic residues in food may affect human health by impairing colonization of normal intestinal flora or by altering metabolic enzyme activity in the normal flora. In an experiment, Oxytetracycline (OTC) was administered to Atlantic salmon and the intestinal microbiota was analyzed. Microbiota from OTC treated ones showed less diversity and was found to account for high resistant bacteria than those without OTC treatment (Navarrete et al. 2008). A similar kind of results were oberved when experiments with oxolinic acid and florfenicol were conducted (Grondel et al. 1985; Samuelsen et al. 1992; Lunden et al. 1998, 1999; Coyne et al. 2001; Lunden and Bylund 2002; Miranda and Zemelman 2002; Buschmann et al. 2012).

In the current situation, development of resistance is much rapid than new drug development by pharmaceutical companies. Some new beta-lactamase inhibitors in combination with antimicrobial agent have been developed for the treatment of Gram negative infections (Zhanel et al. 2013; Tängdén and Giske 2015), however, the emergence of new drug resistant phenotypes are more. Abuse of antibiotics and drastic decline in the development of new antibiotics has been attributed to the rapid evolution of antibiotic resistant pathogens. The applications of antibiotics as

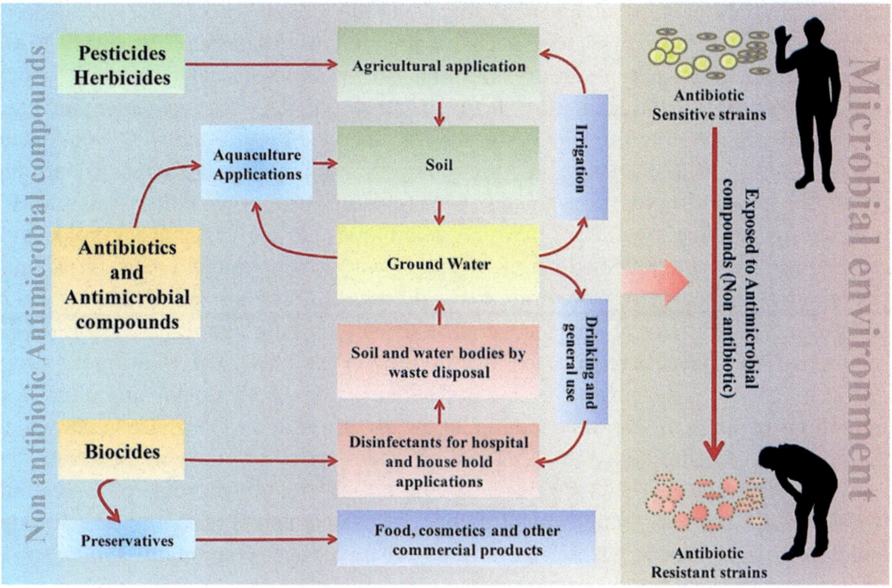

Fig. 5.1 Antibiotic resistance development in clinical isolates due to inappropriate application of non antibiotic antimicrobial compounds

prophylactic agents for plants, food animals and aquaculture growth promotion has raised concerns in recent years for their role in the development of antimicrobial resistance in clinically relevant pathogens. In addition, the use of antimicrobial chemicals such as preservatives and disinfectants and their correlation with antimicrobial resistance in clinical pathogens are also gaining attention in recent years. In this chapter we have discussed the non-medical uses of antibiotics and its impact on AMR in clinical settings (Fig. 5.1).

5.2 Agricultural Practices and Development of AMR

Since 1950's antibiotics are used to protect plants from bacterial blight disease, spot disease and crown gall disease of certain high value crops and ornamental plants. (McManus et al. 2002a).

Earlier streptomycin was used for the treatment of tuberculosis and gram negative infections, later its usage was limited due to the wide spread of streptomycin resistant genes in clinical strains. Then turn into a popular antibacterial agent for agriculture use because of its non-toxic nature to birds, freshwater invertebrates, and honeybees (Vidaver 2002). Streptomycin is primarily used for fire blight management in pear and apple. Few other antibiotics like oxytetracycline, oxolinic acid and gentamicin are also used in some countries for the control of pathogens such as *Erwinia* sp., *Pectobacterium* sp., *Pseudomonas* sp., *Ralstonia* sp., and *Xanthomonas* sp. (Stockwell and Duffy 2012). Oxytetracycline is used as a second line defence against blight disease of apple, pear, peach and nectarine in certain regions where wide spread streptomycin resistant *Erwinia* sp. have been documented (McManus and Stockwell 2000). Streptomycin resistance is due to the mutation in chromosomal gene *rpsL*, which inturn alters the ribosomal binding site of streptomycin. Another mechanism includes the inactivation of streptomycin by the enzyme encoded by *strA* and *strB* (Sundin and Bender 1996). In plant pathogens, the genes *strA* and *strB* usually reside on large (>30 kb), plasmid-borne transposons, accomplishing their own cell-to-cell transfer (McManus and Stockwell 2001). Streptomycin does not found to have an impact on non target soil bacteria (Shade et al. 2013), however, streptomycin resistant isolates were reported even more from non- sprayed orchards (Yashiro and McManus 2012) suggesting the well established concept that bacteria harbours transmissible AMR genes that have never been exposed to antibiotics (Hall 2004). The wide spread streptomycin resistance has made it necessary to investigate whether plant agriculture formulations of streptomycin are contaminated with the producer resistance genes which could accelerate streptomycin resistance in pathogen and non target bacteria. The absence of producer resistance genes suggests that streptomycin resistant gene are found to be a common trait in plant pathogens (Rezzonico et al. 2009). In addition, streptomycin resistance has been reported in fish pathogen *Aeromonas salmonicida*, where streptomycin has never been used (Tolba et al. 2002).

The usage of antibiotics for crop protection and the transfer of antibiotic resistant genes from plant pathogens to medically important human pathogens is the subject of debate and is of great concern (McManus et al. 2002b). This is due to the reason that resistant isolates not only carry streptomycin and oxytetracycline resistant genes but may also carry some other antibiotic resistant genes on a single plasmid that may further get transferred to human pathogens (Schnabel and Jones 1999). When the plasmid of orchard bacteria carrying streptomycin and tetracycline resistant genes were transferred into *E. coli*, it showed resistance to tetracycline and not towards streptomycin and other antibiotics such as cefataxime, pepracilline (Zhang et al. 2011). There are no reports at the molecular levels to correlate the application of antibiotics for crop protection and emergence of AMR in clinical strains

5.2.1 Azole Antifungals

Similarly, the use of azole antifungals in agriculture has also given rise to concerns that azoles may also have an impact on human fungal pathogens, though trizoles for plant and human use are structurally different (Ribas e Ribas et al. 2016). Many genera of fungi such as, *Fusarium* spp., *Aspergillus* spp., *Rhizopus* spp., *Mucor* spp., *Cladosporuim* spp., *Absidia* spp. have been reported as the common causative agents for plant and human fungal diseases. These common fungal pathogens may carry cross-resistance to triazole which is used in human medicine. Resistance to azole antifungal agents was a rare event until the 1980s, however, it became an increasing problem recently. The development of resistance towards Azole drugs are well studied in *C. albicans* and are broadly categorized as (i) drug extrusion by membrane efflux pumps; (ii) mutation of target site; and (iii) development of bypass pathways. A study by

Albrecht Serfling investigated the acquisition of resistance to an agricultural azole (Tebuconazole) and the resulting cross-resistance to various medical antifungal agents. The cross-resistance was investigated in keratitis fungi *Colletotrichum* spp. which occurs primarily after injury by plant material. Tebuconazole adapted *Colletotrichum* spp. strains was found to exhibit cross-resistance to all medical azoles tested (Serfling et al. 2007). In an *in vitro* induction study, upon exposure to agricultural azole, *C. glabrata* developed cross-resistance to fluconazole and posaconazole, suggesting its exposure to agricultural azoles could be associated with the development of cross-resistance. Similarly, prochloraz (agriculture azole) exposure with clinical *A. fumigatus* demonstrated the development of cross-resistance with medical azoles (posaconazole, itraconazole and voriconazole) (Faria-Ramos et al. 2014). Collectively these findings imply the possibilities of triggering the molecular mechanisms in plant pathogens to develop cross-resistance with clinically relevant pathogens

5.2.2 Pesticides and Herbicides

Another area of concern in development of AMR is the wide applications of pesticides and herbicides for agriculture. Recently few studies were conducted to correlate pesticide tolerance and development of antibiotic resistant traits. The pesticides (carbofuran, endosulphan, malathion) tolerant isolates, *Pseudomonas, Azatobacter, Rhizobium* spp. were also found to resist one or more antibiotics tested such as nalidixic acid, cloxacillin, chloramphenicol, tetracycline, amoxycillin, methicillin and doxycycline. Surprisingly, all the pesticide resistant isolates were also found to be resistant to cloxacillin and half of them were resistant to methicillin (Shafiani and Malik 2003). However, this experiment was carried for agricultural soil isolates

Another study has linked herbicides and AMR that demonstrates the toxicity of herbicides towards bacteria. The sub-lethal exposure to dicamba, 2,4-dichlorophenoxyacetic acid, and glyphosate caused changes in susceptibility pattern in *E. coli* and *Salmonella* sp. when tested against five different classes of antibiotics, with an induction of membrane efflux pumps proteins (Kurenbach et al. 2015). This important discovery suggests that herbicides and pesticides which are generally tested for their effectiveness towards killing the pests might underestimate their significance in the emergence of AMR

A solid evidence to correlate the direct relationship between the development of drug resistance in agricultural practice and evocation of drug resistance in clinical practice is yet to be found

5.2.3 Regulatory Policies and Recommendations

United States Environmental Protection Agency (EPA) has set regulations to minimise the exposure of workers to agricultural antibiotics such as by recommending to wear protective clothing during antibiotic applications to crops. EPA also regulates the time period permitted between last spray and crop harvest, ensuring the minimal exposure of humans to antibiotics for plant use. However a suitable alternative such as bio control agents with low toxicity, and less residual problem remains to be an effective method for crop protection (Granatstein 2013). Also, extensive research is needed to find the relevance of antibiotic use in plant agriculture and the resistance problem in clinically relevant antibiotics

5.3 Antibiotics in Animal Agriculture

In animal agriculture, antibiotics are used globally as growth promoters, prevention and treatment of infectious diseases like mastitis, respiratory infections, uterine infections (Ruegg et al. 2012), contagious bovine pleuropneumonia and foot-and-mouth disease, and rinderpest (Chikitsa and Sansthan 2014). The efficacy of growth promoters in animal feed are evidenced by increase in feed intake by 15% in poultry, 10% in lambs, and 15% in swine, 17% in beef cattle (Vishnuraj et al. 2016). However, the mechanisms of antibiotics as growth promoters are unclear. Some proposed mechanisms suggest that antibiotics suppress the activity of intestinal microbiota which are considered to be a major competitor for nutrients within the host (Taylor 2006) and inhibition of sub-clinical infection which further enhance the nutrient uptake via the thinner intestinal wall (Allen and Stanton 2014). However there is lack of experimental proofs for this. Till now there are no clear evidences for the growth promotion activity of antibiotics.

The transmission of antibiotics from animal farms to humans occurs by means of dirct contact and through meat consumption. In addition, animal waste are the major reservoir of antibiotic and their residues and results in certain serious health disorders like hypersensitivity, teratogenicity, carcinogenicity (Darwish et al. 2013), nephropathy, hepatotoxicity, bone marrow toxicity and reproductive disorders (Vishnuraj et al. 2016). In addition, the emerging major area of concern is development and spread of drug resistant pathogens. Drug resistant *Camphylobacter* spp., *Enterococci* spp., *E .coli, Salmonella* spp. were widely reported to be transferred from animals to human.

The three main reasons for development of AMR pathogens in animal farms are:

1. Misuse of antibiotics is a major factor that develops drug resistance. For example, antibiotics mixed with feed or water for sick animals (generally do not take food or water properly), subsequently, the neighboring animals may receive huge quantities of antibiotics without having any disease symptoms.
2. Easy availability of growth promoters in market and often farmers get drugs without any prescription and are then used for extended period of time.
3. Use of sub therapeutic concentrations which may lead to development of AMR. These kinds of practices in animal husbandry promote the emergence of drug resistant isolates. By noticing the case studies in different countries at different time points the spread of antibiotic resistant isolates from animal to human beings are evidenced.

5.3.1 Resistance Development in Clinically Relevant Pathogens

5.3.1.1 *Campylobacter* spp.

Campylobacteriosis, a zoonotic disease is transmitted through uncooked meat products, raw milk and contaminated water (Gorkiewicz et al. 2002). The causative agent *Campylobacter jejuni* was more prevalent in poultry, pigs, cattle, sheep, dogs and cats. Most commonly, fluoroquinolone and macrolides are used to prevent *Campylobacter* infection in poultry (Moore et al. 2006; Payot et al. 2006). High level of tylosin and erythromycin (macrolides) and ciprofloxacin, levofloxaacin (fluoroquinolone) resistant *Campylobacter* spp. were recorded in both animals (broilers, pigs, cattle) and humans. These resistant bacteria might get transferred to human through food chain (EFSA 2015). *Campylobacter* isolates from humans were found to have extremely high resistance to ciprofloxacin and tetracycline (Luber et al. 2003). Hence the treatment options for most common food infections are decreased. World Health Organisation (WHO) recognized *Campylobacter* sp. as one of the major AMR evolving organism in animals and humans. According to CDC, *Campylobacter* spp. causes approximately 1.3 million infections, resulting in 13,000 falling ill, and 120 deaths per annum in United States (WHO 2012). European Food Safety Authority (EFSA) and European Centre for Disease Prevention and Control (ECDC) have taken a step forward to analyse the AMR in food animals and humans, respectively. The latest integrated data of EFSA-ECDC summarise the prevalence of *Campylobacter* resistant isolates among animals and humans. Interestingly, the cross-resistance of *Campylobacter* spp. to cefotaxime, a critically important antibiotic were reported to be low to moderate levels in animals and at low levels in human isolates. A similar kind of low level cross-resistance was noticed in *Salmonella* spp. to cefotaxime (EFSA 2015).

5.3.1.2 *Enterococci* spp.

Enterococci spp. is normal intestinal flora of animals and humans. Few species like *E. cecorum*, *E. faecalis*, *E. durans* are associated with arthritis, spondylitis, osteomyelitis, spondylolisthesis, and femoral head necrosis in broiler and broiler breeder flocks, hepatic granulomas in turkeys, ascites in hens, pulmonary hypertension in broilers (Stępień-Pyśniak et al. 2016). Avopracin an analogue of vancomycin, is widely used as a feed additive in poultry and pigs for production of high meat content. VRE were isolated abundantly during 1994–1995 in animal food products and feacal samples of healthy individuals. Resistant *Enterococci* spp. carries *VanA* and *VanB* gene clusters that produces modified peptidoglycan d-ala-ala that subsequently lose its ability to bind with vancomycin and confers high glycopeptide resistance(Mazel and Davies 1999). The emergence of AMR in *Enterococci* spp. is a major concern, as vancomycin is considered as the last resort for treating gram positive infection. Germany and European countries banned the use of Avopracin during 1996 and 1997. Followed by Avopracin discontinuation at end of 1997s in Germany and European countries there was significant decrease in VRE (Klare et al. 1999) (WHO 2001).

Then linezolid was used as the last resort antibiotic for the treatment of gram positive bacterial infections particularly in VRE (Hammad et al. 2015). In 2009, linezolid resistant gene (*cfr*) was identified in *E. feacalis* that confers resistance to a set of unrelated group of antimicrobials like lincosamides, phenicols, oxazolidinone – linezolid, pleuromutilins and streptogaminA (Liu et al. 2012). *cfr* gene provoked the mutation in 23S rRNA gene by causing methylation of adenine at 2503 position and this prohibited the binding of antibiotics. The dissemination of mobile *cfr* gene resistance was not only limited within the genera, it can be spread to other genera as well (Shen et al. 2013). Recently, other than *cfr* gene, *optrA* gene confers resistance towards medically important antimicrobials like oxazolidinones linezolid and tedizolid (Wang et al. 2015).

A recent study conducted by Stępień-Pyśniak et al. (2016) demonstrates that 88% of *Enterococcus* spp. isolated from poultry displayed resistance to sulphamethoxazole/trimethoprim, 71. 4% of isolates to tylosin, enrofloxacin (69.4%), doxycycline (67.3%), and lincomycin/spectinomycin (56.1%). However, *E. faecalis* and *E. faecium* showed high frequency of sensitivity to amoxicillin, amoxicillin with clavulanic acid and florfenicol. Similarly a study conducted by Rozanska et al. (2015) observed *E. faecalis* strains from poultry meat were susceptible to penicillin. VRE isolated from raw pork, beef, chicken meet and eggs were found to contain *vanA vanB vanC vanD vanE and vanG* genes conferring resistance to pencillin, ampicillin, erythromycin, tetracycline, and ciprofloxacin (Gousia et al. 2015). *E. feacalis* from meat samples carries *tetM, tetL,* and *ermB* showing resistance to tetracyclines and erythromycin (Klibi et al. 2013). The incidence of VRE in poultry reported by authors across the world varied (Werner et al. 2008) as the use of drugs in animal production varied with countries.

5.3.1.3 *E. coli*

E. coli is fast evolving resistant bacteria when compared to other bacteria. Olaquindox and mequindox are commonly used antibiotics in swine farm. High levels of *oqxAB* antibiotic resistant gene carrying *E. coli* was identified in swine animals, swine farm workers and environment. The *oqxAB* is a plasmid mediated quinalone resistance showing resistance to olaquindox, mequindox and other quinolone. The main cause of dissemination of these AMR determinants are due to higher usage of quinolone antibiotics in animal feed (Zhao et al. 2010).

Pig manure used as fertilizer in agriculture provides rich sources of potassium, nitrogen, and phosphorous. However it was associated with risk, when they were carrying drug resistant *E. coli* and *Salmonella* sp. that causes serious illness in human beings. Piggery manure is the major reservoir for antibiotic resistant plasmids like IncP-1 and pHHV216 having *bla-TEM, sul-1, sul-2, sul-3* genes showing resistance towards amoxicillin and sulfadiazine (Binh et al. 2008). Spread of such antibiotic resistance plasmid to animal handlers was described by Levy (2002c). Transfer of *E. coli* resistant plasmid among poultry and animal handlers were reported widely in many countries such as Nigeria, northern India (Singh et al. 1992), Morocco (Amara et al. 1995), and Saudi Arabia (Al-Ghamdi et al. 1999). However, some authors have proved that *E. coli* of human and animal origin has different O serotypes and are unique on their own. Nevertheless, a recent study demonstrated the presence of identical ciprofolaxin resistant clones of *E. coli* in poultry and farmers strongly indicate the transfer of drug resistant bacteria and resistance plasmids of *E. coli* to humans (van den Bogaard et al. 2001). To summarize, antibiotics used for animal agriculture might play a significant role in evoking drug resistance in clinically relevant pathogens.

5.3.2 Regulatory Policies & Recommendations

In 2011, WHO came out with the following recommendations to control AMR in humans and food producing animals (Landers et al. 2012).

- Veterinary prescription is compulsory in live stock for treating microbial diseases
- Providing proper ventilation, maintenance of low live stock densities and good hygienic practice important to regulate the antibiotics use
- Monitoring the use of antibiotics for growth promotion, tracking the use of antibiotics in live stock and providing regulatory policies to control the AMR.
- Conducting of awareness program on proper use of antibiotics in veterinary live stock
- Banning of therapeutically important human antimicrobials in veterinary use
- Monitor the antimicrobial resistance in animal food products.

5.4 Antibiotics in Aquaculture

Aquaculture broadly refers to the farming of aquatic animals such as fish, molluscs, crustaceans etc., and this includes both marine and fresh water organisms. Being a kind of economically important industry, aquaculture involves plenty of human intervention that helps in increasing production of aquaculture animals while limiting input costs. Some of these practices include regular stocking of seeds, feed formulation, predator or disease avoidance etc. (Edwards 1997). With ever dwindling stocks of fishes in the world's oceans, seas and rivers, aqua-farming of animals is expected to contribute to more than half of the sea food that is going to be consumed the world over in the coming years. Antibiotics can be categorized as therapeutic (treatment of established infections) (Eg. chloramphenicol), prophylactic (use of antimicrobials for preventing the onset of infections) (Eg. cerfuroxime, vancomycin) or metaphylactic (dual use of antimicrobials for group medication for treating as well as preventing infections) (Eg. florphenicol, tuluthramycin) (Romero et al. n.d.). Some antibiotics like bacitracin, flavophospholipol, pleuromutilins are used as growth promoters (Wegener et al. 1999). The administration of antibiotics in aquaculture can be done by either mixing them with formulated feed or providing exposure along with water. Some antibiotic classification for veterinary use are beta-lactams (cephalosporins, penicillin, aminopenicillins), macrolides, chloramphenicol, florfenicol, spectinomycin (aminocyclitol), tetracyclins, quinolones, sulphonamides, lincosamides, rifampin, aminoglycosides. According to FDA 2016, the approved drugs for aquaculture are listed in Table 5.1.

As per a 2010 estimate, a total of 63,151 tonnes of antimicrobials was used exclusively for livestocks. In order to reduce such extensive and unregulated usage of antimicrobials in livestocks worldwide, certain criteria should be kept in mind, (a) difference between therapeutic and non-therapeutic application, (b) differences in the livestocks production systems itself and (c) species of animal for production and the role of their natural ecosystem. Antimicrobial agents aim to prevent occurrence and further spread of the infections and diseases. The major issues with exploitation of these agents in aquaculture is that, these are provided along with the feed and placed in water or these are mixed with the water itself, which would affect a broad spectrum of bacteria. The photosynthetic and nitrifying bacteria are essential to maintain ecological balance (Hovanec and DeLong 1992; Hargreaves 1998, 2006; Nagadomi et al. 1999). Antibiotics might affect these beneficial microbial communities, which in turn would affect organic matter degradation, disturb cycling of material and energy flow in the ecological system (UNESCO P. F. 1997). Antibiotics like spectinomycin, enrofloxacin, chloramphenicol, rifampin are banned for usage in animals intended for food production in some countries like USA, Canada, European Union, Japan and Argentina. These antibiotics are known to readily develop bacterial resistance (Spectinomycin, Enrofloxacin), cause aplastic anaemia (Chloramphenicol) and could be potential teratogens in animals (Rifampin). Development of resistant strains is one of the major risk factors associated with indiscriminate use of antibiotics. Resistance transfer can occur from aquaculture to the non-target species (humans), and then onwards to microbes infecting humans. This can occur through consumption (when trace amount of antibiotic drug remains in the flesh, promotes drug resistance within the individual human's bacterial community) (Barton 2002; Froehlicher et al. 2009; Baquero 2009) or even handling of products containing the residues. This can be toxic, mutagenic or allergenic to the organs (Hayse-Gregson and Diana 2011).

5.4.1 Regulatory Policies

There are set regulations by governments that prescribe obligatory Maximum Residue Levels for a variety of aquaculture products. However the actual quantity of antimicrobial residues consumed or encountered by people determines the public health risk associated with these compounds (Muñoz et al. 2010). Moreover, as per the WHO and FAO reports, it is now known that it is the AMR bacteria in food rather than antimicrobial residues in food that poses a grave risk for human health worldwide (World Health Organization. and Food and Agriculture Organization of the United Nations. 2003). After the second joint FAO/OIE (World Organisation for Animal Health- Office international des epizooties) /WHO workshop, 2004, it was recommended to follow Good Agricultural Practices (GAP), as defined by Codex (Recommended International Code of Practice – General Principles of Food

Table 5.1 List of approved drugs for aquaculture by FDA (2016)

Mode of administration	Antibiotics	Effect	Commercial products
Immersion	Chloramine- T	Bacterial gill disease	HALAMID®AQUA-NADA 141-423
	Formalin	Control of Protozoa	Formalin-F™ – NADA 137-687
			Formacide-B- ANADA 200-414
			Parasite-S®- NADA 141-255
	Hydrogen peroxide	Ectoparasites (Sea lice- *Lepeophtherius salmonis*), bacterial gill disease	35% PEROX-AID®-NADA 141-255
	Oxytetracycline hydrochloride	External columnaris, bacterial hemorrhagic septicemia	Oxymarine™ – NADA 130-435
			Oxytetracycline HCl Soluble Powder-343-ANADA 200-247
			PENNOX 343- ANADA 200-026
			TERRAMYCIN 343 Soluble Powder- NADA 008-622
			TETROXY Aquatic-ANADA 200-460
	Tricaine methanesulfonate	Anaesthetic	Tricaine S- ANADA 200-226
Injectable	Chorionic gonadotropin	Spawning function	Chorulon®- NADA 140-297
Medicated Articles/ Feeds	Florfenicol	Enteric septicemia, cold water disease, furunculosis, columnaris	Aquaflor® -NADA 141-246
	Oxatetracycline dehydrate	Control mortality due to cold water disease and columnaris	Terramycin® 200 for Fish-NADA 038-439
	Sulfadimethoxine/ Ormetroprim	Furunculosis, enteric septicemia	Romet-30®- NADA 125-933
			Sulfamerazine- NADA 033-950

Jacobsen and Berglind (1988), Samuelsen (1992), Björklund et al. (1991), Coyne et al. (1994), Samuelsen et al. (1994), Capone et al. (1996), Weston (2000), Rach et al. (2008), Heuer et al. (2009), and Yanong and Francis-Floyd (2016)

Hygiene, CAC/RCP (Codex Alementarius Commission/Recommended international Code of Practise) 1–1969, Rev. 4 (2003), with an annex on "Hazard Analysis Critical Control Point (HACCP) system and guidelines for its application" (1969), in order to minimize or reduce the use of antimicrobials in both agriculture and aquaculture. Subsequently, as per the consultation meeting of 2004 it was concluded that development and spread of resistant bacteria and resistant gene together with antibiotic residues in aquaculture products constitute part of human health hazards due to antibiotic use in aquaculture. Antimicrobial resistance is a growing problem everywhere and in all fields where antimicrobials are used. Due to globalization resistant microbes and genes are easily transferred all over different geographical locations. Based on a workshop on Antimicrobials and AMR in Aquaculture in Seoul in 2006 (Seoul 2006), the following guidelines were proposed for regulating the indiscriminate use of antimicrobial substances for both food and non-food animals: Use of extra-label could be an important part of antimicrobial control especially in those countries where there is a limited availability of registered veterinary medicines for aquaculture use.

(i) Decisions on registration of veterinary antibiotics should consider the possibility to reduce the impact on human health of AMR developing in aquatic animals and environment.
(ii) Need to follow appropriate use of evaluated antimicrobial agents for aquaculture following established regulations.
(iii) The importance of educating veterinarians and veterinary paraprofessionals (e.g. animal health professionals), on the proper and need-based use of antimicrobials in aquaculture.

Setting up and development of extension services for antimicrobial application in aquaculture in countries where they are used under limited or no expert supervision. Infection control and prevention with improvement in diagnostic practices, disease surveillance and management health plans, monitoring health status of ornamental fish and providing optimal environment, proper feeding, sanitation, vaccination, selective breeding and immunostimulants are some other proposals for prevention of diseases and infection control and to minimize antimicrobial usage. Owing to weak or inadequate legislation, regulatory surveillance and monitoring systems, FAO insists on a strict and controlled use of antimicrobial agents so as to reduce the risk of spread of AMR. They have identified four main focus areas and five objectives of the Global Action Plan for 2016–2020 on AMR (FAO 2016).

5.4.2 Recommendations

Due to the rapid increase in antibiotic resistance and their residues, there is a urgent need for identifying and developing alternative strategies against bacterial infections in animals and this is especially true for aquaculture industry. One such alternate strategy could be the use of bacteriophages for specific killing of pathogenic bacteria (Nakai and Park 2002). This strategy will not be deleterious to non-target bacterial species and also there is a lower chance of resistance development as the pool of bacteria available is small (Defoirdt et al. 2006b). The ability to infect a wide range of target bacterial species should be the primary criteria for selection of phages as biocontrol agents in aquaculture. A study tested the lytic activity against *Vibrio harveyi* by isolating phage from shrimp farm water and its ability to protect *Penaeus monodon* from a natural infection of luminescent vibriosis was evaluated. The phage was seen to increase the survival from 17% to 86% after 17 days when compared to daily addition of antibiotics (shrimp survival only 40%) (Steen et al. 2004). But the only problem with this approach is that bacteriophage may transfer virulence factors (Vázquez et al. 2005) and bacteria might develop resistance in phage attack which can be overcome by use of mixture of phages or their components (Anderson and Dawes 1990; Halet et al. 2007). A study by Imbeault et al. (2006) demonstrated the efficacy of a mixture of bacteriophages in brook trout against *Aeromonas salmonicida* infection. However, it was realized that the resistant bacteria, when compared to the original strain, had a shorter generation time and moreover the success rate when the same was repeated was low (Park and Nakai 2003). Probiotics are live microorganisms that when administered in adequate amounts, confer a health benefit on the host (Reid et al. 2003). Bacteria like lactic acid bacteria (Villamil et al. 2002; Vázquez et al. 2005; Balcázar and Rojas-Luna 2007; Balcázar et al. 2008; Pérez-Sánchez et al. 2011), *Bacillus* spp.(Ochoa-Solano and Olmos-Soto 2006; Balcázar et al. 2007; Bandyopadhyay and Das Mohapatra 2009; Liu et al. 2009; Nayak 2010; Ai et al. 2011; Antony et al. 2011), *Vibrio* spp. (Vaseeharan and Ramasamy 2003; Fjellheim et al. 2007; Thompson et al. 2010), *Aeromonas* spp. (Irianto et al. 2003; Lategan et al. 2004a; Lategan et al. 2004b; Lategan et al. 2006), *Actinobacteria* (You et al. 2007; Das et al. 2010) have been studied for use as probiotics in aquatic organisms like prawns (Van Hai et al. 2009), shrimps (Farzanfar 2006; Ninawe and Selvin 2009) and bivalve molluscs (Kesarcodi-Watson et al. 2008; Prado et al. 2010). Some of the mechanism of action of these probiotics may involve inhibitory compound production, enhancement of disease resistance, immunemodulation (Verschuere et al. 2000; Ochoa-Solano and Olmos-Soto 2006; Geovanny et al. 2007; Tinh et al. 2008b; Sahu et al. 2008; Flesche 2011; Palaty 2011; Abellán 2014). Growth inhibition is another important strategy to control pathogenic bacteria and short chain fatty acids are well known for this effect (Van Immerseel et al. 2002). Acidophilic bacteria such as lactic acid bacteria are important symbionts and can also withstand intestinal pH due to the presence of short chain fatty acids, when compared to neutrophilic bacteria (Vázquez et al. 2005; Defoirdta et al. 2006). Polyhydroxylactonates, such as poly-β-hydroxybutyrate are frequently found to be degraded in the gastrointestinal tracts, and these were found to give similar effects like short chain fatty acids (Defoirdt et al. 2009; Liu et al. 2010). Anti-virulence therapy, regulation of virulence factor expression or specifically inhibiting virulence factor, can also serve as an alternative (Clatworthy et al. 2007). These can be achieved by disrupting bacterial communication (quorum sensing) system (Defoirdt et al. 2004; Milton 2006; Defoirdt et al. 2008; Sahu et al. 2008; Merrifield et al. 2010; Defoirdt et al. 2011; Natrah et al. 2011), by interfering with (Steinberg et al. 2002; Rasch et al. 2004; Vine et al. 2006; Defoirdt et al. 2007) or inactivation (Defoirdt et al. 2006a; Dong et al. 2007; Tinh et al. 2008a; Nhan et al. 2010) of signal molecule detection. Essential oils, that contribute to plant aroma can also be used for exploiting the phenolic compounds that is part of the majority of plant antimicrobial components (Cosentino et al. 1999; Ultee et al. 2002). Some of their effect on bacterial membranes causes increase in membrane permeability resulting in disruption of proton motive force, electron flow and active transport (Hammer et al. 1999; Cosentino et al. 1999; Lambert et al. 2001; Cross et al. 2007). These are also found to inhibit quorum sensing, induce heat shock proteins and prevent flagella movement (Mahmoud et al. 2004; Maenner et al. 2011).

Aquatic organisms grown in farms have essentially required the increased use of antibiotics, which in turn concerns public health. Proper surveillance and strict regulations for the use of licensed antibiotics is necessary for eradicating pathogenic bacteria without addition of ill effects to the human and environment. Alternatives to antibiotics in aquaculture have also emerged which have been proved promising in reducing the negativities faced due to antimicrobials as such. Further studies on these alternatives can be more helpful in understanding the actual mechanisms and evaluating their long term impact on environment and host microbiota. Need of antibiotics itself can be reduced by application of good husbandry practices, thus replacing prophylactic usage and reinforcing immunization programs, reduce the stress on organisms and proper hygienic maintenance.

5.5 Biocides

Biocides are chemicals commonly used in food industries, pharmaceuticals, and cosmetics and to protect the product from spoilage. Biocides are classified into two types: (i) Preservatives and (ii) Disinfectants.

5.5.1 Preservatives

Worldwide there are demands to produce market products with increased shelf life, which can be achieved economically by using chemical preservatives like chlorhexidine, phenolics, ammonium compounds, organic acids, sulphites and carbon dioxide, which targets mainly microbe's cell wall and cell membrane, (Table 5.2) and which are similar to the action of antibiotics. Continuous and inappropriate use of these biocides can increase cross-resistance to antimicrobial compounds, which are an important and emerging concern in clinical setup (Condell et al. 2012). *E. coli* ATCC 12806, a widely used strain in food industry, has recently been shown to possess antibiotic resistance, raising concerns for humans. The adapted *E. coli* (trisodium phosphate, sodium nitrite, and sodium hypochlorite) showed resistance towards aminoglycosides, cephalosporins, and

Table 5.2 Mode of action of preservatives

Mode of action	Targets bacteria	Preservatives	References
Cell wall inhibitors	Gram negative (*E. coli* and *Salmonella*) and positive bacteria	Phenolic compound, Phenol, Aryl alkyl acid, Organo mercurials, EDTA, Chlorhexidine, Cetrimide, Glutaraldehyde, Anionic surfactants, weak acids	Puupponen-Pimia et al. (2001) and Stratford et al. (2013)
Cell membrane inhibitor	Gram negative, positive bacteria and yeast	2-phenoxyethanol, weak acids	Komatsu et al. (2002) and Stratford et al. (2013)
Cytoplasmic membrane inhibition	Gram negative and positive bacteria	2-phenoxyethanol, Parabens, Organo mercurials, Hexachlorophene, Formaldehyde donators, Bronopol, Imidurea and Benzalkonium chloride	Denyer (1990), Denyer (1995), and Maillard (2002)

quinolones (Capita et al. 2014). Similarly, benzalkonium chloride, chlorhexidine, and triclosan induced resistance by permeability alteration or upregulation of efflux pump in *Salmonella* spp. and other gram negative bacteria (Condell et al. 2012). A similar kind of resistance profile was observed in triclosan treated *E. coli*, with an enhanced susceptibility to multiple clinically relevant antibiotics such as ampicillin, fluoroquinolones, chloramphenicol, tetracycline. The clinical strains of *E. coli* were found to have over-expression of *marA*, *soxS* or *acrAB* (Mcmurry 1998). Though many laboratory investigations suggested the linkage between triclosan usage and AMR, some lab reported that triclosan tolerant bacterial species did not increase emerging AMR (Lear et al. 2006; Birosova and Mikulasova 2009; Cottell et al. 2009). Recently, *P. aeruginosa*, *E. coli* and *S. aureus* isolated from different commercially available cosmetics were found to exhibit resistance against all the preservatives (methyl paraben, propyl paraben, imidazolidinyl urea and dimethyl dimethylol hydantoin) and some commonly used antibiotics (amoxacillin, tetracycline, cotrimoxazol, gentamycin, norfloxacin, nalidixic acid, tobramycin) (Abu Shaqra et al. 2014). The cross-resistance might be triggered between preservatives and antibiotics because of similar target site, transport mechanism, evolution of resistance mechanism and similar kind of mobile resistance genes (Gilbert and McBain 2003; Condell et al. 2012). One of the commonly used resistance mechanism against biocides and antibiotics in microbes is the use of membrane situated multidrug efflux pump. These pumps are important to gram negative bacteria for the establishment of intrinsic or acquired resistance against a wide variety of antibiotics as well as biocides, especially to the resistance-nodulation-division family (RND family) of gram negative bacteria (Piddock 2006; Hinchliffe et al. 2013). This mechanism showed resistance to both amphiphilic and lipophilic compounds, including detergents (bile acids, sodium dodecyl sulfate), dyes (crystal violet, ethidium bromide), and antimicrobial agents (β-lactams, chloramphenicol, tetracycline, fluoroquinolones, erythromycin) (Piddock 2006). Thus, further studies are needed to understand and correlate the role of antimicrobial agents for different applications and its role in development of AMR.

5.5.2 Disinfectants

Antibiotics and disinfectants are two different categories of antimicrobial agents. Antibiotics has specific drug target site whereas disinfectants has multiple target sites (Table 5.3) and some of the antibiotics and disinfectants has common target sites (Chopra 1998; Hugo 1999). For example *Mycobacterium* sp. can be inhibited by multiple agents such as isoniazid, triclosan, cationic biocides, chlorhexidine salts and Quaternary Ammonium Compounds (QAC), wherein these agents target mycobacterial enoyl reductase. Hence the probability of developing AMR in disinfectant-resistant bacteria is high. Various resistance mechanisms are responsible for cross-resistance between biocides and antibiotics, specifically over expression of efflux pumps and horizontal transfer of mobile genes carrying resistant determinants. Fluoroquinolone resistance is mainly related to antibiotic abuse. Cross-resistance of fluoroquinolone and QAC is observed because of over-expression of RND type AcrAB efflux pump, which is primarily responsible for displaying resistance against FQ and QAC in *E. coli.* (Levy 2002a). Mutated AcrAB efflux pump did not show resistance against FQ (Buffet-Bataillon et al. 2016). Biocide resistant clinical isolates possess plasmid mediated QAC resistant genes such as qacA/B/C/D/E. A single plasmid carrying multiple numbers of resistant genes against various antimicrobial agents are evidenced in various reports, for example *S.aureus* pSK-01carries resistant genes of both antibiotics and biocides (Lyon and Skurray 1987; Paulsen et al. 1993; Lucey et al. 2000). Extensive use of disinfectants showed resistance towards disinfectants and antibiotics. Nosocomial pathogens (*Proteus mirabilis, P. aeruginosa, Serratia marcescens and Providencia stuartii)* showed resistance against chlorohexidine and antibiotics (Stickler and Chawla 1988; Russell 2002). However, this cross-resistance can be circumstantial also.

5.6 Conclusion

In conclusion, extensive uses of antimicrobial agents under non-clinical settings such as in agriculture, animal husbandry, aquaculture, and other food and non-food industries has raised concerns in recent years and are highly suspected for their role of development of AMR in health care settings. Though various laboratories have demonstrated the decrease in susceptibility to antibiotics in bacteria that are resistant to biocides and others, evidences such as transfer of resistant genes and plasmids to clinically relevant pathogens are required to correlate non-medical use of antibiotics and evocation of AMR mechanisms. However, antimicrobial usage for non-medical purposes should be minimised by way of developing suitable alternatives like biocontrol agents, use of plant based products such as essential oils, ban on inappropriate use of antibiotics for growth promotion in animals, use of

Table 5.3 Mode of action of disinfectants

Group	Disinfectant	Application	Mode of action	References
Halogen	Chlorine-releasing agents (CRAs) N-chloro compounds Chlorine dioxide Sodium hypochlorite	Water treatment, Deodrization Antiseptics	DNA synthesis Protein synthesis	Mcdonnell and Russell (1999)
	Iodine	Fungicidal Bactericidal Virucidal Sporicidal Tuberculocidal To treat wound	Cellular Protein, Fatty Acids, Nucleotides	Kinnaman (1908), Dakin et al. (1916), Lawrence et al. (1957), and Mcdonnell and Russell (1999)
Heavy metal ions	Mercury Mercuric chloride Organic synthesized mercurials	Dental amalgam Wound cleaning Vaccine preservative, seed preservation, medicines and paint	Cytoplasmic membrane	Leistevuo et al. (2000). Block (2001), and Maillard (2002)
	Copper Copper oxide nano particals	Against gram negative and gram positive bacteria Wood preservation	Cell wall	Theivasanthi and Alagar (2011), Matsunaga et al. (2010)
Alcohols	*n*-propanol Isopropyl alcohol Ethyl alcohol	broad-spectrum antimicrobial activity against virus, fungus and bacteria	protein and membrane damage	Atiyeh et al. (2009)
	Isopropyl alcohol	Antibacterial for surface disinfectant and skin antiseptic because of lipophilic nature		
	Ethyl alcohol	Antiviral because of hydrophilic nature		
Aldehydes	Glutaraldehyde Formaldehyde Anilides Biguanides (Chlorhexidine, Alexidine, Polymeric biguanides), Diamidines, Peroxygens, Phenols, Halophenols	Broad-spectrum antimicrobial	Cell Wall	GORMAN et al. (1980), Ganz et al. (1985), and Denyer (1995)

probiotics in animal farms and aquaculture, control in providing and usage of antibiotics in all sectors including health care.

References

Abellán FG (2014) Inmunotoxicología producida por metales pesados y caracterización del moco de piel en peces= Heavy metal immunotoxicology and skin mucus in fish

Abu Shaqra QM, Al-Momani W, Al-Groom RM (2014) Susceptibility of some bacterial contaminants recovered from commercial cosmetics in Jordan to preservatives and antibiotics. Trop J Pharm Res 13:255–259. https://doi.org/10.4314/tjpr.v13i2.14

Ai Q, Xu H, Mai K, et al (2011) Effects of dietary supplementation of *Bacillus subtilis* and fructooligosaccharide on growth performance, survival, non-specific immune response and disease. Aquaculture 317:155–161

Al-Ghamdi MS, El-Morsy F, Al-Mustafa ZH et al (1999) Antibiotic resistance of Escherichia coli isolated from poultry workers, patients and chicken in the eastern province of Saudi Arabia. Trop Med Int Heal 4:278–283. https://doi.org/10.1046/j.1365-3156.1999.00392.x

Allen HK, Stanton TB (2014) Altered egos : antibiotic effects on food animal microbiomes. Annu Rev Microbiol 68:297–315. https://doi.org/10.1146/annurev-micro-091213-113052

Amara A, Ziani Z, Bouzoubaa K (1995) Antibioresistance of Escherichia coli strains isolated in Morocco from chickens with colibacillosis. Vet Microbiol 43:325–330. https://doi.org/10.1016/0378-1135(94)00101-2

Anderson A, Dawes E (1990) Occurrence, metabolism, metabolic role, and industrial uses of bacterial polyhydroxyalkanoates. Microbiol Rev 54:450–472

Antony S, Singh I, Jose R et al (2011) Antimicrobial peptide gene expression in tiger shrimp, Penaeus monodon in response to gram-positive bacterial probionts and white spot virus challenge Aquaculture 316:6–12

Atiyeh BS, Dibo SA, Hayek SN (2009) Wound cleansing, topical antiseptics and wound healing. Int Wound J 6:420–430. https://doi.org/10.1111/j.1742-481X.2009.00639.x

Balcázar J, Vendrell D, de Blas I , Ruiz-Zarzuela I (2008) Characterization of probiotic properties of lactic acid bacteria isolated from intestinal microbiota of fish. Aquaculture 278:188–191

Balcázar J, Vendrell D, Blas I De, Ruiz-Zarzuela I (2007) In vitro competitive adhesion and production of antagonistic compounds by lactic acid bacteria against fish pathogens. Vet Microbiol 122:373–380

Balcázar JL, Rojas-Luna T (2007) Inhibitory activity of probiotic *Bacillus subtilis* UTM 126 Against Vibrio Species Confers Protection Against Vibriosis in Juvenile Shrimp (Litopenaeus vannamei). Curr Microbiol 55:409–412. https://doi.org/10.1007/s00284-007-9000-0

Bandyopadhyay P, Das Mohapatra PK (2009) Effect of a probiotic bacterium Bacillus circulans PB7 in the formulated diets: on growth, nutritional quality and immunity of *Catla catla* (Ham.). Fish Physiol Biochem 35:467–478. https://doi.org/10.1007/s10695-008-9272-8

Baquero F (2009) Environmental stress and evolvability in microbial systems. Clin Microbiol Infect 15:5–10. https://doi.org/10.1111/j.1469-0691.2008.02677.x

Barton BA (2002) Stress in Fishes: A Diversity of Responses with Particular Reference to Changes in Circulating Corticosteroids. Integr Comp Biol 42:517–525. https://doi.org/10.1093/icb/42.3.517

Binh CTT, Heuer H, Kaupenjohann M, Smalla K (2008) Piggery manure used for soil fertilization is a reservoir for transferable antibiotic resistance plasmids. FEMS Microbiol Ecol 66:25–37. https://doi.org/10.1111/j.1574-6941.2008.00526.x

Birosova L, Mikulasova M (2009) Development of triclosan and antibiotic resistance in Salmonella enterica serovar Typhimurium. J Med Microbiol 58:436–441. https://doi.org/10.1099/jmm.0.003657-0

Björklund HV, Råbergh CMI, Bylund G (1991) Residues of oxolinic acid and oxytetracycline in fish and sediments from fish farms. Aquaculture 97:85–96. https://doi.org/10.1016/0044-8486(91)90281-B

Block SS (2001) Disinfection, sterilization, and preservation – Google books. Lippincott, Philadelphia

Buffet-Bataillon S, Tattevin P, Maillard J-Y et al (2016) Efflux pump induction by quaternary ammonium compounds and fluoroquinolone resistance in bacteria. Future Microbiol 11:81–92. https://doi.org/10.2217/fmb.15.131

Buschmann AH, Tomova A, López A et al (2012) Salmon aquaculture and antimicrobial resistance in the marine environment. PLoS One 7:e42724. https://doi.org/10.1371/journal.pone.0042724

Capita R, Riesco-Peláez F, Alonso-Hernando A, Alonso-Calleja C (2014) Exposure of *Escherichia coli* ATCC 12806 to sublethal concentrations of food-grade biocides influences its ability to form biofilm, resistance to antimicrobials, and ultrastructure. Appl Environ Microbiol 80:1268–1280. https://doi.org/10.1128/AEM.02283-13

Capone DG, Weston DP, Miller V, Shoemaker C (1996) Antibacterial residues in marine sediments and invertebrates following chemotherapy in aquaculture. Aquaculture 145:55–75. https://doi.org/10.1016/S0044-8486(96)01330-0

Chikitsa P, Sansthan V (2014) Strategies for combating and eradicating important infectious diseases of animals with particular reference to India: Present and future perspectives. Asian J Anim Vet Adv 9:77–106

Chopra I (1998) Research and development of antibacterial agents. Curr Opin Microbiol 1:495–501

Clatworthy A, Pierson E, Hung D (2007) Targeting virulence: a new paradigm for antimicrobial therapy. Nat Chem Biol 3:541–548

Condell O, Iversen C, Cooney S et al (2012) Efficacy of biocides used in the modern food industry to control salmonella enterica, and links between biocide tolerance and resistance to clinically relevant antimicrobial compounds. Appl Environ Microbiol 78:3087–3097. https://doi.org/10.1128/AEM.07534-11

Cosentino S, Tuberoso CIG, Pisano B et al (1999) In-vitro antimicrobial activity and chemical composition of Sardinian Thymus essential oils. Lett Appl Microbiol 29:130–135. https://doi.org/10.1046/j.1472-765X.1999.00605.x

da Costa P, Loureiro L, Matos A (2013) Transfer of multidrug-resistant bacteria between intermingled ecological niches: the interface between humans, animals and the environment. Intl J Environ Res Pub Health 10:278–294

Cottell A, Denyer SP, Hanlon GW et al (2009) Triclosan-tolerant bacteria: changes in susceptibility to antibiotics. J Hosp Infect 72:71–76. https://doi.org/10.1016/j.jhin.2009.01.014

Coyne R, Hiney M, O'Connor B et al (1994) Concentration and persistence of oxytetracycline in sediments under a marine salmon farm. Aquaculture 123:31–42. https://doi.org/10.1016/0044-8486(94)90117-1

Coyne R, Smith P, Moriarty C (2001) The fate of oxytetracycline in the marine environment of a salmon cage farm

Cross DE, McDevitt RM, Hillman K, Acamovic T (2007) The effect of herbs and their associated essential oils on performance, dietary digestibility and gut microflora in chickens from 7 to 28 days of age. Br Poult Sci 48:496–506. https://doi.org/10.1080/00071660701463221

Dakin HD, Cohen JB, Daufresne M, Kenyon J (1916) The antiseptic action of substances of the chloramine group. R Soc 89:232–251

Darwish WS, Eldaly EA, El-Abbasy MT, Ishizu M (2013) Antibiotic residues in food: the African scenario: HUSCAP. Jpn J Vet Res. http://133.87.26.249/dspace/handle/2115/52350. Accessed 7 Jan 2017

Das S, Ward L, Burke C (2010) Screening of marine Streptomyces spp. for potential use as probiotics in aquaculture. Aquaculture 305:32–41

Defoirdt T, Boon N, Bossier P, Verstraete W (2004) Disruption of bacterial quorum sensing: an unexplored strategy to fight infections in aquaculture. Aquaculture 240:69

Defoirdt T, Boon N, Sorgeloos P, Verstraete W (2009) Short-chain fatty acids and poly-β-hydroxyalkanoates:(New) biocontrol agents for a sustainable animal production. Biotechnol Adv 27:680–685

Defoirdt T, Boon N, Sorgeloos P, Verstraete W (2008) Quorum sensing and quorum quenching in Vibrio harveyi: lessons learned from in vivo work. ISME J 2:19–26

Defoirdt T, Crab R, Wood T, Sorgeloos P (2006a) Quorum sensing-disrupting brominated furanones protect the gnotobiotic brine shrimp Artemia franciscana from pathogenic Vibrio harveyi, Vibrio campbellii, and vibrio parahaemolyticus isolates, Appl Environ Microl 72:6419–6423

Defoirdt T, Halet D, Sorgeloos P, Bossier P (2006b) Short-chain fatty acids protect gnotobiotic *Artemia franciscana* from pathogenic Vibrio campbellii. Aquaculture 261:804–808

Defoirdt T, Miyamoto CM, Wood TK et al (2007) The natural furanone (5Z)-4-bromo-5-(bromomethylene)-3-butyl-2(5H)-furanone disrupts quorum sensing-regulated gene expression in Vibrio harveyi by decreasing the DNA-binding activity of the transcriptional regulator protein luxR. Environ Microbiol 9:2486–2495. https://doi.org/10.1111/j.1462-2920.2007.01367.x

Defoirdt T, Sorgeloos P, Bossier P (2011) Alternatives to antibiotics for the control of bacterial disease in aquaculture. Curr Opin Microbiol 14:251–258

Defoirdta T, Haleta D, Sorgeloosb P et al (2006) Short-chain fatty acids protect gnotobiotic *Artemia franciscana* from pathogenic Vibrio campbellii. Aquaculture 261:804–404

Denyer SP (1990) Mechanisms of action of biocides. Int Biodeterior 26:89–100. https://doi.org/10.1016/0265-3036(90)90050-H

Denyer SP (1995) Mechanisms of action of antibacterial biocides. Int Biodeterior Biodegradation 36:227–245. https://doi.org/10.1016/0964-8305(96)00015-7

Dong Y, Wang L, Zhang L (2007) Quorum-quenching microbial infections: mechanisms and implications. Philos Trans R Soc B Biol Sci 362:1201–1211

Edwards P (1997) Rural aquaculture: overview and framework for country reviews. FAO Regional Office for Asia and the Pacific

EFSA (2015) EU Summary Report on antimicrobial resistance in zoonotic and indicator bacteria from humans, animals and food in 2013. EFSA J 13:1–178. https://doi.org/10.2903/j.efsa.2015.4036

FAO (2016) The FAO action plan on antimicrobial resistance supporting the food and agriculture sectors in implementing the Global Action Plan on Antimicrobial Resistance to minimize the impact of antimicrobial resistance

Faria-Ramos I, Farinha S, Neves-Maia J et al (2014) Development of cross-resistance by Aspergillus fumigatus to clinical azoles following exposure to prochloraz, an agricultural azole. BMC Microbiol 14:155. https://doi.org/10.1186/1471-2180-14-155

Farzanfar A (2006) The use of probiotics in shrimp aquaculture: Table 1. FEMS Immunol Med Microbiol 48:149–158. https://doi.org/10.1111/j.1574-695X.2006.00116.x

Fjellheim A, Playfoot K, Skjermo J, Vadstein O (2007) Vibrionaceae dominates the microflora antagonistic towards Listonella anguillarum in the intestine of cultured Atlantic cod (Gadus morhua L.) larvae

Flesche K (2011) Pro-og prebiotika for torskelarver

Froehlicher M, Liedtke A, Groh K, Neuhauss S (2009) Zebrafish (Danio rerio) neuromast: promising biological endpoint linking developmental and toxicological studies. Aquat Toxicol 95:307–319

Ganz T, Selsted ME, Szklarek D et al (1985) Defensins. Natural peptide antibiotics of human neutrophils. J Clin Invest 76:1427–1435. https://doi.org/10.1172/JCI112120

Geovanny DGR, Balcázar JL, Ma S (2007) Probiotics as control agents in aquaculture. J Ocean Univ China 6:76–79. https://doi.org/10.1007/s11802-007-0076-8

Gilbert P, McBain A (2003) Potential impact of increased use of biocides in consumer products on prevalence of antibiotic resistance. Clin Microbiol Rev 16:189–208

Gorkiewicz G, Feierl G, Zechner R, Zechner EL (2002) Transmission of Campylobacter hyointestinalis from a pig to a human. J Clin Microbiol 40:2601–2605. https://doi.org/10.1128/JCM.40.7.2601

Gorman SP, Scott EM, Russell AD (1980) Antimicrobial activity, uses and mechanism of action of glutaraldehyde. J Appl Bacteriol 48:161–190. https://doi.org/10.1111/j.1365-2672.1980.tb01217.x

Gousia P, Economou V, Bozidis P, Papadopoulou C (2015) Vancomycin-Resistance Phenotypes, Vancomycin-Resistance Genes, and Resistance to Antibiotics of Enterococci Isolated from Food of Animal Origin. Foodborne Pathog Dis 12:214–220. https://doi.org/10.1089/fpd.2014.1832

Granatstein, David, Stockwell V (2013) Report: lack of evidence for direct linkage of plant agriculture use of oxytetracycline to antibiotic resistance in human pathogens

Grondel J, Angenent G, Egberts E (1985) The influence of antibiotics on the immune system III. Investigations on the cellular functions of chicken leukocytes in vitro. Vet Immunol Immunopathol 10(4):307–316

Van Hai N , Buller N, Fotedar R (2009) The use of customised probiotics in the cultivation of western king prawns (*Penaeus latisulcatus* Kishinouye, 1896). Fish Shellf Immunol 27:100–104

Halet D, Defoirdt T, Damme P Van (2007) Poly-β-hydroxybutyrate-accumulating bacteria protect gnotobiotic Artemia franciscana from pathogenic Vibrio campbellii. FEMS Microbiol Ecol 60:363–369

Hall BG (2004) Innovation: Predicting the evolution of antibiotic resistance genes. Nat Rev Microbiol 2:430–435. https://doi.org/10.1038/nrmicro888

Hammad AM, Hassan HA, Shimamoto T (2015) Prevalence, antibiotic resistance and virulence of Enterococcus spp. in Egyptian fresh raw milk cheese. Food Control 50:815–820. https://doi.org/10.1016/j.foodcont.2014.10.020

Hammer KA, Carson CF, Riley TV (1999) Antimicrobial activity of essential oils and other plant extracts. J Appl Microbiol 86:985–990. https://doi.org/10.1046/j.1365-2672.1999.00780.x

Hargreaves JA (1998) Nitrogen biogeochemistry of aquaculture ponds. Aquaculture 166:181–212. https://doi.org/10.1016/S0044-8486(98)00298-1

Hargreaves JA (2006) Photosynthetic suspended-growth systems in aquaculture. Aquac Eng 34:344–363. https://doi.org/10.1016/j.aquaeng.2005.08.009

Hayse-Gregson K, Diana J (2011) The problematic relationship between aquaculture and antibiotics | Human World | EarthSky. Hum WORLD

Heuer OE, Kruse H, Grave K et al (2009) Human health consequences of use of antimicrobial agents in aquaculture. Clin Infect Dis 49:1248–1253. https://doi.org/10.1086/605667

Hinchliffe P, Symmons MF, Hughes C, Koronakis V (2013) Structure and operation of bacterial tripartite pumps. Annu Rev Microbiol 67:221–242. https://doi.org/10.1146/annurev-micro-092412-155718

Hovanec TA, DeLong EF (1992) Deep-sea research. Part B, oceanographic literature review. Pergamon Press, New York

Hugo W (1999) Disinfection mechanisms

Imbeault S, Parent S, Lagacé M et al (2006) Using bacteriophages to prevent Furunculosis caused by *Aeromonas salmonicida* in farmed Brook Trout. J Aquat Anim Health 18:203–214. https://doi.org/10.1577/H06-019.1

Van Immerseel F, Cauwerts K, Devriese LA et al (2002) Feed additives to control Salmonella in poultry. Sci J 58:501–513

Irianto A, Robertson PAW, Austin B (2003) Oral administration of formalin-inactivated cells of *Aeromonas hydrophila* A3-51 controls infection by atypical A. salmonicida in goldfish, *Carassius auratus* (L.). J Fish Dis 26:117–120. https://doi.org/10.1046/j.1365-2761.2003.00439.x

Jacobsen P, Berglind L (1988) Persistence of oxytetracycline in sediments from fish farms. Aquaculture 70:365–370. https://doi.org/10.1016/0044-8486(88)90120-2

Kesarcodi-Watson A, Kaspar H, Lategan M, Gibson L (2008) Probiotics in aquaculture: the need, principles and mechanisms of action and screening processes. Elsevier Science, Amsterdam

Kinnaman GC (1908) The antimicrobic action of bromin. J Am Med Assoc L:345. https://doi.org/10.1001/jama.1908.25310310021001e

Klare I, Badstübner D, Konstabel C et al (1999) Decreased incidence of VanA-type vancomycin-resistant enterococci isolated from poultry meat and from fecal samples of humans in the

community after discontinuation of avoparcin usage in animal husbandry. Microb Drug Resist 5:45–52. https://doi.org/10.1089/mdr.1999.5.45
Klibi N, Ben SL, Jouini A et al (2013) Species distribution, antibiotic resistance and virulence traits in enterococci from meat in Tunisia. Meat Sci 93:675–680. https://doi.org/10.1016/j.meatsci.2012.11.020
Komatsu E, Yamazaki H, Abe K et al (2002) Influence of Temperature on the Efficacy of 2-Phenoxyethanol as a Preservative for Adsorbed Diphtheria-Purified Pertussis-Tetanus Combined Vaccine. J Heal Sci 48:89–92. https://doi.org/10.1248/jhs.48.89
Kumarasamy K, Toleman M, Walsh T (2010) Emergence of a new antibiotic resistance mechanism in India, Pakistan, and the UK: a molecular, biological, and epidemiological study. Lancet Infect Dis 10:597–602
Kurenbach B, Marjoshi D, Amábile-cuevas CF et al (2015) Sublethal exposure to commercial formulations of the herbicides changes in antibiotic susceptibility in Escherichia coli and Salmonella enterica serovar. Typhimurium MbioAsmOrg 6:1–9. https://doi.org/10.1128/mBio.00009-15.Editor
Lambert RJW, Skandamis PN, Coote PJ, Nychas G-JE (2001) A study of the minimum inhibitory concentration and mode of action of oregano essential oil, thymol and carvacrol. J Appl Microbiol 91:453–462. https://doi.org/10.1046/j.1365-2672.2001.01428.x
Landers TF, Cohen B, Wittum TE, Larson EL (2012) A review of antibiotic use in food animals: perspective, policy, and potential. Public Health Rep 127:4–22. https://doi.org/10.1177/003335491212700103
Lategan M, Booth W, Shimmon R, Gibson L (2006) An inhibitory substance produced by Aeromonas media A199, an aquatic probiotic. Aquaculture 254:115–124
Lategan M, Torpy F, Gibson L (2004a) Biocontrol of saprolegniosis in silver perch Bidyanus bidyanus (Mitchell) by Aeromonas media strain A199. Aquaculture 235:77–88
Lategan M, Torpy F, Gibson L (2004b) Control of saprolegniosis in the eel Anguilla australis Richardson, by Aeromonas media strain A199. Aquaculture 240: 19
Lawrence CA, Carpenter CM, Naylor-Foote AWC (1957) Iodophors as Disinfectants**Health Department, County of Los Angeles, Los Angeles 12, Calif. J Am Pharm Assoc (Scientific ed) 46:500–505. https://doi.org/10.1002/jps.3030460813
Lawson AJ, Dassama MU, Ward LR, Threlfall EJ (2002) Multiply resistant (MR) *Salmonella enterica* serotype Typhimurium DT 12 and DT 120: a case. Emerging Infect Dis 8(4):434–436
Lear JC, Maillard J-Y, Dettmar PW et al (2006) Chloroxylenol- and triclosan-tolerant bacteria from industrial sources—susceptibility to antibiotics and other biocides. Int Biodeterior Biodegradation 57:51–56. https://doi.org/10.1016/j.ibiod.2005.11.002
Leistevuo J, Järvinen H, Osterblad M et al (2000) Resistance to mercury and antimicrobial agents in Streptococcus mutans isolates from human subjects in relation to exposure to dental amalgam fillings. Antimicrob Agents Chemother 44:456–457. https://doi.org/10.1128/AAC.44.2.456-457.2000
Levy SB (2002a) Active efflux, a common mechanism for biocide and antibiotic resistance. Symp Ser Soc Appl Microbiol 92:65S–71S. https://doi.org/10.1046/j.1365-2672.92.5s1.4.x
Levy SB (2002b) The 2000 Garrod lecture. Factors impacting on the problem of antibiotic resistance. J Antimicrob Chemother 49:25–30. https://doi.org/10.1093/jac/49.1.25
Levy SB (2002c) The antibiotic paradox: how the misuse of antibiotics destroys their curative powers, 2nd edn. Int Microbiol 5:155–156. https://doi.org/10.1007/s10123-002-0082-z
Liu C-H, Chiu C-S, Ho P-L, Wang S-W (2009) Improvement in the growth performance of white shrimp, *Litopenaeus vannamei* , by a protease-producing probiotic, *Bacillus subtilis* E20, from natto. J Appl Microbiol 107:1031–1041. https://doi.org/10.1111/j.1365-2672.2009.04284.x
Liu Y, De Schryver P, Van Delsen B et al (2010) PHB-degrading bacteria isolated from the gastrointestinal tract of aquatic animals as protective actors against luminescent vibriosis. FEMS Microbiol Ecol 74:196–204. https://doi.org/10.1111/j.1574-6941.2010.00926.x
Liu Y, Wang Y, Wu C et al (2012) First report of the multidrug resistance gene cfr in Enterococcus faecalis of animal origin. Antimicrob Agents Chemother 56:1650–1654. https://doi.org/10.1128/AAC.06091-11

Luber P, Wagner J, Hahn H, Bartelt E (2003) Antimicrobial resistance in Campylobacter jejuni and Campylobacter coli strains isolated in 1991 and 2001–2002 from poultry and humans in Berlin, Germany. Antimicrob Agents Chemother 47:3825–3830. https://doi.org/10.1128/AAC.47.12.3825-3830.2003

Lucey B, Crowley D, Moloney P et al (2000) Integronlike structures in Campylobacter spp. of human and animal origin. Emerg Infect 6:50–55

Lunden T, Bylund G (2002) Effect of sulphadiazine and trimethoprim on the immune response of rainbow trout (Oncorhynchus mykiss). Vet Immunol Immunopathol 85:99–108

Lunden T, Miettinen S, Lönnström L, Lilius E (1998) Influence of oxytetracycline and oxolinic acid on the immune response of rainbow trout (Oncorhynchus mykiss). Fish Shellf Immunol 8:217–230

Lunden T, Miettinen S, Lönnström L, Lilius E (1999) Effect of florfenicol on the immune response of rainbow trout (Oncorhynchus mykiss). Vet Immunol Immunopathol 67:317

Lyon B, Skurray R (1987) Antimicrobial resistance of Staphylococcus aureus: genetic basis

Maenner K, Vahjen W, Simon O (2011) Studies on the effects of essential-oil-based feed additives on performance, ileal nutrient digestibility, and selected bacterial groups in the gastrointestinal tract of piglets, J Anim Sci 89:2106–2112

Mahmoud B, Yamazaki K, Miyashita K, Il-Shik S (2004) Bacterial microflora of carp (Cyprinus carpio) and its shelf-life extension by essential oil compounds. Food Microbiol 21:657–666

Maillard J-Y (2002) Bacterial target sites for biocide action. J Appl Microbiol 92:16S–27S. https://doi.org/10.1046/j.1365-2672.92.5s1.3.x

Matsunaga H, Kataoka Y, Kiguchi M (2010) Copper nanoparticles in southern pine wood treated with a micronised preservative: can nanoparticles penetrate the cell walls of tracheids and ray parenchyma?

Mazel D, Davies J (1999) Antibiotic resistance in microbes. Cell Mol Life Sci 56:742–754

Mcdonnell G, Russell AD (1999) Antiseptics and disinfectants: activity, action, and resistance. Clin Microbiol Rev 12:147–179

McManus P, Stockwell V (2000) Antibiotics for plant diseases control: Silver Bullets or Rusty Sabers. APSnet Feat. https://doi.org/10.1094/APSnetFeature-2000-0600

McManus PS, Stockwell VO (2001) Antibiotic use for plant disease management in the united states. Plant Heal Prog 2:2–900. https://doi.org/10.1094/PHP-2001-0327-01-RV

McManus PS, Stockwell VO, Sundin GW, Jones AL (2002a) Antibiotic use in plant agriculture. Annu Rev Phytopathol 40:443–465. https://doi.org/10.1146/annurev.phyto.40.120301.093927

McManus PS, Stockwell VO, Sundin GW, Jones AL (2002b) A NTIBIOTIC U SE IN P LANT A GRICULTURE. Annu Rev Phytopathol 40:443–465. https://doi.org/10.1146/annurev.phyto.40.120301.093927

Mcmurry L (1998) Overexpression of marA, soxS, or acrAB produces resistance to triclosan in laboratory and clinical strains of Escherichia coli. FEMS Microbiol Lett 166:305–309. https://doi.org/10.1016/S0378-1097(98)00347-4

Merrifield D, Dimitroglou A, Foey A, Davies S (2010) The current status and future focus of probiotic and prebiotic applications for salmonids. Aquaculture 302:1–18

Milton D (2006) Quorum sensing in vibrios: complexity for diversification

Miranda C, Zemelman R (2002) Bacterial resistance to oxytetracycline in Chilean salmon farming. Aquaculture 212:31–47

Moore JE, Barton MD, Blair IS et al (2006) The epidemiology of antibiotic resistance in Campylobacter. Microbes Infect 8:1955–1966. https://doi.org/10.1016/j.micinf.2005.12.030

Muñoz I, Martínez Bueno MJ, Agüera A, Fernández-Alba AR (2010) Environmental and human health risk assessment of organic micro-pollutants occurring in a Spanish marine fish farm. Environ Pollut 158:1809–1816. https://doi.org/10.1016/j.envpol.2009.11.006

Nagadomi H, Hiromitsu T, Takeno K et al (1999) Treatment of aquarium water by denitrifying photosynthetic bacteria using immobilized polyvinyl alcohol beads. J Biosci Bioeng 87:189–193. https://doi.org/10.1016/S1389-1723(99)89011-2

Nakai T, Park S (2002) Bacteriophage therapy of infectious diseases in aquaculture. Res Microbiol 153:13–18

Natrah FMI, Defoirdt T, Sorgeloos P, Bossier P (2011) Disruption of Bacterial Cell-to-Cell Communication by Marine Organisms and its Relevance to Aquaculture. Mar Biotechnol 13:109–126. https://doi.org/10.1007/s10126-010-9346-3

Navarrete P, Mardones P, Opazo R et al (2008) Oxytetracycline Treatment Reduces Bacterial Diversity of Intestinal Microbiota of Atlantic Salmon. J Aquat Anim Health 20:177–183. https://doi.org/10.1577/H07-043.1

Nayak S (2010) Probiotics and immunity: a fish perspective. Fish Shellf Immunol 29:2–14

Nhan DT, Cam DTV, Wille M et al (2010) Quorum quenching bacteria protect Macrobrachium rosenbergii larvae from Vibrio harveyi infection. J Appl Microbiol 109:1007–1016. https://doi.org/10.1111/j.1365-2672.2010.04728.x

Ninawe AS, Selvin J (2009) Probiotics in shrimp aquaculture: Avenues and challenges. Crit Rev Microbiol 35:43–66. https://doi.org/10.1080/10408410802667202

O'Neill J (2015) Review on antimicrobial resistance: antimicrobial resistance: tackling a crisis for the health and wealth of nations chaired by Jim O'Neill, December 2014

Ochoa-Solano J, Olmos-Soto J (2006) The functional property of Bacillus for shrimp feeds. Food Microbiol 23:519–25

Palaty C (2011) Health effects of aquacultured seafood consumption: evidence and intervention strategies

Park S, Nakai T (2003) Bacteriophage control of Pseudomonas plecoglossicida infection in ayu Plecoglossus altivelis

Paulsen I, Littlejohn T, Rådström P (1993) The 3'conserved segment of integrons contains a gene associated with multidrug resistance to antiseptics and disinfectants, vol 37, pp 761–768

Payot S, Bolla JM, Corcoran D et al (2006) Mechanisms of fluoroquinolone and macrolide resistance in Campylobacter spp. Microbes Infect 8:1967–1971. https://doi.org/10.1016/j.micinf.2005.12.032

Pérez-Sánchez T, Balcázar JL, García Y et al (2011) Identification and characterization of lactic acid bacteria isolated from rainbow trout, Oncorhynchus mykiss (Walbaum), with inhibitory activity against Lactococcus garvieae. J Fish Dis 34:499–507. https://doi.org/10.1111/j.1365-2761.2011.01260.x

Piddock LJV (2006) Multidrug-resistance efflux pumps — not just for resistance. Nat Rev Microbiol 4:629–636. https://doi.org/10.1038/nrmicro1464

Prado S, Romalde J, Barja J (2010) Review of probiotics for use in bivalve hatcheries

Puupponen-Pimia R, Nohynek L, Meier C et al (2001) Antimicrobial properties of phenolic compounds from berries. J Appl Microbiol 90:494–507. https://doi.org/10.1046/j.1365-2672.2001.01271.x

Rach JJ, Johnson A, Rudacille JB, Schleis SM (2008) Efficacy of oxytetracycline hydrochloride bath immersion to control external columnaris disease on Walleye and Channel Catfish Fingerlings. N Am J Aquac 70:459–465. https://doi.org/10.1577/A07-062.1

Rasch M, Buch C, Austin B, Slierendrecht W (2004) An inhibitor of bacterial quorum sensing reduces mortalities caused by vibriosis in rainbow trout (Oncorhynchus mykiss, Walbaum)

Reid G, Jass J, Sebulsky MT, McCormick JK (2003) Potential uses of probiotics in clinical practice. Clin Microbiol 16:658–672

Rezzonico F, Stockwell VO, Duffy B (2009) Plant agricultural streptomycin formulations do not carry antibiotic resistance genes. Antimicrob Agents Chemother 53:3173–3177. https://doi.org/10.1128/AAC.00036-09

Ribas e Ribas AD, Spolti P, Del Ponte EM et al (2016) Is the emergence of fungal resistance to medical triazoles related to their use in the agroecosystems? A mini review. Brazilian J Microbiol 47:793–799. https://doi.org/10.1016/j.bjm.2016.06.006

Romero J, Feijoó C, Navarrete P (2012) Antibiotics in aquaculture-use, abuse and alternatives

Romero J, Feijoó CG, Navarrete P (n.d.) Antibiotics in aquaculture – use, abuse and alternatives

Różańska H, Lewtak-Piłat A, Osek J (2015) Antimicrobial resistance of *Enterococcus faecalis* isolated from meat. Bull Vet Inst Pulawy 59:229–233. https://doi.org/10.1515/bvip-2015-0034

Ruegg PL, Science D, Wi M (2012) Antimicrobial residues and resistance: understanding and managing drug usage on dairy farms

Russell AD (2002) Introduction of biocides into clinical practice and the impact on antibiotic-resistant bacteria. J Appl Microbiol 92:121S–135S. https://doi.org/10.1046/j.1365-2672.92.5s1.12.x

Sahu MK, Swarnakumar NS, Sivakumar K et al (2008) Probiotics in aquaculture: importance and future perspectives. Indian J Microbiol 48:299–308. https://doi.org/10.1007/s12088-008-0024-3

Samuelsen O (1992) The fate of antibiotics/chemotherapeutics in marine aquaculture sediments. Probl Chemother Aquac Theory:87–95

Samuelsen O, Torsvik V, Ervik A (1992) Long-range changes in oxytetracycline concentration and bacterial resistance towards oxytetracycline in a fish farm sediment after medication. Sci Total Environ 114:25–36

Samuelsen OB (1989) Degradation of oxytetracycline in seawater at two different temperatures and light intensities, and the persistence of oxytetracycline in the sediment from a fish farm. Aquaculture 83:7–16. https://doi.org/10.1016/0044-8486(89)90056-2

Samuelsen OB, Lunestad BT, Ervik A, Fjelde S (1994) Stability of antibacterial agents in an artificial marine aquaculture sediment studied under laboratory conditions. Aquaculture 126:283–290. https://doi.org/10.1016/0044-8486(94)90044-2

Sanco D (2002) European Commission, Health & Consumer Protection Directorate-General

Schnabel EL, Jones AL (1999) Distribution of tetracycline resistance genes and transposons among phylloplane bacteria in Michigan Apple Orchards distribution of tetracycline resistance genes and transposons among Phylloplane bacteria in Michigan Apple Orchards. Appl Environ Microbiol 65:4898–4907

Sekizaki T, Otani Y, Osaki M et al (2001) Evidence for horizontal transfer of SsuDAT1I restriction-modification genes to the Streptococcus suis genome. J Bacteriol 183:500–511. https://doi.org/10.1128/JB.183.2.500-511.2001

Seoul R of K (2006) Antimicrobial use in aquaculture and antimicrobial resistance. Report of a joint FAO/OIE/WHO expert consultation on antimicrobial use in aquaculture and antimicrobial resistance. In: Seoul

Serfling A, Wohlrab J, Deising HB (2007) Treatment of a clinically relevant plant-pathogenic fungus with an agricultural azole causes cross-resistance to medical azoles and potentiates caspofungin efficacy. Antimicrob Agents Chemother 51:3672–3676. https://doi.org/10.1128/AAC.00654-07

Shade A, Klimowicz AK, Spear RN et al (2013) Streptomycin application has no detectable effect on bacterial community structure in apple orchard soil. Appl Environ Microbiol 79:6617–6625. https://doi.org/10.1128/AEM.02017-13

Shafiani S, Malik A (2003) Tolerance of pesticides and antibiotic resistance in bacteria isolated from wastewater- irrigated soil. World J Microbiol Biotechnol 19:897–901. https://doi.org/10.1023/B:WIBI.0000007290.94694.4f

Shen J, Wang Y, Schwarz S (2013) Presence and dissemination of the multiresistance gene CFR in Gram-positive and Gram-negative bacteria. J Antimicrob Chemother 68:1697–1706. https://doi.org/10.1093/jac/dkt092

Singh M, Chaudhry MA, Yadava JNS, Sanyal SC (1992) The spectrum of antibiotic resistance in human and veterinary isolates of *Escherichia coli* collected from 1984–86 in Northern India. J Antimicrob Chemother 29:159–168. https://doi.org/10.1093/jac/29.2.159

Smith RD, Coast J (2002) Antimicrobial resistance: a global response. Bull World Health Organ 80:126–133. https://doi.org/10.1590/S0042-96862002000200008

Steen RJCA, Ariese F, van HB et al (2004) Monitoring and evaluation of the environmental dissipation of the marine antifoulant 4,5-dichloro-2-n-octyl-4-isothiazolin-3-one (DCOIT) in a Danish Harbor. Chemosphere 57:513–521. https://doi.org/10.1016/j.chemosphere.2004.06.043

Steinberg P, Rasmussen TB, Henzter M et al (2002) Halogenated furanones inhibit quorum sensing through accelerated LuxR turnover. Microbiology 148:1119–1127. https://doi.org/10.1099/00221287-148-4-1119

Stępień-Pyśniak D, Marek A, Banach T et al (2016) Prevalence and antibiotic resistance of Enterococcus strains isolated from poultry. Acta Vet Hung 64:148–163. https://doi.org/10.1556/004.2016.016

Stickler D, Chawla J (1988) Antiseptics and long-term bladder catheterization. J Hosp Infect 12:337–338
Stockwell VO, Duffy B (2012) Use of antibiotics in plant agriculture. Rev Sci Tech 31:199–210
Stratford M, Steels H, Nebe-von-Caron G et al (2013) Extreme resistance to weak-acid preservatives in the spoilage yeast Zygosaccharomyces bailii. Int J Food Microbiol 166:126–134. https://doi.org/10.1016/j.ijfoodmicro.2013.06.025
Sun J, Deng Z, Yan A (2014) Bacterial multidrug efflux pumps: mechanisms, physiology and pharmacological exploitations. Biochem Biophys Res Commun 453(2): 254–267
Sundin GW, Bender CL (1996) Dissemination of the strA-strB streptomycin-resistance genes among commensal and pathogenic bacteria from humans, animals, and plants. Mol Ecol 5:133–143
Tängdén T, Giske C (2015) Global dissemination of extensively drug-resistant carbapenemase-producing Enterobacteriaceae: clinical perspectives on detection, treatment and infection control. J Int Med 277(5):501–512
Taylor P (2006) Antibiotics as growth promotants : mode of action. 37–41. https://doi.org/10.1081/ABIO-120005768
Theivasanthi T, Alagar M (2011) Studies of Copper Nanoparticles Effects on Micro-organisms
Thompson J, Gregory S, Plummer S et al (2010) An in vitro and in vivo assessment of the potential of Vibrio spp. as probiotics for the Pacific White shrimp, Litopenaeus vannamei. J Appl Microbiol 109:1177–1187. https://doi.org/10.1111/j.1365-2672.2010.04743.x
Tinh N, Yen V, Dierckens K, Sorgeloos P (2008a) An acyl homoserine lactone-degrading microbial community improves the survival of first-feeding turbot larvae (Scophthalmus maximus L.). Aquaculture 285:56–62
Tinh NTN, Dierckens K, Sorgeloos P, Bossier P (2008b) A Review of the Functionality of Probiotics in the Larviculture Food Chain. Mar Biotechnol 10:1–12. https://doi.org/10.1007/s10126-007-9054-9
Tolba S, Egan S, Kallifidas D, Wellington EMH (2002) Distribution of streptomycin resistance and biosynthesis genes in streptomycetes recovered from different soil sites. FEMS Microbiol Ecol 42:269–276. https://doi.org/10.1111/j.1574-6941.2002.tb01017.x
Ultee A, Bennik M, Moezelaar R (2002) The phenolic hydroxyl group of carvacrol is essential for action against the food-borne pathogen Bacillus cereus. Appl Environ Microbiol 68:4
UNESCO PFW (1997) Towards safe and effective use of chemicals in coastal aquaculture. Reports Stud 65:1–38
van den Bogaard AE, London N, Driessen C et al (2001) Antibiotic resistance of faecal *Escherichia coli* in poultry, poultry farmers and poultry slaughterers. J Antimicrob Chemother 47:763–771. https://doi.org/10.1093/jac/47.6.763
Vaseeharan B, Ramasamy P (2003) Control of pathogenic Vibrio spp. by *Bacillus subtilis* BT23, a possible probiotic treatment for black tiger shrimp Penaeus monodon. Lett Appl Microbiol 36:83–87. https://doi.org/10.1046/j.1472-765X.2003.01255.x
Vázquez J, González M, Murado M (2005) Effects of lactic acid bacteria cultures on pathogenic microbiota from fish. Aquaculture 245:149–161
Verschuere L, Rombaut G, Sorgeloos P (2000) Probiotic bacteria as biological control agents in aquaculture. Microbiol Mol Biol Rev 64:655–671
Vidaver AK (2002) Uses of antimicrobials in plant agriculture. Clin Infect Dis 34(Suppl 3):S107–S110. https://doi.org/10.1086/340247
Villamil L, Tafalla C, Figueras A (2002) Evaluation of immunomodulatory effects of lactic acid bacteria in turbot (Scophthalmus maximus)
Vine N, Leukes W, Kaiser H (2006) Probiotics in marine larviculture, FEMS Microbiol Rev 30:404–427
Vishnuraj MR, Kandeepan G, Rao KH et al (2016) Occurrence , public health hazards and detection methods of antibiotic residues in foods of animal origin : A comprehensive review. Cogent Food Agric 486:1–8. https://doi.org/10.1080/23311932.2016.1235458
Wang Y, Lv Y, Cai J et al (2015) A novel gene, optrA, that confers transferable resistance to oxazolidinones and phenicols and its presence in Enterococcus faecalis and *Enterococcus faecium*

of human and animal origin. J Antimicrob Chemother 70:2182–2190. https://doi.org/10.1093/jac/dkv116
Wegener HC, Aarestrup FM, Jensen LB et al (1999) Use of antimicrobial growth promoters in food animals and *Enterococcus faecium* resistance to therapeutic antimicrobial drugs in Europe. Emerg Infect Dis 5:329–335. https://doi.org/10.3201/eid0503.990303
Werner G, Coque TM, Hammerum AM et al (2008) Emergence and spread of vancomycin resistance among enterococci in Europe. Eur Secur 13:1–11
Weston D (2000) Ecological effects of the use of chemicals in aquaculture
WHO (2001) Original: English WHO global strategy for containment of antimicrobial strategy for containment of antimicrobial resistance. World Health WHO/CDS/CS:105. WHO/CDS/CSR/DRS/2001.2, 99 p
WHO (2012) The global view of Campylobacteriosis
Wilson G (2004) A day on the fish farm: FDA and the regulation of aquaculture. Virginia Environ Law J 23:351–395
World Health Organization., Food and Agriculture Organization of the United Nations. (2003) Hazard characterization for pathogens in food and water: guidelines. World Health Organization, Geneva
Yanong RPE, Francis-Floyd R (2016) Therapeutic considerations in aquaculture – exotic and laboratory animals – veterinary manual. In: Merck Sharp & Dohme Corp, Kenilworth
Yashiro E, McManus PS (2012) Effect of streptomycin treatment on bacterial community structure in the apple phyllosphere. PLoS One 7:e37131. https://doi.org/10.1371/journal.pone.0037131
You J, Xue X, Cao L et al (2007) Inhibition of Vibrio biofilm formation by a marine actinomycete strain A66. Appl Microbiol Biotechnol 76:1137–1144. https://doi.org/10.1007/s00253-007-1074-x
Zhanel GG, Lawson CD, Adam H et al (2013) Ceftazidime-avibactam: a novel cephalosporin/??-lactamase inhibitor combination. Drugs 73:159–177
Zhang M, C a P, Zhou L et al (2011) Chemical compounds effective against the Citrus Huanglongbing bacterium "Candidatus Liberibacter asiaticus" in planta. Phytopathology 101:1097–1103. https://doi.org/10.1094/PHYTO-09-10-0262
Zhao J, Chen Z, Chen S et al (2010) Prevalence and dissemination of oqxAB in Escherichia coli isolates from animals, farmworkers, and the environment. Antimicrob Agents Chemother 54:4219–4224. https://doi.org/10.1128/AAC.00139-10
(1969) General principles of food hygiene
(2016) WHO | FAO/WHO Second International Conference on Nutrition (ICN2)
(2015) Status Report on Antimicrobial Resistance
USP Dietary Supplement Standards and USP Verified.

Biofilms in Antimicrobial Activity and Drug Resistance

6

Timsy Bhando, Vineet Dubey, and Ranjana Pathania

Abstract

Biofilms are communistic and complex network of microorganisms concealed in an autogenic polymeric matrix made up of polysaccharides, proteins and extracellular DNA. These surface attached microbial communities are responsible for more than 65% of human infections and have emerged as a major public health concern. Owing to their high population densities and cellular proximity, biofilms act as a barrier to antibiotic diffusion and are notoriously difficult to eradicate. Hence, high resistance of biofilm-associated infections to antibiotic therapy is one of the biggest clinical challenges. Yet our understanding about them needs further research and strategies for their control remain to be elucidated. This chapter is dedicated to gain insight into biofilm architecture and to study the mechanisms for their recalcitrance to antimicrobial therapy. Given the serious and pervasive clinical impact of biofilm-related infections, most recent strategies for their treatment have also been discussed.

Keywords

Bacterial biofilm · Antimicrobial resistance · Biofilm resistance mechanisms · Biofilm treatment strategies

T. Bhando · V. Dubey · R. Pathania (✉)
Department of Biotechnology, Indian Institute of Technology Roorkee, Roorkee, Uttarakhand, India
e-mail: rpathfbs@iitr.ac.in

S. M. Mandal, D. Paul (eds.), *Bacterial Adaptation to Co-resistance*,
https://doi.org/10.1007/978-981-13-8503-2_6

6.1 Introduction

Charles Darwin established the evolutionary theory and gave the phrase "Survival of the fittest" to describe the mechanism of natural selection (Murray 1860). The biological concept of fitness is defined as reproductive success and any action that improves the reproductive fitness endures within a species. Genetic adaptation forms the basis for survival and occurs via mutation, genetic recombination, acquisition of novel genes or regulation of gene expression. In order to acclimatize to a plethora of host niches and survive harsh environmental conditions, bacteria have evolved the ability to adhere to biotic or abiotic surfaces and form architecturally complex communities termed as biofilms. In the 1930s, Arthur Henrici and Claude Zobell first introduced the concept of bacterial adhesion to submerged surfaces (Henrici 1933; Zobell and Allen 1935). Later in 1978, William J. Costerton coined and described the term "biofilm" as surface-associated microbial aggregations and paved the way for understanding the significance of biofilm infection in the field of medicine (McCoy et al. 1981).

In the last few years, scientists have become keenly interested to study the process of bacterial biofilm formation and carried out extensive research to find answers for two major questions: what forces drive formation of biofilms and how do these confer a selective fitness advantage. Interestingly after years of research, it has now been suggested that the motivation behind the biofilm mode of growth is a strategy for defence against hostile environment and a mechanism to abide by the favourable niche of human host (Costerton et al. 1978). Further, molecular studies on biofilm formation have also begun to elucidate the driving forces responsible for the shift towards biofilm approach for sustenance (Joo and Otto 2012).

Biofilm formation is a dynamic, intricate and multifaceted process whereby the single-celled bacteria assume an interim multicellular lifestyle and establish communication amongst them (Hall-Stoodley et al. 2004). An extracellular matrix of polymeric substances predominantly composed of exopolysaccharides, proteins and nucleic acids support the biofilm communities (López et al. 2010). This protective milieu of biofilm structure can withstand high concentrations of antimicrobial agents and is extraordinarily resilient to host immune responses, thus making them particularly difficult to eradicate. Biofilms also catalyse the exchange of genetic material among bacterial species serving as reservoirs of antibiotic resistance.

Biofilm population contributes to majority of microbial infections such as endocarditis, urinary tract infections, dental caries, medical device-related infections and chronic lung infections in cystic fibrosis patients (Joo and Otto 2012). Moreover, infections by opportunistic biofilm forming pathogens in immunocompromised patients could be calamitous and sometimes even fatal. Besides serving as a source of various infections, biofilms also play an imperative role in pathogen physiology and persistence, which is responsible for recalcitrance of biofilm infections in clinical settings (Hall-Stoodley et al. 2004). Biofilm infections are not only associated with increased morbidity and mortality but also contribute to the emergence and dissemination of antibiotic resistance traits in the nosocomial setting (Okshevsky and Meyer 2015; Percival et al. 2015). The devastating and inescapable clinical impact of bacterial biofilms has inspired many researchers, in recent years to investigate the regulatory mechanisms of biofilm formation, with the ultimate goal of finding novel targets for chemotherapeutic agents and to device novel strategies for inhibition of biofilms.

6.2 Biofilms: Structure and Function

Biofilms are ubiquitous in nature comprising of either homogenous or heterogeneous bacterial populations in close proximity within the extracellular matrix. Formation of the biofilm matrix induces a unique environment for the bacteria that leads to a dynamic biofilm approach of life. It is an important trait that fortifies biofilm resistance and accounts for 90% of the total biofilm biomass. The Extracellular Polymeric Substances (EPS) matrix mainly comprises of exopolysaccharides that immobilize the biofilm cells, keeping them in long-term close proximity. Apart from EPS, the matrix is composed of carbohydrate-binding proteins followed by pili, flagella, adhesive fibers and extracellular DNA that provides a stabilizing platform for complex three dimensional biofilm architecture (Flemming and Wingender 2010; Kostakioti et al. 2013).

6.2.1 Exopolysaccharides

The extracellular matrix protects the cells within from external aggressions and acts as a barrier to the diffusion of various small molecules. It traps nutrients for metabolic utilizations by the resident bacteria and retains water through H-bond interactions with hydrophilic polysaccharides (López et al. 2010). Mannose, glucose and galactose are the most abundant carbohydrates in the matrix, followed by *N*-acetylglucosamine, galacturonic acid, fucose, arabinose, rhamnose and xylose. The composition and contribution of exopolysaccharides in the matrix vary depending on growth conditions, substrates and could be even species specific or strain specific. For example, mucoid strains of *Pseudomonas aeruginosa* synthesise alginate which is a critically important virulence factor during chronic infections such as cystic fibrosis (Rabin et al. 2015b). Alginate protects *P. aeruginosa* cells from the inhibitory action of antibiotics ciprofloxacin, ceftazidime etc. and suppresses the host immune response (Hodges and Gordon 1991; Leid et al. 2005). *P. aeruginosa* strains also produce two other exopolysaccharides namely, Pel (Glucose-rich) and Psl (mannose-rich) which serve as the primary biofilm matrix polysaccharides (Colvin et al. 2012).

6.2.2 Extracellular Proteins

Extracellular proteins are another major component of the EPS matrix. These adhesive proteins attach to cell surface and polysaccharides and help in biofilm formation and stabilization. For instance, *Staphylococcus aureus* matrix harbors Biofilm Associated Proteins (Bap) that are required for biofilm formation and infection processes (Lasa and Penadés 2006). *Bacillus subtilis* expresses amyloid-like protein TasA, which associates with the extracellular matrix and mutants lacking TasA are known to be deficient in biofilm formation (Branda et al. 2006). Other examples include Glucan binding proteins (Gbps) in *Streptococcus mutans* biofilms and Esp protein from *Enterococcus faecalis* (Lynch et al. 2007; Toledo-Arana et al. 2001). Additionally, matrix associated lectin proteins are also known to facilitate cell–matrix or cell–cell interactions within the biofilms. *P. aeruginosa* has two such lectin binding proteins, LecA and LecB that influence the biofilm architecture (Diggle et al. 2006; Tielker et al. 2005).

6.2.3 Extracellular DNA

In addition to the exopolysaccharides and proteins, extracellular DNA (eDNA) also provides.

structural integrity and stability to the biofilm (Okshevsky and Meyer 2015). These extracellular DNAs are actively secreted by the cells and are produced by cell lysis. eDNAs are critical for biofilm attachment since they interact with receptors on substratum to facilitate adhesion (Rabin et al. 2015a, b). Young *Pseudomonas* biofilms are vulnerable to DNase treatment as compared to mature biofilm, thus emphasing the stabilizing role of eDNA during the initial stages of biofilm formation (Whitchurch et al. 2002). Also, eDNA are known to coordinate cellular movement in the twitching motility mediated expansion of *P. aeruginosa* biofilms (Gloag et al. 2013). Being anionic in nature, eDNA can chelate Mg^{2+} ions and activate the PhoPQ/PmrAB two component system, leading to antibiotic resistance in *P. aeruginosa* and other Gram-negative bacteria (Mulcahy et al. 2008). eDNA also represents an important mechanism for horizontal gene transfer in bacteria (Montanaro et al. 2011; Vorkapic et al. 2016).

Thus, the structural components of the matrix give rise to a hydrated, robust structure having high tensile strength that keeps bacteria in close proximity, enabling cell-to-cell communication and genetic exchange (Flemming and Wingender 2010). The biofilm structure also protects the cells from wide range of environmental challenges such as dessication, metal toxicity, oxidative stress, UV exposure, antimicrobials etc. (Flemming and Wingender 2010; Hall-Stoodley et al. 2004; Walters et al. 2003). Ecologically, cooperation and competition in the presence of limited nutrients within the enclosed system of EPS matrix leads to constant adaptation of population fitness (Flemming and Wingender 2010).

6.3 Process of Biofilm Formation: Stages in a Biofilm Infection

Biofilm formation is a multi-step process in which microorganisms transform from planktonic to sessile mode of growth. The complex process of biofilm formation incorporates a series of events leading to adaptation under diverse nutritional and environmental conditions. Bacterial cells within biofilm communities elicit specific mechanisms and undergo changes upon surface adhesion. The biofilm communities are built in specific, defined steps that allow intense cellular interactions such as cell–cell communication and horizontal gene transfer. The developmental stages of biofilm formation have been well scrutinized and concluded to consist of the following key stages (a) initial reversible attachment to a surface (b) irreversible attachment and micro-colony formation (c) formation of a three dimensional biofilm structure and (d) biofilm maturation followed by detachment and dispersal (Fig. 6.1).

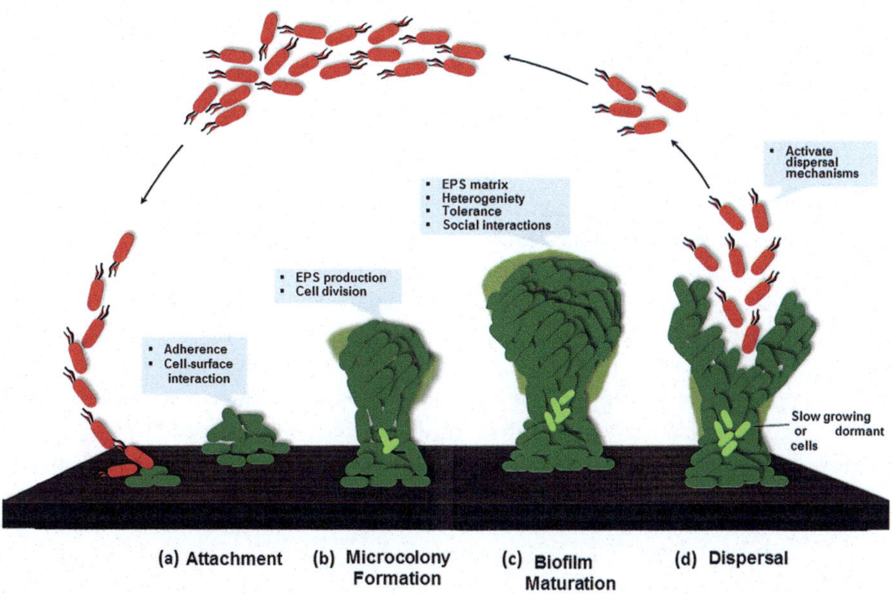

Fig. 6.1 Stages of Biofilm formation: Biofilm formation is a complex multi-step process. The different stages of biofilm formation include (**a**) reversible attachment to the surface followed by (**b**) irreversible attachment (**c**) formation of microcolonies through extracellular matrix production and (**d**) formation of a mature three-dimensional biofilm architecture. Mature biofilms then disintegrate and distribute planktonic cells to new sites of infection in the human host

6.3.1 Attachment

The first stage of initial attachment of biofilm to the surface is critical for the survival of all cells within a biofilm structure. In this stage, planktonic microbial cells loosely adhere to the surface either by physical forces or by virtue of bacterial appendages such as flagella, fimbriae, pili etc. (Marić and Vraneš 2007). The initial attachment is reversible and dynamic and involves Van der Waals forces, electrostatic forces and hydrophobic interactions (Rijnaarts et al. 1995). Different factors like surface functionality, temperature and pressure also influence the bacterial adhesion significantly. During this initial contact, bacteria exhibit Brownian motion and could be washed away by fluid shear forces (Palmer et al. 2007).

In the second stage, when the attractive forces are greater than repulsive forces, some of the reversibly attached cells maintain a grip on the surface and tend to become irreversibly adhered (Garrett et al. 2008). With the aid of physical appendages like flagella, fimbriae and pili bacteria can weather the shear repulsive forces and maintain strong interactions with the surface (Kumar and Anand 1998). For example, Uropathogenic *Escherichia coli* and other *E. coli* pathotypes rely on type 1 pili which are multisubunit adhesive organelles assembled by the chaperone usher pathway. The hydrophobicity of bacterial cell surface also plays a critical role in biofilm formation to help bacteria adhere to a hydrophobic nonpolar surface (Tribedi and Sil 2014).

6.3.2 Microcolonies and Formation of Three Dimensional Biofilm Structure

Further upon irreversible attachment, micro-colony formation occurs where microbial cells communicate among each other by the production of autoinducers that trigger signal transduction cascades for the expression of biofilm-specific genes (López et al. 2010). In this stage, bacterial cells secrete a matrix of extracellular polysaccharide substances which is critical for stability of biofilm network. As discussed previously, in *P. aeruginosa,* the three polysaccharides, namely alginate, Pel and Psl provide stability to the biofilm. Also eDNAs are known to be responsible for cellular communication and stabilization of *P. aeruginosa* biofilms (Gloag et al. 2013). Twitching motility is speculated to be involved in the formation of microcolonies (O'Toole et al. 2000; Stoodley et al. 2002). At this stage, the biofilm becomes multi-layered and its thickness increases up to 10 μm. Matrix formation is followed by formation of water-filled channels for transportation of nutrients throughout the biofilm and removal of waste materials from within the biofilm communities. Thus, in the first and second stages of biofilm development, bacterial cells initially loosely associate with the substratum, followed by strong adhesion and microcolony formation (Hall-Stoodley et al. 2004).

Adhered cells within the microcolonies mature into large cellular aggregates which attain a thickness of about 100 μm. These macrocolonies manipulate their physiology, metabolism and are eventually encased into a matrix so as to withstand harsh external conditions. In many Gram-negative species, cyclic di-GMP (c-di-GMP) is a key intracellular signalling molecule involved in this process that promotes production of biofilm matrix and has been extensively implicated in the shift between sessility and motility in bacteria (Kostakioti et al. 2013). Microcolonies in biofilms often consist of diverse and complex microbial communities. These multi-species micro-consortia act synergistically, enhancing substrate exchange, distribution of metabolic products and removal of toxic end products in close proximity. Thus, biofilms also encourage the establishment of synergistic microconsortium where two or more metabolically distinct bacteria depend on each other in order to utilize certain substrates as energy source (Davey and O'toole 2000). Cell to cell communication system known as quorum sensing (QS) also plays critical role in the maturation stage of biofilm formation. It regulates cell differentiation and development of biofilm structures. Cells in a biofilm produce and release QS signaling molecules that are used for intraspecies as well as interspecies communication. Several major types of QS systems such as N-acyl-homoserine lactone (AHL) systems and AI2/LuxS systems have been characterised in bacteria (Bassler 2002).

6.3.3 Detachment and Dispersal

In the fifth stage of biofilm dispersion, sessile cells commit themselves to return back to motile form and recommence the process of biofilm formation, on encountering a favourable environment (Hall-Stoodley et al. 2004). Biofilm dispersal can occur in response to several cues, such as alterations in nutrient availability, oxygen fluctuations and accumulation of toxic products, or other stress-inducing conditions (Herrmann et al. 2010; Karatan and Watnick 2009; Tielker et al. 2005). For example, Uropathogenic *E. coli* (UPEC) responds to extracellular iron whereas increased amounts of carbon and nitrogen sources induce dispersal of *P. aeruginosa* biofilms (Karatan and Watnick 2009; Sauer et al. 2004). For dispersion and colonisation of new surfaces, microbial community produces saccharolytic enzymes that destabilize the matrix polysaccharides and release bacteria residing within the complex biofilm structure. *E. coli* release *N*-acetyl-heparosan lyase while *P. aeruginosa* is known to release alginate lyase for the breakdown of the biofilm matrix (Sutherland 1999). For translocation of the biofilm infection to a new site, microorganisms upregulate the expression of the flagellar proteins to induce bacterial motility (Sauer et al. 2004). Non-motile bacteria such as *S. aureus* gain benefit from the swimming motility of flagellated *P. aeruginosa* for translocation and distribution to new ecological niches (Samad et al. 2017).

6.4 Biofilms in Infectious Diseases

Microorganisms growing as aggregates encased in a matrix are extremely resistant to antibiotics and host immune response and this has serious implications for the host. It is estimated that about 65% of all bacterial infections and 80% of all chronic infections are associated with bacterial biofilms (Jamal et al. 2018). These could be either tissue associated or device associated. Due to their high tolerance toward antibiotics, these chronic tissue-related and device-related infections are difficult to treat and are responsible for recurrence of infections.

6.4.1 Tissue Associated Infections

Biofilm-induced pathogenesis in the host have been characterised on the basis of four major criteria that define its etiology: the bacteria must adhere to a surface; occur as clusters within the matrix of host components; the biofilm infection be constraint to a site; and resistant to action of antibiotics even though the constituent planktonic organisms remain susceptible (Parsek and Singh 2003). Cystic fibrosis (CF) is a genetic disorder that impairs the normal functioning of the lungs. The viscous mucus produced on the epithelium causes difficulty in breathing and harbours bacterial infections in patients. Increased mucosal secretions within the CF lung provides a favourable environment conducive to *P. aeruginosa* colonisation and biofilm formation (Moreau-Marquis et al. 2008). Other bacteria found to be associated with lung infections in cystic fibrosis patients are *S. aureus* and *Haemophilus influenza* (Høiby 1988). Infective endocarditis (IE) is an infection of the heart endothelium and is associated with substantial morbidity and mortality. The microorganisms associated with this serious condition are species of *Staphylococcus, Candida, Pneumococci, Streptococcus* and few other Gram-negative bacteria. The condition occurs when these organisms enter into the blood stream through oropharynx, gastrointestinal tract or genitourinary tract and attach to the injured heart valves (Holland et al. 2016). Periodontitis is a serious infection of the gums that damages soft tissue and the bone supporting the teeth. The bacteria associated within the plaque biofilm are mainly anaerobes such as *Pseudomonas gingivalis, Staphylococcus intermedius, Bacteroides forsythus, Campylobacter rectus, Aggregatibacter actinomycetemcomitans, Eubacterium nodatum, Peptostreptococcus micros,* and *Treponema* sp. (Popova et al. 2013). Chronic rhinosinusitis is another inflammatory disorder of the nasal sinuses. The condition causes mucosal build up that may harbour mixed species biofilms of *S. aureus, Streptococcus pneumoniae, Haemophilus influenza* and *Moraxella catarrhalis* (Fastenberg et al., 2016; Stevens et al. 2015). Chronic wounds such as diabetic foot ulcers or non-healing surgical sites are an ideal environment for biofilm formation. Almost 88–98% of wound infections are found to be *S. aureus* positive. *P. aeruginosa* is the next most common bacteria at sites of chronic wounds (Zhao et al. 2013). Other biofilm associated infections include urinary tract infections caused by uropathogenic *E. coli* and middle ear infections mediated by *M. catarrhalis, H. influenza* and *S. pneumoniae* (Hall-Stoodley and Stoodley 2009).

6.4.2 Device Associated Infections

Medical devices are responsible for a majority of nosocomial infections, particularly in critically ill individuals. The first evidence of the involvement of biofilms in device-related infections was provided in 1982 by an electron microscopy study of a pacemaker lead in a patient with recurrent *S. aureus* bloodstream infection (Lebeaux et al. 2014). Since then, there has been an upsurge in biofilm-related infections because of the widespread use of indwelling medical devices in health-care settings. Device-associated infections are an increasing cause of morbidity and mortality in clinics (Darouiche and Darouiche 2001). Biofilm populations usually occur on or within medical devices such as contact lenses, central venous catheters, mechanical heart valves, peritoneal dialysis catheters, prosthetic joints, pacemakers, urinary catheters and voice prostheses. Staphylococci (particularly *S. epidermidis* and *S. aureus*) are the most common microorganisms infecting medical devices followed by multidrug-resistant Gram-negative bacteria, particularly *Acinetobacter baumannii, E. coli, Klebsiella pneumoniae, P. aeruginosa* and other bacteria that can opportunistically infect immunocompromised host (Donlan 2001).

6.5 Mechanism of Resistance to Antimicrobial Agents by Biofilms

Resistance of bacterial biofilms to various classes of antibiotics and other antimicrobial agents is an acquired property of the bacterial cells in which the genetic makeup of cells changes either by mutations or integration of foreign DNA (Høiby et al. 2010). This bestows upon the cells, the ability to withstand high concentration of antibiotics (otherwise lethal for sensitive cells) even after the cells are dispersed from the EPS matrix. Tolerance, on the other hand, is the characteristic specific to biofilm, and is lost once the bacterial cells detach from EPS matrix. The fact that pathogenic bacteria within the biofilm milieu utilize both tolerance and resistance mechanisms to survive antimicrobial challenges intriguing. Based upon the evidence gained from the recent studies, the mechanisms by which biofilm associated bacteria survive in the presence of antibiotics and biocides are described as below (Fig. 6.2).

6.5.1 Extracellular Polymeric Substance (EPS) Matrix as Diffusion Barrier

The components secreted by bacterial cells as extracellular polymeric substance (EPS) matrix forms a plausible barrier for the penetration of antimicrobial agents and is known as the diffusion inhibition reaction in biofilms (Oubekka et al. 2012). It could be attained via antibiotic degradation, complex formation, chelation or sacrificial oxidation mediated inactivation of antibiotics. These processes either alone or together lower the effective concentration of the antibiotics to sub lethal levels,

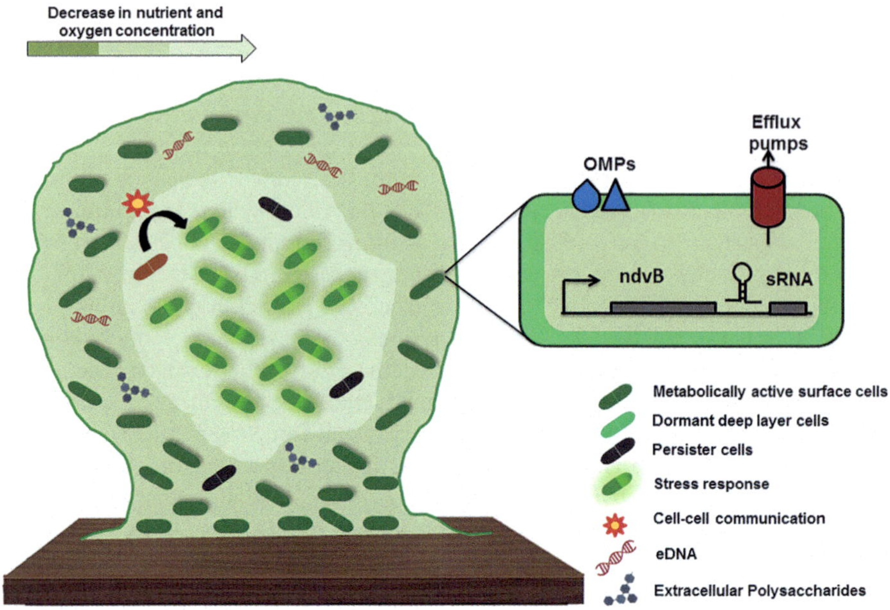

Fig. 6.2 Major antimicrobial resistance and tolerance mechanisms employed by bacterial biofilms. Biofilm cells (dark green) embed in a mushroom-shaped matrix (shown in light green). The biofilm attaches to a biotic or abiotic surface (brown rectangle). The various resistant mechanisms depicted in the figure are as follows: (1) nutrient gradient (demonstrated here as a colour-intensity gradient) with less nutrient availability in the core of the biofilm, (2) matrix exopolysaccharides, (3) extracellular DNA, (4) stress responses (oxidative stress response, stringent response etc.), (5) multidrug efflux pumps and outer membrane proteins, (6) intercellular interactions (horizontal gene transfer, quorum sensing, multispecies communication etc.), (7) expression of biofilm specific genes and non-coding small RNAs and (8) persister cells. Figure adapted and modified from Hall and Mah (2017)

thereby promoting the selection of resistant cells within the biofilm that could evade the antibiotic stress.

Glycocalyx, the extracellular polysaccharide containing structure is an integral part of biofilms and serves as one of the distinguishing features in antibiotic tolerance. It varies in thickness from 0.2-1 μm and can accumulate antibiotic molecules up to 25% of its weight. The exoenzymes on the glycocalyx surface also inhibit diffusion and inhibition by antibiotics. For instance, the accumulation of β-lactamases in the glycocalyx of *P. aeruginosa* results in increase in resistance to β-lactam class of antibiotics (Giwercman et al. 1991). *Erwinia amylovora* biofilms can detoxify and degrade various aromatics, phenolics and heavy metals by reduction of these ions (Ordax et al. 2010). One study highlighted the integration of numerous processes such as extracellular signaling, metal immobililisation, phenotypic variation and gene mutation in mediating tolerance to antibiotics (Harrison et al. 2007). Alginate, a major constituent of mucoid biofilm produced by *Pseudomonas sp.,* is the primary antibiotic binding entity that limits the availability of aminoglycosides (tobramycin) to bacterial cells inside the biofilm (Khan et al. 2010). In another study, aminoglycoside tolerance of *S. aureus* and *P. aeruginosa* were compared in monoculture and coculture wound like *in vitro* model of biofilm infection, supplemented with or without coagulated plasma (host-derived factors). The cells showed increased tolerance to gentamycin and tetracycline when cocultured and supplemented with plasma (Fux et al. 2005). Thus, the EPS matrix along with host derived factors is responsible for inhibiting diffusion of antibiotics.

6.5.2 Horizontal Gene Transfer Mediated Resistance Gene Dissemination

Increased cell density and genetic competency, a physical environment that favours direct cell to cell contact, accumulation of extracellular DNA enhances the uptake of antibiotics resistance markers between cells by means of horizontal gene transfer (Hall and Mah 2017; Mah 2012). Transfer of antibiotic resistance marks via plasmid conjugation is a common mechanism of horizontal gene transfer as reported between the cells of *E. coli* and *Pseudomonas putida.* Conjugation has been shown to be hundreds of fold more favourable in biofilms as compared to planktonic bacterial cells (Merod and Wuertz 2014). Often these resistance genes have been found to be localized on mobile genetic elements (MGEs) in the genome. The conjugative transposon Tn*916* is an important example of a MGE that confers resistance to tetracycline and minocycline in the oral microflora (Connell et al. 2003). Type VI secretion system (T6SS) is also responsible for horizontal gene transfer in *Vibrio cholerae* biofilms. The T6SS machinery causes lysis of the neighbouring cells and uptake of DNA by natural competence (Borgeaud et al. 2015). The uptake of eDNA by transformation also serves as the source of transfer of a plethora of resistance genes among biofilm populations as reported in case of *Streptococcus gordonii* (Wang et al. 2002). The potency of EPS matrix in biofilm of *Acinetobacter baylyi* to stabilize extracellular plasmid for uptake by bacterial cells in biofilms (Merod and Wuertz 2014).

6.5.3 Slow Growth Rate and Dormancy

Limitation of nutrient in the biofilm leads to slow growth rate, starvation of bacterial cells and metabolic dormancy that contributes to antimicrobial tolerance (Brown et al. Brown et al. 1988). In biofilms, the number of bacterial cells in stationary phase significantly increases and so does their tolerance to antibiotics which rely majorly on metabolic activity of the cell for their action (Amato et al. 2014). For instance, mature biofilms are less susceptible to antibiotic vancomycin (Monzón et al. 2002). Slow growth rate leads to one of the possible dormancy states of non-sporulating bacteria that is, viable-but-non-culturable state (VBNC state). VBNC is a state of dormancy in which the organism fails to grow in rich media, while retaining viability parameters such as respiration, presence of rRNA and plasma membrane integrity (Li et al. 2014). The VBNC state of several pathogens such as *E. faecalis, S. epidermidis, Helicobacter pylori, Mycobacterium tuberculosis* and *H. influenza* has been found to be resistant to several antimicrobials (Ramamurthy et al. 2014). In biofilms, *S. aureus* can enter the VBNC state in the presence of vancomycin/quinupristin and can cause recurrent infections (Irie et al. 2010).

6.5.4 Physiological Dormancy of Persisters

Biofilms contain a small fraction of bacterial population exhibiting increased tolerance to biocides and antimicrobials (Conlon et al. 2015). These so called "Persister cells" easily survive an antibiotic attack by shutting down their cellular targets. Multidrug tolerance in persisters is a phenotypic phenomenon thus discriminating them from resistant mutants that undergo genetic modifications to acquire resistance. As the antibiotic pressure within the biofilms drop, persister cells give rise to a population that is as susceptible as the original one and again possess a subset of tolerant cells. Thus, persister cells are a major culprit that account for the recalcitrance of biofilm associated infections *in vivo* and for their relapsing nature (Ayrapetyan et al. 2015; Helaine and Kugelberg 2014; Fauvart et al. 2011). Glycocalyx matrix also helps the biofilm persisters evade the immune response of the host. Formation of persister cells in response to the released toxin has been linked in many reports, for example RelE and MazF, from toxin-antitoxin (TA) system (Keren et al. 2011). Toxins blocks an essential processes such as protein synthesis to establish a state of dormancy, thereby leading to a state of tolerance since antibiotics cannot target inactive metabolic targets. *E. coli* RelBE is one of the best studied TA module and cells expressing RelE from an inducible promoter display high tolerance to antibiotics – ofloxacin, cefotaxime and tobramycin (Lewis 2005). Ectopic expression of other toxins such as MazF and HipA are also known to produce persistent cells in biofilms (Lewis 2008). Other than the TA modules, genes for glycerol-3-phosphate dehydrogenase (glpD) and glycerol-3-phosphate acetyltransferase (plsB) have been identified as persister genes and are potential targets for antipersister therapy (Fauvart et al. 2011).

6.5.5 Antimicrobial Resistance and Quorum Sensing

Quorum sensing (QS) is a communication process amongst bacterial cells that enables bacteria to regulate the expression of certain genes in response to changes in population density conferring them a beneficial phenotype by forming biofilms and expressing virulence factors. An autoinducers such as acyl homoserine lactone (AHL) acts as signalling molecule which is detected by cell population. The role of AHL signaling molecules in biofilm formation has been shown in many model organisms. Quorum sensing facilitates proper development of EPS matrix architecture, as quorum sensing deficiency is correlated with thinner biofilm formation and lower EPS production, and such a mutant or deficient biofilm is susceptible to kanamycin. In another example, the expression of superoxide dismutase and catalase genes are regulated by AHL in *P. aeruginosa*, which makes the cells resistant to oxidative stress (Singh et al. 2017).

6.5.6 Stress Mediated Change in Cellular Morphology

Cells in biofilm continually face stress in the form of nutrition deficiency, pH or accumulation of toxic products. Bacteria have the innate ability to cope stress by changing its cellular physiology and morphology, which decreases its sensitivity to antimicrobial agents. Stress response also controls the composition and arrangement of cellular envelope. The gene rpoE which encodes for the sigma factor sigma-24 (σ24, sigma E, or RpoE) plays crucial role in exocytoplasmic stress response in bacterial biofilms. RpoE also known as AlgU (or AlgT) in *P. aeruginosa* has been reported to control alginate biosynthesis in the opportunistic pathogen (Flores-Kim and Darwin 2014). Antibiotic resistance in biofilms is also mediated by several genetic mechanisms that are specific to the biofilm phenotype and are not expressed in planktonic cells. One such mechanism is the expression of gene *ndvB*, which encodes for a glucosyltransferase enzyme that catalyzes the synthesis of periplasmic β-(1 $\rightarrow$ 3)-cyclic glucans. Glucans promote resistance by sequestering antibiotics in the periplasm away from their cellular targets. The transcription of *ndvB* has been shown to be dependent on the stationary-phase sigma factor RpoS (Flores-Kim and Darwin 2014). RpoS is also known to regulate the expression of 50 other genes responsible for stress tolerance and physiological or metabolic rearrangements (Vijayakumar et al. 2004). Such rearrangements lead to recalcitrance of biofilms to antibiotics. Bacteria within biofilms also encounter conditions of increased osmolarity and greater oxygen limitation (Sauer 2003). Under condition of high-osmolality, the nosocomial pathogen *A. baumannii* releases outer membrane porins such as OMP33–36 and CarO from their membrane which indirectly render the bacteria resistant to carbapenem class of antibiotics (Novović et al. 2018). Also, OmpR response regulator of the two-component regulatory system helps *Burkholderia cepacia* survive high osmolarity condition in cystic fibrosis lung (Silva et al. 2018).

6.5.7 Efflux Pumps

Drug efflux is a major mechanism of resistance in both Gram-positive and Gram-negative bacteria. Resistant microbes maintain their normal physiology by effluxing out toxins, salts, heavy metal, antibiotics and biocides (Pearson et al. 1999). Efflux pumps may be single component or multicomponent, the latter being more prevalent in Gram-negative (Lee et al. 2000). There are six major superfamily of efflux pump namely, (i).Resistance-Nodulation-Division (RND) superfamily; (ii).the Major Facilitator Superfamily (MFS); (iii).the Multidrug And Toxic compound Extrusion (MATE); (iv). ATP-Binding Cassette (ABC) superfamily; (v). Small

Multidrug Resistance (SMR) family; (vi) the Drug Metabolite Transporter (DMT) superfamily. Amongst clinical strains, RND pumps are most common in Gram-negative and are multipartite pumps typically composed of a transmembrane and periplasmic subunits along with an outer membrane protein. In Gram- positive bacteria, MFS efflux pumps are more prevalent than single component pumps.

It is known that QS plays important role in the formation of biofilms and expression of many virulence genes. Efflux pumps transport Homoserine lactone (HSL) across the cytoplasmic membrane. Studies in *P. aeruginosa* and *A. baumannii* involving ~~the~~ efflux pump mutant strains, showed a decrease in biofilm formation, elucidating the role of efflux pumps for extrusion of biofilm material (Pamp et al. 2008; Sharma et al. 2017). Biofilm constitutes both metabolically active and inactive population. Since the action of many antibiotics require cells in metabolically active state, the latter population shows tolerance to most classes of antibiotics. Multidrug resistance efflux pump are involved in resistance to metabolic active cells in biofilm (Pamp et al. 2008). In entero- and uropathogenic strains of *E. coli*, AcrAB-TolC is the most commonly found Multidrug Resistance (MDR) pump (Fernandes et al. 2003). The substrates for this pump are chloramphenicol, fluoroquinolones, rifampicin, SDS, ethidium bromide, etc. (Piddock 2006). The protein TolC that act as outer membrane protein for AcrAB pump, aids in adherence of bacteria onto human epithelial cells (HEp-2) and in formation of biofilms (Wakimoto et al. 2004). AcrAB-TolC are also involved in bacterial colonization and persistence (Piddock 2006). A study reported the relationship between biofilm formation and efflux pumps in *E. coli*, wherein deletion of six genes for proton motive force pumps, *i.e. emrD*, *emrE*, *emrK*, *acrD*, *acrE* and *mdtE* displayed decreased in biofilm formation (Matsumura et al. 2011).

P. aeruginosa biofilms exhibit high resistance to tetracycline, chloramphenicol, β-lactams and this resistance pattern closely resembles the drug efflux profile by the MexAB-OprM pump. It has been demonstrated that MexAB-OprM plays a role in the resistance of aztreonam, gentamicin, tetracycline and tobramycin in biofilm structures (De Kievit et al. 2001).

6.6 Small RNA Based Regulation of Biofilm Formation

The choice that bacteria make to attach to a surface and form biofilms or to stay in planktonic state involves the circuit of regulatory genes. Recent findings in the area of post-transcriptional regulation suggest that such precision in gene regulation is governed by non-coding small RNAs (sRNA). Although sRNA based regulation was initially described as stress based response but it is now evident that sRNA are also involved in multicellular processes, and biofilm formation is one such phenomenon. The sRNA based regulation involving biofilm formation occurs by two mechanisms (1) Base paring of sRNA with target mRNAs or (2) sequesteration of target protein by secondary structures of sRNA. Based upon their location in genome and their target mRNA, sRNA in bacteria could be either *cis* acting or *trans* acting (Thomason and Storz 2010). *Cis* acting sRNAs show substantial complementarity with its targets while *trans*-acting sRNA show limited complementarity and need a RNA chaperone Hfq for interaction with target mRNAs (Chao and Vogel 2010). In *E. coli* and *Salmonella sp.*, CsgD is a key biofilm regulator that facilitates attachment and biofilm formation by activating the synthesis of curli fimbrae and repressing the flagellar biosynthesis genes (Ogasawara et al. 2011; Römling et al. 1998). The synthesis of *bis*-(3′, 5′)-cyclic-diguanosine monophosphate (c-di-GMP), a second messenger for production of cellulose, in biofilms is also regulated by CsgD (Beloin et al. 2008). Five trans-acting sRNAs namely OmrA/OmrB, AcrZ,, McaS, RprA and GcvB have been reported to regulate CsgD (Chambers and Sauer 2013).

In biofilms, bacterial cells encounter delayed growth and various stress conditions. Four sRNAs namely OxyS, ArcZ, DsrA, and RprA regulate the expression of RpoS, a master regulator in stress conditions (Hengge-Aronis 2002). The expression of DsrA small RNA is also under the control of *E. coli* AI2-based Quorum sensing system, that regulates biosynthesis of capsule associated polysaccharides and MDR efflux pumps (Li et al. 2007; Nishino et al. 2011). In *P. aeruginosa* PAO1, RpoS positively regulates *psl* gene expression, which synthesises matrix polysaccharide in biofilms (Irie et al. 2010). An RNA binding protein CsrA, negatively regulates RpoS by activating the genes that promotes growth, and represses bacterial biofilm formation. CsrA activity is known to be stalled by two sRNAs – CsrB and CsrC (Bak et al. 2015; Mika and Hengge 2014).

In *Vibrio harveyi*, sRNAs have been demonstrated to regulate mechanisms such as quorum sensing and biofilm formation. In quorum-sensing circuits of *Vibrio harveyi*, several *trans*-encoded quorum regulatory RNAs, (Qrr) sRNAs function at the center of QS pathways. These sRNAs positively control the production of the AphA, while repressing the assembly of LuxR, which are master regulators of low and high-cell-density in biofilm mode of growth respectively (Rutherford et al. 2011). sRNAs such as ArcZ, DsrA, RprA, McaS, OmrA/OmrB, and GcvB have been reported to control the expression of *flhDC, rpoS* or *csgD* (encoding the master regulator biofilm formation). Bak et al. identified 33 sRNAs in *E. coli* that significantly affect biofilm formation and related phenotypes of swimming and swarming motilities, type I fimbriae or curli fimbriae formation (Bak et al. 2015).

6.7 Strategies for Treatment of Biofilm Associated Infections

The physiological properties of biofilm confer the bacterial communities within them resistance to traditional antibiotic therapies and increased resistance to antimicrobial therapy. Hence, there is a dire need for the discovery of novel molecules or strategies that target both planktonic and biofilm associated cells. Various innovative anti-biofilm remedies have been developed that aim to limit the bacterial adhesion to biotic or abiotic surfaces, inhibit signalling mechanisms that regulate the switch to the biofilm means of growth and dislodges cells from mature biofilms. Numerous effective antimicrobial agents to repress drug resistance in biofilms have been developed lately, a few of which have been discussed below (Fig. 6.3).

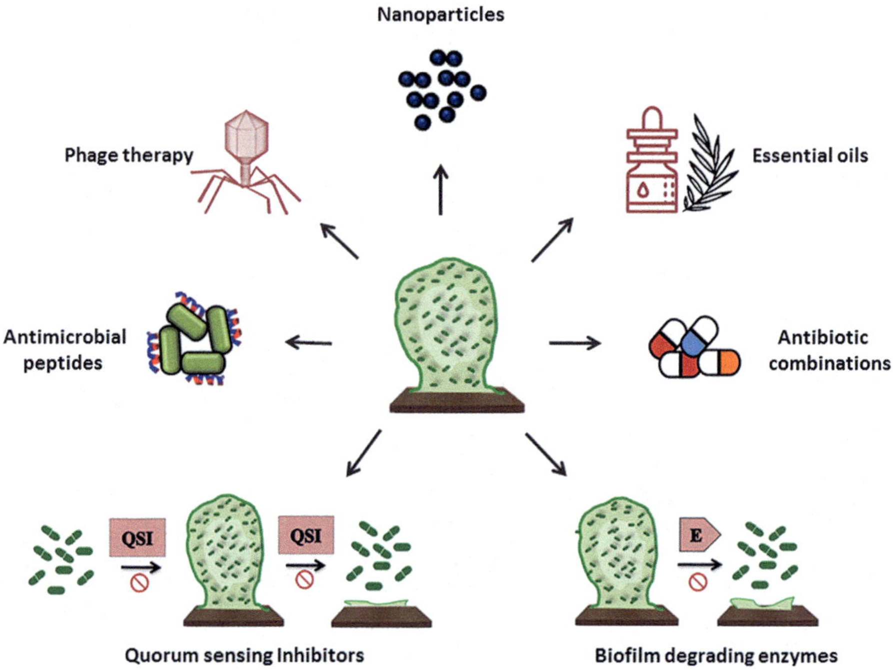

Fig. 6.3 Strategies to tackle antibiotic resistance within biofilms. QSI: Quorum Sensing Inhibitors, E: Enzymes

6.7.1 Aggressive Antibiotic Treatment Regime

Biofilm infections are extremely difficult to handle in the clinical settings due to the restricted penetration of bacteria through the matrix. Antibiotics like rifampicin, tetracycline, fusidic acid and those belonging to classes- quinolones, macrolides have higher penetration ability than β-lactams, aminoglycosides, glycoproteins and colistin. During a chronic biofilm wound infection, high cell density and oxygen limitation leads to increased glycolysis and activation of acidosis. As a result, most antibiotics fail to act and rifampicin could be the antibiotic of choice for treatment at low pH (Laub et al. 1989). However, biofilm embedded slow growing or stationary phase bacteria are several times less susceptible to antibiotics as compared to the planktonic cells rendering antibiotic monotherapy mostly inappropriate (Hengzhuang et al. 2012; Ciofu et al. 2017). Therefore favourable treatment of biofilm infections requires the use of antibiotic combinations, ideally including an agent that can penetrate and act upon adherent bacterial cells (Hengzhuang et al. 2012; Herrmann et al. 2010). Combination of antibiotics rifampicin with fusidic acid and ciprofloxacin with vancomycin were reported to be bactericidal against Methicillin Resistant *S. aureus* (MRSA) biofilms (Saginur et al. 2006). Also, combination of azithromycin (AZM) and ciprofloxacin (CIP) has been reported to be synergistic against biofilm cells of *P. aeruginosa* (Saini et al. 2015). Apart from improving the therapeutic efficacy, combinatorial therapy in biofilm infections can also delay the emergence of resistant populations. Optimisation of the treatment dose to achieve maximum bactericidal effect and to determine concentrations that lie within the safety parameters of hepatic and renal function is extremely important. Minimum biofilm inhibition concentration (MBIC) and minimum biofilm eradication concentration (MBEC) are two criterion used to determine a suitable dose for drug combinations (Hengzhuang et al. 2012). Biofilm formation on indwelling medical devices such as central venous catheters (CVCs) can also be prevented by coating them with topical antibiotics. For instance, minoclycine/rifampin-coated CVCs are implanted in case of *S. aureus* catheter-related bloodstream infections (Ciofu et al. 2017). Another approach may be the use of sequential antibiotic treatments since sequential therapy with inhaled antibiotics tobramycin and aztreonam was found to be superior for treatment of patients with cystic fibrosis (Ciofu et al. 2017).

6.7.2 Biofilm Killing by Essential Oils/Natural Products

Essential oils are plant secondary metabolites with broad spectrum antibacterial activity (Hammer et al. 1999). Essential oils exert their antibacterial activity by multiple mechanisms such as inhibiting ATP synthesis and compromising membranes leading to leakage of metabolites and ions. They have also been reported to act as efflux pumps inhibitors and quenchers for quorum signaling molecules like AHLs (Nazzaro et al. 2013). Essential oils extracted from *Origanum vulgare*, carvacrol and thymol inhibit formation in *S. aureus* (Nostro et al. 2007). Interestingly, Cassia, Peru balsam and red thymol have been found to be more effective in killing biofilm associated bacteria than antibiotics ofloxacin and gentamycin (Kavanaugh and Ribbeck 2012).

Combination between essential oils and conventional antibiotics is another strategy where essential oils are known to modify the uptake and tolerance of biofilm cells to antibiotics. Three essential oils cinnamon (*Cinnamomum zeylanicum*), tea tree oil (*Melaleuca alternifolia*), and palmarosa (*Cymbopogon martini*) have been proved synergistic with the fluoroquinolone antibiotic, ciprofloxacin against established *P. aeruginosa* biofilms (Coelho and Pereira 2013). Carvacrol and thymol have also been identified to possess efflux pump inhibitory activity thus rendering the biofilm cells susceptible to otherwise ineffective antibiotics (Cirino et al. 2014). Since essential oils are multi-component in nature, they are believed to be more promising in preventing bacterial resistance. Further, the ease of extraction, non-toxicity and potential health benefits warrant the use of essential oils as candidate antimicrobial agents in biofilm associated infections (Yap et al. 2014). Coumarins represent a class of natural compounds reported to act as broad spectrum antibiofilm agents. Several synthetic coumarins have also been synthesised to develop more effective and pharmacologically active molecules (Reen et al. 2018). Ginger extract has also been reported to inhibit biofilm formation through reduction of cellular c-di-GMP, which is an important signalling molecule during the biofilm mode of growth (Kim and Park 2013). Resveratrol is another natural compound found in red grapes, peanuts that has been reported to have antibacterial and antibiofilm properties against *E. coli* O157:H7 (Rabin et al. 2015a).

6.7.3 Nanoparticles and Their Combination with Antibiotics

Nanoparticles (NPs) are materials of size less than 100 nm in one dimension and are increasingly being used against bacteria as an alternative to antibiotics. Smaller size, larger surface area and extremely reactive nature makes nanoparticles a preferred choice in comparison to other antibacterial agents (Wang et al. 2017). NPs have been reported to possess broad-spectrum properties against both Gram-positive and Gram-negative bacterial pathogens. On the basis of several studies, it has been established that nanoparticles exert their antibacterial activity by virtue of the following mechanisms: (1) disruption and penetration of the bacterial cell membrane; (2) generation of Reactive oxygen species; (3) ATP depletion and (4) interaction with cellular DNA and proteins (Wang et al. 2017).

Nanoparticles coupled with metals or natural product extracts have been shown to possess inhibitory activity against both planktonic and biofilm associated cells. NPs compromise the biofilm integrity by interacting with its polysaccharides, extracellular DNA, proteins, and lipids. Silver and Zinc oxide nanoparticles are reported to inhibit approximately 95% of *P. aeruginosa* biofilms (Martinez-Gutierrez et al. 2013; Lee et al. 2014). NPs and antibiotics, in combination have also been reported to possess substantial antibiofilm activity. Silver nanoparticles are known to potentiate the activity of antibiotics chloramphenicol, gentamycin, ampicillin, tobramycin and vancomycin by membrane damage and rapid antibiotic uptake by biofilm associated cells (Peulen and Wilkinson 2011; Su et al. 2009; Habash et al. 2017). Biofunctionalized polymyxin B-capped silver nanoparticles (PBSNPs) were sythesised where PBSNPs displayed antibiofilm activity against multiple drug-resistant *Vibrio fluvialis* and *P. aeruginosa* (Lambadi et al. 2015). Other metal nanoparticles such as copper NPs (CuNPs), Iron NPs (FeNPs) and magnesium NPs (MgNPs) are also known to inhibit biofilms. Combination of selenite nanoparticles antibiotics such as oxacillin have been proved effective to dislodge MRSA biofilms (Cihalova et al. 2015). Non-metallic inorganic NPs also represent an important category of antibiofilm agents. The antibacterial activity of Nitric oxide (NO) in many NO-releasing NPs has been exploited for their ability to disperse *P. aeruginosa* biofilms. For example, Methylaminopropyl trimethoxysilane (MAP3) NPs have been reported to exhibit excellent ability to disperse biofilms (Qayyum and Khan 2014).

Combination of leaf extract from *Allophyus cobbe* with Ag nanoparticles have shown potential antibiofilm activities against important bacterial pathogens (Gurunathan et al. 2014). For use in combination therapy, safety levels for consumption of nanoparticle must be evaluated before their use in pharmaceutical formulations. The use of NPs for functionalization of biomedical devices such as catheters etc. is increasingly being sought. For example, AgNP-coated catheters showed enhanced *in vitro* activity preventing the formation by *E. coli* and *Enterococcus sp.* biofilms (Qayyum and Khan 2014).

6.7.4 Quorum Sensing Quenchers/Inhibitors

Cell to cell signaling has been reported one of the key processes during biofilm formation and maturation. Quorum sensing allows bacteria to establish communication within the population, thus aiding its survival and virulence genes expression. Therefore, use of quorum sensing inhibitors or quencher molecules that weaken the biofilm architecture or its formation altogether is another promising approach (Uroz et al. 2009). The two chemical auto-inducer molecules namely N-acyl homoserine lactone (AHL) and Autoinducer-2 (AI-2) in Gram-negative and Gram-positive bacteria respectively regulate the bacterial cell signaling system. The inhibitors could be either an enzyme which degrades the establishment of biofilm into its constituents or a molecule which subsequently decreases or shuts off the expression of genes important for biofilm formation. Efforts to develop QS inhibitors have included screens of natural products, small molecule libraries, virtual screening, and used synthetic libraries derived from native AI structures (Paczkowski et al. 2017). The enzymes lactonase and paraoxonase are also known to degrade QS signals and display potential anti-biofilm strategy (Chen et al. Chen et al. 2009). RNAIII-activating protein (RAP) and *arg* quorum sensing systems play important role for formation of *S. aureus* biofilms. In one study, rat graft infected with MRSA when administered with RNAIII-Inhibiting Peptide (RIP) repressed RAP and *arg* system (Balaban et al. 2007). A secondary metabolite in lichens, Usnic acid is also known to interfere with QS in *S. aureus* (Donelli et al. 2007). Several natural compounds including penicillic acid, solenopsin A, catechin, ellagic acid derivatives, and curcumin have been reported to exhibit QS-inhibiting effects. Flavonoids are a group of natural products that have been reported to inhibit *P. aeruginosa* biofilm formation by interacting with the QS receptors, LasR and RhlR, thus significantly reducing their ability to bind to DNA encoding QS-regulated promoters (Paczkowski et al. 2017). Also, ginger oil extract containing 6-gingerol has been demonstrated to adhere to QS receptors in *P. aeruginosa* and inhibit QS processes altogether (Kim et al. 2015). The use of QS inhibitors in combination with various antibiotics is another strategy to tackle the problem of biofilms. In one report, an analog of AI-2, phenyl-DPD (phenyl-4,5-dihydroxy-2,3-pentanedione) in combination with gentamycin resulted in clearance of *P. aeruginosa* biofilms (Roy et al. 2013). AHL degrading lactonase is a promising candidate in reducing *S. aureus* biofilm formation. The activity displayed promising efficacy when lactone was used in combination with ciprofloxacin and gentamycin (Kiran et al. 2011). Baicalein, 14-α-lipoyl andrographolide and ajoene are ancient medicines which could be promising Quorum Sensing Inhibitors (QSI) in *P. aeruginosa* biofilms (Zeng et al. 2011).

6.7.5 Antimicrobial Peptides (AMP)

Antimicrobial peptides (AMPs) are 15–30 amino acids long peptides produced by the innate immune response and represent another important approach to treat biofilms (Kostakioti et al. 2013). AMPs are positively charged and show their antibacterial activity by targeting the negatively charged cell membrane and biofilm surface (Melo et al. 2009). Peptides from the class cathelicidins such as BMAP-27, BMAP-28, SMAP-29 exhibit promising activity in eradication of preformed biofilms and killing of biofilm associated cells in MDR strain of *P. aeruginosa* (Pompilio et al. 2011). Milk lactoferrins are metal chelating AMPs reported to chelate iron and inhibit biofilm formation in *P. aeruginosa* (Singh et al. 2002). Few AMPs are known to bind eDNA, resulting in the weakening of biofilm structure in *P. aeruginosa.* Lytic peptides are a class of AMPs that are known to cause cell membrane lysis and inhibition of biofilm formation. Studies in *S. aureus* showed that lytic peptide PTP-7 could diffuse deep into the *S. aureus* biofilms killing over 95% of bacteria. Interestingly, PTP-7 can retain its activity in high acidic condition of biofilm (Kharidia and Liang 2011). The most pursued and successful strategy for the use of AMPs so far has been their surface coating onto biomaterials such as prosthetic implants since it confers AMPs with long term stability and low toxicity (Di Luca et al. 2014). For example, coating of Tet-20 peptides tethered onto titanium implant surface exhibited broad antimicrobial activities against *P. aeruginosa* and *S. aureus,* both *in vivo* and *in vitro* and appeared to be have no associated toxicity (Gao et al. 2011).

Synergistic combinations of AMPs with conventional antibiotics offer an alternative approach to target multi-drug resistant pathogens. Several *in vivo* studies have shown that combinations of nisin with daptomycin, indolicidin with teicoplanin and cecropin (1–7)-melittin-A (2–9) amide with ciprofloxacin potentiated the activity of antibiotics against biofilms (Dosler and Mataraci 2013). Antimicrobial peptide AMP 1018 has broad spectrum anti-biofilm activity and degrades the secondary messenger, guanosine pentaphosphate [(p)ppGpp] which is essential for biofilm formation in most bacteria. At low concentration, peptide 1018 inhibits biofilm formation, but it can eradicate preformed biofilms at higher concentrations (de la Fuente-Núñez et al. 2014). Because of their ability to permeabilise cell membranes, AMPs can also act against metabolically inactive cells within the biofilms. For example, Several Trp/Arg containing AMPs were able to disperse and kill persister cells in preformed biofilms (Koo et al. 2017). However, AMPs have a few limitations such as their ability to bind to components of EPS matrix and other host molecules, which drastically reduces their effectiveness. Also, AMPS are prone to degradation by microbial proteases which further diminish their potency. Therefore, extensive preclinical efficacy studies are required before AMPs could be used for treatment. Another major drawback of AMPs is their synthesis cost that acts as a barrier for clinical development and commercialization (Koo et al. 2017).

6.7.6 Phage Therapy

Another promising approach to control and eradicate biofilms is the use of bacteriophages. Phages could be isolated from varying environmental conditions and are harmless to human host since they specifically target and kill bacteria. They have a high mutation accumulation rate to match with the magnitude of mutations in the genome of bacteria within biofilms. Phage therapy takes advantage of the lytic cycle of virulent viruses that cause cell lysis, which confers them their antimicrobial activity. Staphylococcal phage K combined with another staphylococcal phage, DRA88 (MOI 10) could completely remove preformed biofilms of three *S. aureus* isolates after 48 h of treatment. Recent advances in biotechnology have enabled the development of engineered phages to improve their activity in biofilm eradication. For instance, Lu and Collins genetically engineered T7 phage to express enzyme dispersin B that can degrade biofilms (Merril et al. 1996) . Also, a filamentous phage overexpressing a repressor of the SOS DNA repair system in *E. coli* has been engineered to target metabolically inactive persister cells within biofilms (Lu and Collins 2007).

Bacteriophages also secrete EPS degrading enzymes (Sutherland et al. 2004). Phage-encoded lytic proteins such as endolysins and virion-associated peptidoglycan hydrolases (VAPGHs) have been assessed as antimicrobial agents against pathogens such as *S. aureus* (Nelson et al. 2012; Rodríguez-Rubio et al. 2013). Several groups have also exploited the combination of sub-lethal concentrations of antibiotics with virulent phages called as phage-antibiotic synergy (PAS). Ryan et al. reported the efficacy of T4 phage-cefotaxime combination in the eradication of *E. coli* ATCC 11303 biofilms such that 10^7 PFU mL^{-1}could reduce the minimum biofilm eradication concentration (MBEC) of cefotaxime against *E. coli* biofilms by eightfolds (Pires et al. 2017). Combinations of phages and antibiotics have also been tested against *S. aureus* and *K. pneumoniae* biofilms (Pires et al. 2017).

6.7.7 Matrix Degrading Enzymes

EPS matrix in biofilms facilitates the adhesion and protection of biofilm associated cells from antimicrobial activity of antibiotics. Therefore use of enzymes that could inhibit and disrupt the EPS matrix formation and facilitates detachment is another approach. Various classes of enzymes, specifically proteases, deoxyribonucleases, and glycoside hydrolase have been exploited for the dispersal of medical biofilms. In *S. aureus*, ten secreted proteases have been identified till date, four of which namely SspA, SspB, ScpA, and aureolysin (Aur) have been shown to be involved in biofilm disruption (Koo et al. 2017). Fungal strains of *Aspergillus clavatus* have also been isolated that produce enzymes- proteases, amylases and pectinases that can degrade *P. aeruginosa*, *B. subtilis* and *S. aureus* biofilms (Singh et al. 2015).

The EPS mass could also be reduced by DNase-1, dispersin B (DspB), α-amylase, while the cell number within biofilms could be controlled by the combination of enzymes with antibacterial agents (Izano et al. 2007; Kalpana et al. 2012). Older biofilms and good quality biofilms of *P. aeruginosa* can be controlled by using recombinant DNase-1 derivative (DNase1L2), isolated from human stratum corneum. *B. subtilis* S8–18 α-amylase was evaluated against biofilms of a clinical MRSA strain and *P. aeruginosa* ATCC 10145 (Eckhart et al. 2007). Biofilm-degrading enzymes, such as lysostaphin and alginate lyase, showed antibiofilm activities against various pathogenic bacteria (Algburi et al. 2017). Although these enzymes destroy and detach biofilms, biofilm reestablishment is not guaranteed. DspB when use in combination with cefamandolenfate, hydrolyzed the EPS of *S. aureus* biofilm. Despite the high cost of production, biofilm eradicating enzymes could possibly be used as an alternative or as a synergistic helper to antibiotics in the treatment of persistent infections (Donelli et al. 2007).

6.7.8 Anti-adhesion Agents

Apart from their role in matrix stabilization and energy storage in biofilms, bacterial exopolysaccharides have also been reported to perform functions that inhibit or destabilize the biofilm. Capsular polysaccharide secreted by uropthaogenic *E. coli* CFT073 (UPEC) was the first antibacterial polysaccharide to be reported in 2006. It displayed a broad-spectrum activity against biofilm formation by *E. coli*, *P. aeruginosa*, *K. pneumoniae*, *S. aureus*, *S. epidermidis* and *E. faecalis* (Valle et al. 2006). Scott Hultgren and colleagues reported that a mannose analogue, mannoside M4284, inhibits the adhesive function of type 1 pili and effectively reduces intestinal colonization of UPEC. M4284 can be used to treat urinary tract infections (UTI) and has no negative impact on the composition of the gut microbiota (Spaulding et al. 2017). Mannosides also enhanced the antimicrobial effects of trimethoprim-sulfamethoxazole (TMP-SMZ) preventing infection by PBC-1, a UPEC isolate that was resistant to TMP-SMZ treatment in the clinical settings (Cusumano et al. 2011).

Type-1 pili is an adhesive pili assembled by the chaperone/usher pathway (CUP) that plays a critical role in biofilm formation both on the host surface and for colonization of catheters and other surfaces in nosocomial settings. Pilicide compounds have been designed that bind to the chaperone and block critical functions thus preventing pilus assembly. They have been shown to inhibit UPEC biofilm formation *in vitro* by 50%, at concentrations as low as 3 mM (Kostakioti et al. 2013). Curli are functional extracellular amyloid fibers produced by uropathogenic *E. coli* (UPEC) and other Enterobacteriaceae. Ring-fused 2-pyridone compounds such as FN075 and BibC6 are examples of curlicide compounds that inhibit curli biogenesis and polymerization in UPEC strains (Cegelski et al. 2009).

6.8 Conclusions

The recalcitrance of biofilms to the action of antibiotics poses a major impediment in the treatment of biofilm-related infections. Over the past few decades, several *in vitro* studies have revealed the importance of biofilm components and mechanisms regulating biofilm formation. This knowledge has led to the development of novel strategies directed at inhibiting biofilm formation and inducing dispersal of preformed biofilms. However, *in vitro* biofilm research alone is insignificant as compared with the complexity of *in vivo* biofilm-associated infections. Therefore, biofilm researchers need to re-evaluate their experimental approaches thoroughly and chose appropriate animal models for representation of biofilm-associated infections. In the near future, this would help in developing lead therapeutics for the treatment of biofilm-related infections in the clinic.

References

Algburi A et al (2017) Control of biofilm formation: antibiotics and beyond. Appl Environ Microbiol Edited by M. J. Pettinari 83(3). https://doi.org/10.1128/AEM.02508-16

Amato SM et al (2014) The role of metabolism in bacterial persistence. Front Microbiol Frontiers Media SA 5:70. https://doi.org/10.3389/fmicb.2014.00070

Ayrapetyan M, Williams TC, Oliver JD (2015) Bridging the gap between viable but non-culturable and antibiotic persistent bacteria. Trends Microbiol 23(1):7–13. https://doi.org/10.1016/j.tim.2014.09.004

Bak G et al (2015) Identification of novel sRNAs involved in biofilm formation, motility and fimbriae formation in *Escherichia coli*. Sci Rep Nature Publishing Group 5(1):15287. https://doi.org/10.1038/srep15287

Balaban N et al (2007) Treatment of *Staphylococcus aureus* biofilm infection by the quorum-sensing inhibitor RIP. Antimicrob Agents Chemother American Society for Microbiology Journals 51(6):2226–2229. https://doi.org/10.1128/AAC.01097-06

Bassler BL (2002) Small talk: cell-to-cell communication in bacteria. Cell Cell Press 109(4):421–424. https://doi.org/10.1016/S0092-8674(02)00749-3

Beloin C, Roux A, Ghigo JM (2008) *Escherichia coli* biofilms. Curr Top Microbiol Immunol 322:249–289. Available at: http://www.ncbi.nlm.nih.gov/pubmed/18453280. Accessed 9 Sept 2018

Borgeaud S et al (2015) The type VI secretion system of *Vibrio cholerae* fosters horizontal gene transfer. Science 347(6217):63–67. https://doi.org/10.1126/science.1260064

Branda SS et al (2006) A major protein component of the Bacillus subtilis biofilm matrix. Mol Microbiol 59(4):1229–1238. https://doi.org/10.1111/j.1365-2958.2005.05020.x

Brown MR, Allison DG, GilbertP (1988) Resistance of bacterial biofilms to antibiotics: a growth-rate related effect?. J Antimicrob Chemother 22(6):777–780. Available at: http://www.ncbi.nlm.nih.gov/pubmed/3072331. Accessed 26 Aug 2018

Cegelski L et al (2009) Small-molecule inhibitors target *Escherichia coli* amyloid biogenesis and biofilm formation. Nat Chem Biol NIH Public Access 5(12):913–919. https://doi.org/10.1038/nchembio.242

Chambers JR, Sauer K (2013) Small RNAs and their role in biofilm formation. Trends Microbiol NIH Public Access 21(1):39–49. https://doi.org/10.1016/j.tim.2012.10.008

Chao Y, Vogel J (2010) The role of Hfq in bacterial pathogens. Curr Opin Microbiol 13(1):24–33. https://doi.org/10.1016/j.mib.2010.01.001

Chen T, Li F, Chen B-S (2009) Cross-talks of sensory transcription networks in response to various environmental stresses. Interdiscipl Sci Comput Life Sci 1(1):46–54. https://doi.org/10.1007/s12539-008-0018-1
Cihalova K et al (2015) Staphylococcus aureus and MRSA growth and biofilm formation after treatment with antibiotics and SeNPs. Intl J Mol Sci Multidisciplinary Digital Publishing Institute 16(10):24656–24672. https://doi.org/10.3390/ijms161024656
Ciofu O et al (2017) Antibiotic treatment of biofilm infections. APMIS. Wiley/Blackwell (10.1111) 125(4):304–319. https://doi.org/10.1111/apm.12673
Cirino ICS et al (2014) The essential oil from *Origanum vulgare* L. and its individual constituents carvacrol and thymol enhance the effect of tetracycline against Staphylococcus aureus. Chemotherapy. Karger Publishers 60(5–6):290–293. https://doi.org/10.1159/000381175
Coelho FL, Pereira MO (2013) Exploring new treatment strategies for *Pseudomonas aeruginosa* biofilm infections based on plant essential oils. Formatex:83–89
Colvin KM et al (2012) The Pel and Psl polysaccharides provide *Pseudomonas aeruginosa* structural redundancy within the biofilm matrix. Environ Microbiol NIH Public Access 14(8):1913–1928. https://doi.org/10.1111/j.1462-2920.2011.02657.x
Conlon BP, Rowe SE, Lewis K (2015) Persister cells in biofilm associated infections. In: Adv Exp Med Biol, vol 831, pp 1–9. https://doi.org/10.1007/978-3-319-09782-4_1
Connell SR et al (2003) Ribosomal protection proteins and their mechanism of tetracycline resistance.', Antimicrob Agents Chemother, 47(12), 3675–3681. Available at: http://www.ncbi.nlm.nih.gov/pubmed/14638464. Accessed 24 Aug 2018
Costerton JW, Geesey GG, Cheng KJ (1978) How bacteria stick. Sci Am 238(1):86–95. Available at: http://www.ncbi.nlm.nih.gov/pubmed/635520. Accessed 12 Aug 2018
Cusumano CK et al (2011) Treatment and prevention of urinary tract infection with orally active FimH inhibitors. Sci Transl Med 3(109):109ra115–109ra115. https://doi.org/10.1126/scitranslmed.3003021
Darouiche RO, Darouiche RO (2001) Device-associated infections: a macroproblem that starts with microadherence. Clin Infect Dis Oxford University Press 33(9):1567–1572. https://doi.org/10.1086/323130
Davey ME, O'toole GA (2000) Microbial biofilms: from ecology to molecular genetics. Microbiol Mol Biol Rev MMBR 64(4):847–867. Available at http://www.ncbi.nlm.nih.gov/pubmed/11104821. Accessed 12 Aug 2018
De Kievit TR et al (2001) Multidrug efflux pumps: expression patterns and contribution to antibiotic resistance in *Pseudomonas aeruginosa* biofilms. Antimicrob Agents Chemother 45(6):1761–1770. https://doi.org/10.1128/AAC.45.6.1761-1770.2001
de la Fuente-Núñez C et al (2014) Broad-spectrum anti-biofilm peptide that targets a cellular stress response. PLoS Pathogens Edited by M R Parsek Public Library of Science 10(5):e1004152. https://doi.org/10.1371/journal.ppat.1004152
Di Luca M, Maccari G, Nifosì R (2014) Treatment of microbial biofilms in the post-antibiotic era: prophylactic and therapeutic use of antimicrobial peptides and their design by bioinformatics tools. Pathog Dis Oxford University Press 70(3):257–270. https://doi.org/10.1111/2049-632X.12151
Diggle SP et al (2006) The galactophilic lectin, LecA, contributes to biofilm development in *Pseudomonas aeruginosa*. Environ Microbiol 8(6):1095–1104. https://doi.org/10.1111/j.1462-2920.2006.001001.x
Donelli, G. et al (2007) Synergistic activity of dispersin B and cefamandole nafate in inhibition of staphylococcal biofilm growth on polyurethanes. Antimicrob Agents Chemother American Society for Microbiology Journals 51(8):2733–2740. https://doi.org/10.1128/AAC.01249-06
Donlan RM (2001) Biofilms and device-associated infections. Emerg Infect Dis Centers for Disease Control and Prevention 7(2):277–281. https://doi.org/10.3201/eid0702.700277
Dosler S, Mataraci E (2013) In vitro pharmacokinetics of antimicrobial cationic peptides alone and in combination with antibiotics against methicillin resistant *Staphylococcus aureus* biofilms. Peptides 49:53–58. https://doi.org/10.1016/j.peptides.2013.08.008

Eckhart L et al (2007) DNase1L2 suppresses biofilm formation by Pseudomonas aeruginosa and *Staphylococcus aureus*. Br J Dermatol 156(6):1342–1345. https://doi.org/10.1111/j.1365-2133.2007.07886.x

Fauvart M, De Groote VN, Michiels J (2011) Role of persister cells in chronic infections: clinical relevance and perspectives on anti-persister therapies. J Med Microbiol 60(6):699–709. https://doi.org/10.1099/jmm.0.030932-0

Fernandes P, Ferreira BS, Cabral JMS (2003) Solvent tolerance in bacteria: role of efflux pumps and cross-resistance with antibiotics. Int J Antimicrob Agents 22(3):211–216. Available at: http://www.ncbi.nlm.nih.gov/pubmed/13678823. Accessed: 26 Aug 2018

Flemming H-C, Wingender J (2010) The biofilm matrix. Nat Rev Microbiol 8(9):623–633. https://doi.org/10.1038/nrmicro2415

Flores-Kim J, Darwin AJ (2014) Regulation of bacterial virulence gene expression by cell envelope stress responses. Virulence Taylor & Francis 5(8):835–851. https://doi.org/10.4161/21505594.2014.965580

Fux CA et al (2005) Survival strategies of infectious biofilms. Trends Microbiol 13(1):34–40. https://doi.org/10.1016/j.tim.2004.11.010

Gao G et al (2011) The biocompatibility and biofilm resistance of implant coatings based on hydrophilic polymer brushes conjugated with antimicrobial peptides. Biomaterials Elsevier 32(16):3899–3909. https://doi.org/10.1016/J.BIOMATERIALS.2011.02.013

Garrett TR, Bhakoo M, Zhang Z (2008) Bacterial adhesion and biofilms on surfaces. Prog Nat Sci Elsevier 18(9):1049–1056. https://doi.org/10.1016/J.PNSC.2008.04.001

Giwercman B et al (1991) Induction of beta-lactamase production in *Pseudomonas aeruginosa* biofilm. Antimicrob Agents Chemother American Society for Microbiology (ASM) 35(5):1008–1010. Available at: http://www.ncbi.nlm.nih.gov/pubmed/1906694. Accessed: 23 Aug 2018

Gloag ES et al (2013) Self-organization of bacterial biofilms is facilitated by extracellular DNA. Proc Natl Acad Sci 110(28):11541–11546. https://doi.org/10.1073/pnas.1218898110

Gurunathan S et al (2014) Enhanced antibacterial and anti-biofilm activities of silver nanoparticles against Gram-negative and Gram-positive bacteria. Nanoscale Res Lett 9(1):373. https://doi.org/10.1186/1556-276X-9-373

Habash MB et al (2017) Potentiation of tobramycin by silver nanoparticles against Pseudomonas aeruginosa biofilms. Antimicrob Agents Chemother 61(11). https://doi.org/10.1128/AAC.00415-17

Hall CW, Mah T-F (2017) Molecular mechanisms of biofilm-based antibiotic resistance and tolerance in pathogenic bacteria. FEMS Microbiol Rev 41(3):276–301. https://doi.org/10.1093/femsre/fux010

Hall-Stoodley L, Stoodley P (2009) Evolving concepts in biofilm infections. Cell Microbiol Wiley/Blackwell (10.1111) 11(7):1034–1043. https://doi.org/10.1111/j.1462-5822.2009.01323.x

Hall-Stoodley L, Costerton JW, Stoodley P (2004) Bacterial biofilms: from the natural environment to infectious diseases. Nat Rev Microbiol 2(2):95–108. https://doi.org/10.1038/nrmicro821

Hammer KA, Carson CF, Riley TV (1999) Antimicrobial activity of essential oils and other plant extracts. J Appl Microbiol Wiley/Blackwell (10.1111) 86(6):985–990. https://doi.org/10.1046/j.1365-2672.1999.00780.x

Harrison JJ, Ceri H, Turner RJ (2007) Multimetal resistance and tolerance in microbial biofilms. Nat Rev Microbiol 5(12):928–938. https://doi.org/10.1038/nrmicro1774

Helaine S, Kugelberg E (2014) Bacterial persisters: formation, eradication, and experimental systems. Trends Microbiol 22(7):417–424. https://doi.org/10.1016/j.tim.2014.03.008

Hengge-Aronis R (2002) Signal transduction and regulatory mechanisms involved in control of the sigma(S) (RpoS) subunit of RNA polymerase. Microbiol Mol Biol Rev MMBR, 66(3), p. 373–395, table of contents. Available at: http://www.ncbi.nlm.nih.gov/pubmed/12208995. Accessed 9 Sept 2018

Hengzhuang W et al (2012) In vivo pharmacokinetics/pharmacodynamics of colistin and imipenem in Pseudomonas aeruginosa biofilm infection. Antimicrob Agents Chemother American Society for Microbiology Journals 56(5):2683–2690. https://doi.org/10.1128/AAC.06486-11

Henrici AT (1933) Studies of freshwater bacteria: I. A direct microscopic technique. J Bacteriol American Society for Microbiology (ASM) 25(3):277–287. Available at: http://www.ncbi.nlm.nih.gov/pubmed/16559616. Accessed 12 Aug 2018

Herrmann G et al (2010) Colistin-Tobramycin combinations are superior to monotherapy concerning the killing of biofilm *Pseudomonas aeruginosa*. J Infect Dis Oxford University Press 202(10):1585–1592. https://doi.org/10.1086/656788

Hodges NA, Gordon CA (1991) Protection of *Pseudomonas aeruginosa* against ciprofloxacin and beta-lactams by homologous alginate. Antimicrob Agents Chemother 35(11):2450–2452. Available at: http://www.ncbi.nlm.nih.gov/pubmed/1804025. Accessed 12 Aug 2018

Høiby, N. (1988) 'Hemophilus influenzae, *Staphylococcus aureus*, Pseudomonas cepacia, and *Pseudomonas aeruginosa* in patients with cystic fibrosis.', Chest, 94(2 Suppl), p. 97S–103S. Available at: http://www.ncbi.nlm.nih.gov/pubmed/3293941. Accessed 24 Aug 2018

Høiby N et al (2010) Antibiotic resistance of bacterial biofilms. Intl J Antimicrob Agents Elsevier 35(4):322–332. https://doi.org/10.1016/J.IJANTIMICAG.2009.12.011

Holland TL et al (2016) Infective endocarditis. Nature Rev Dis Primers NIH Public Access 2:16059. https://doi.org/10.1038/nrdp.2016.59

Irie Y et al (2010) *Pseudomonas aeruginosa* biofilm matrix polysaccharide Psl is regulated transcriptionally by RpoS and post-transcriptionally by RsmA. Mol Microbiol Wiley-Blackwell 78(1):158–172. https://doi.org/10.1111/j.1365-2958.2010.07320.x

Izano EA et al (2007) Detachment and killing of *Aggregatibacter actinomycetemcomitans* biofilms by Dispersin B and SDS. J Dent Res 86(7):618–622. https://doi.org/10.1177/154405910708600707

Jamal M et al (2018) Bacterial biofilm and associated infections. J Chin Med Assoc 81(1):7–11. https://doi.org/10.1016/j.jcma.2017.07.012

Joo H-S, Otto M (2012) Molecular basis of in vivo biofilm formation by bacterial pathogens. Chem Biol 19(12):1503–1513. https://doi.org/10.1016/j.chembiol.2012.10.022

Kalpana BJ, Aarthy S, Pandian SK (2012) Antibiofilm Activity of α-Amylase from *Bacillus subtilis* S8–18 Against Biofilm Forming Human Bacterial Pathogens. Appl Biochem Biotechnol Springer-Verlag 167(6):1778–1794. https://doi.org/10.1007/s12010-011-9526-2

Karatan E, Watnick P (2009) Signals, regulatory networks, and materials that build and break bacterial biofilms. Microbiol Mol Biol Rev 73(2):310–347. https://doi.org/10.1128/MMBR.00041-08

Kavanaugh NL, Ribbeck K (2012) Selected antimicrobial essential oils eradicate Pseudomonas spp. and *Staphylococcus aureus* biofilms. Appl Environ Microbiol. American Society for Microbiology 78(11):4057–4061. https://doi.org/10.1128/AEM.07499-11

Keren I et al (2011) Characterization and transcriptome analysis of *Mycobacterium tuberculosis* persisters. mBio American Society for Microbiology 2(3):e00100-11. https://doi.org/10.1128/mBio.00100-11

Khan W et al (2010) Aminoglycoside resistance of *Pseudomonas aeruginosa* biofilms modulated by extracellular polysaccharide. Intl Microbiol 13(4):207–212. https://doi.org/10.2436/20.1501.01.127

Kharidia R, Liang JF (2011) The activity of a small lytic peptide PTP-7 on Staphylococcus aureus biofilms. J Microbiol 49(4):663–668. https://doi.org/10.1007/s12275-011-1013-5

Kim H-S, Park H-D (2013) Ginger extract inhibits biofilm formation by *Pseudomonas aeruginosa* PA14. PLoS ONE Edited by J Vadivelu Public Library of Science 8(9):e76106. https://doi.org/10.1371/journal.pone.0076106

Kim H-S et al (2015) 6-Gingerol reduces *Pseudomonas aeruginosa* biofilm formation and virulence via quorum sensing inhibition. Sci Rep Nature Publishing Group 5(1):8656. https://doi.org/10.1038/srep08656

Kiran, S. et al. (2011) 'Enzymatic quorum quenching increases antibiotic susceptibility of multidrug resistant *Pseudomonas aeruginosa*.', Iranian J Microbiol Tehran University of Medical Sciences, 3(1), pp. 1–12. Available at: http://www.ncbi.nlm.nih.gov/pubmed/22347576. Accessed 10 Sept 2018

Koo H et al (2017) Targeting microbial biofilms: current and prospective therapeutic strategies. Nat Rev Microbiol 15(12):740–755. https://doi.org/10.1038/nrmicro.2017.99

Kostakioti M, Hadjifrangiskou M, Hultgren SJ (2013) Bacterial biofilms: development, dispersal, and therapeutic strategies in the Dawn of the Postantibiotic era. Cold Spring Harb Perspect Med 3(4):a010306–a010306. https://doi.org/10.1101/cshperspect.a010306

Kumar, C. G. and Anand, S. K. (1998) 'Significance of microbial biofilms in food industry: a review.', Int J Food Microbiol, 42(1–2), pp. 9–27. Available at: http://www.ncbi.nlm.nih.gov/pubmed/9706794. Accessed 10 Sept 2018

Lasa I, Penadés JR (2006) Bap: a family of surface proteins involved in biofilm formation. Res Microbiol 157(2):99–107. https://doi.org/10.1016/j.resmic.2005.11.003

Laub R, Schneider YJ, Trouet A (1989) Antibiotic susceptibility of Salmonella spp. at different pH values. J Gen Microbiol 135(6):1407–1406. https://doi.org/10.1099/00221287-135-6-1407

Lebeaux D, Ghigo J-M, Beloin C (2014) Biofilm-related infections: bridging the gap between clinical management and fundamental aspects of recalcitrance toward antibiotics. Microbiol Mol Biol Rev MMBR American Society for Microbiology (ASM) 78(3):510–543. https://doi.org/10.1128/MMBR.00013-14

Lee, A. et al. (2000) 'Interplay between efflux pumps may provide either additive or multiplicative effects on drug resistance.', J Bacteriol, 182(11), pp. 3142–3150. Available at: http://www.ncbi.nlm.nih.gov/pubmed/10809693. Accessed 26 Aug 2018

Lee J-H et al (2014) ZnO nanoparticles inhibit *Pseudomonas aeruginosa* biofilm formation and virulence factor production. Microbiol Res Urban & Fischer 169(12):888–896. https://doi.org/10.1016/J.MICRES.2014.05.005

Leid, J. G. et al. (2005) 'The exopolysaccharide alginate protects *Pseudomonas aeruginosa* biofilm bacteria from IFN-gamma-mediated macrophage killing.', J Immunol, 175(11), pp. 7512–7518. Available at: http://www.ncbi.nlm.nih.gov/pubmed/16301659. Accessed 12 Aug 2018

Lewis, K. (2005) 'Persister cells and the riddle of biofilm survival.', Biochemistry Biokhimiia, 70(2), pp. 267–274. Available at: http://www.ncbi.nlm.nih.gov/pubmed/15807669. Accessed 9 Sept 2018

Lewis K (2008) Multidrug tolerance of biofilms and persister cells. In Current topics in microbiology and immunology, vol 322, pp 107–131. Available at: http://www.ncbi.nlm.nih.gov/pubmed/18453274. Accessed 9 Sept 2018

Li J et al (2007) Quorum sensing in *Escherichia coli* is signaled by AI-2/LsrR: effects on small RNA and biofilm architecture. J Bacteriol 189(16):6011–6020. https://doi.org/10.1128/JB.00014-07

Li L et al (2014) The importance of the viable but non-culturable state in human bacterial pathogens. Front Microbiol 5:258. https://doi.org/10.3389/fmicb.2014.00258

López D, Vlamakis H, Kolter R (2010) Biofilms. Cold Spring Harbor Perspect Biol Cold Spring Harbor Laboratory Press 2(7):a000398. https://doi.org/10.1101/cshperspect.a000398.

Lu TK, Collins JJ (2007) Dispersing biofilms with engineered enzymatic bacteriophage. Proc Natl Acad Sci 104(27):11197–11202. https://doi.org/10.1073/pnas.0704624104

Lynch DJ et al (2007) Glucan-binding proteins are essential for shaping Streptococcus mutans biofilm architecture. FEMS Microbiol Lett NIH Public Access 268(2):158–165. https://doi.org/10.1111/j.1574-6968.2006.00576.x.

Mah T-F (2012) Biofilm-specific antibiotic resistance. Future Microbiol 7(9):1061–1072. https://doi.org/10.2217/fmb.12.76

Marić S, Vraneš J (2007) Characteristics and significance of microbial biofilm formation. Period Biol. Available at: https://bib.irb.hr/datoteka/314210.9._Maric_Vranes-Periodicum.pdf. Accessed 12 Aug 2018

Martinez-Gutierrez F et al (2013) Anti-biofilm activity of silver nanoparticles against different microorganisms. Biofouling Routledge 29(6):651–660. https://doi.org/10.1080/08927014.2013.794225

Matsumura, K. et al. (2011) 'Roles of multidrug efflux pumps on the biofilm formation of Escherichia coli K-12.', Biocontrol Sci, 16(2), pp. 69–72. Available at: http://www.ncbi.nlm.nih.gov/pubmed/21719992. Accessed: 26 Aug 2018

McCoy WF et al (1981) Observations of fouling biofilm formation. Can J Microbiol NRC Research Press Ottawa, Canada 27(9):910–917. https://doi.org/10.1139/m81-143

Melo MN, Ferre R, Castanho MARB (2009) Antimicrobial peptides: linking partition, activity and high membrane-bound concentrations. Nat Rev Microbiol 7(3):245–250. https://doi.org/10.1038/nrmicro2095

Merod RT, Wuertz S (2014) Extracellular polymeric substance architecture influences natural genetic transformation of Acinetobacter baylyi in biofilms. Appl Environ Microbiol. American Society for Microbiology 80(24):7752–7757. https://doi.org/10.1128/AEM.01984-14

Merril CR et al (1996) Long-circulating bacteriophage as antibacterial agents. Proc Natl Acad Sci National Academy of Sciences 93(8):3188–3192. https://doi.org/10.1073/pnas.93.8.3188

Mika F, Hengge R (2014) Small RNAs in the control of RpoS, CsgD, and biofilm architecture of *Escherichia coli*. RNA Biol Taylor & Francis 11(5):494–507. https://doi.org/10.4161/rna.28867

Montanaro L et al (2011) Extracellular DNA in biofilms. Int J Artif Organs 34(9):824–831. https://doi.org/10.5301/ijao.5000051

Monzón M et al (2002) Biofilm testing of *Staphylococcus epidermidis* clinical isolates: low performance of vancomycin in relation to other antibiotics. Diag Microbiol Infect Dis 44(4):319–324. Available at.: http://www.ncbi.nlm.nih.gov/pubmed/12543535. Accessed 26 Aug 2018

Moreau-Marquis S, Stanton BA, O'Toole GA (2008) Pseudomonas aeruginosa biofilm formation in the cystic fibrosis airway. Pulm Pharmacol Therap NIH Public Access 21(4):595–599. https://doi.org/10.1016/j.pupt.2007.12.001

Mulcahy H, Charron-Mazenod L, Lewenza S (2008) Extracellular DNA chelates cations and induces antibiotic resistance in *Pseudomonas aeruginosa* Biofilms. PLoS Pathog Edited by M. S. Gilmore. Public Library of Science 4(11):e1000213. https://doi.org/10.1371/journal.ppat.1000213

Murray J (1860) On the origin of species by means of natural selection, or the preservation of favoured races in the struggle for life. Br Foreig Medico-Chirurgical Rev Biomedical Journal Digitization Project 25(50):367–404. Available at: http://www.ncbi.nlm.nih.gov/pubmed/30164232. Accessed 10 Sept 2018.

Lambadi et al (2015) Facile biofunctionalization of silver nanoparticles for enhanced antibacterial properties, endotoxin removal, and biofilm control. Int J Nanomedicine 10:2155. https://doi.org/10.2147/IJN.S72923

Nazzaro F et al (2013) Effect of essential oils on pathogenic bacteria. Pharmaceuticals (Basel, Switzerland) Multidisciplinary Digital Publishing Institute (MDPI) 6(12):1451–1474. https://doi.org/10.3390/ph6121451

Nelson DC et al (2012) Endolysins as antimicrobials. In: Advances in virus research, vol 83. Academic, New York, pp 299–365. https://doi.org/10.1016/B978-0-12-394438-2.00007-4

Nishino K et al (2011) Effect of overexpression of small non-coding DsrA RNA on multidrug efflux in Escherichia coli. J Antimicrob Chemother 66(2):291–296. https://doi.org/10.1093/jac/dkq420

Nostro A et al (2007) Effects of oregano, carvacrol and thymol on Staphylococcus aureus and *Staphylococcus epidermidis* biofilms. J Med Microbiol Microbiology Society 56(4):519–523. https://doi.org/10.1099/jmm.0.46804-0

Novović K et al (2018) Acinetobacter spp. porin Omp33-36: classification and transcriptional response to carbapenems and host cells. PLOS ONE Edited by Z Ruan 13(8):e0201608. https://doi.org/10.1371/journal.pone.0201608

O'Toole G, Kaplan HB, Kolter R (2000) Biofilm formation as microbial development. Annu Rev Microbiol 54(1):49–79. https://doi.org/10.1146/annurev.micro.54.1.49

Ogasawara H, Yamamoto K, Ishihama A (2011) Role of the biofilm master regulator CsgD in cross-regulation between biofilm formation and flagellar synthesis. J Bacteriol 193(10):2587–2597. https://doi.org/10.1128/JB.01468-10

Okshevsky M, Meyer RL (2015) The role of extracellular DNA in the establishment, maintenance and perpetuation of bacterial biofilms. Crit Rev Microbiol Informa Healthcare 41(3):341–352. https://doi.org/10.3109/1040841X.2013.841639

Ordax M et al (2010) Exopolysaccharides favor the survival of Erwinia amylovora under copper stress through different strategies. Res Microbiol 161(7):549–555. https://doi.org/10.1016/j.resmic.2010.05.003

Oubekka SD, Briandet R, Fontaine-Aupart MP, Steenkeste K (2012) Correlative time-resolved fluorescence microscopy to assess antibiotic diffusionreaction in biofilms. Antimicrob Agents Chemother 56(6):3349–3358

Paczkowski JE et al (2017) Flavonoids suppress *Pseudomonas aeruginosa* virulence through allosteric inhibition of quorum-sensing receptors. J Biol Chem American Society for Biochemistry and Molecular Biology 292(10):4064–4076. https://doi.org/10.1074/jbc.M116.770552

Palmer J, Flint S, Brooks J (2007) Bacterial cell attachment, the beginning of a biofilm. J Ind Microbiol Biotechnol Springer-Verlag 34(9):577–588. https://doi.org/10.1007/s10295-007-0234-4

Pamp SJ et al (2008) Tolerance to the antimicrobial peptide colistin in Pseudomonas aeruginosa biofilms is linked to metabolically active cells, and depends on the pmr and mexAB-oprM genes. Mol Microbiol 68(1):223–240. https://doi.org/10.1111/j.1365-2958.2008.06152.x

Parsek MR, Singh PK (2003) Bacterial biofilms: an emerging link to disease pathogenesis. Annu Rev Microbiol 57(1):677–701. https://doi.org/10.1146/annurev.micro.57.030502.090720

Pearson JP, Van Delden C, Iglewski BH (1999) Active efflux and diffusion are involved in transport of *Pseudomonas aeruginosa* cell-to-cell signals. J Bacteriol 181(4):1203–1210. Available at: http://www.ncbi.nlm.nih.gov/pubmed/9973347. Accessed 26 Aug 2018

Percival SL et al (2015) Healthcare-associated infections, medical devices and biofilms: risk, tolerance and control. https://doi.org/10.1099/jmm.0.000032

Peulen T-O, Wilkinson KJ (2011) Diffusion of nanoparticles in a biofilm. Environ Sci Technol American Chemical Society 45(8):3367–3373. https://doi.org/10.1021/es103450g

Piddock LJV (2006) Clinically relevant chromosomally encoded multidrug resistance efflux pumps in Bacteria. Clin Microbiol Rev 19(2):382–402. https://doi.org/10.1128/CMR.19.2.382-402.2006

Pires D et al (2017) Phage therapy as an alternative or complementary strategy to prevent and control biofilm-related infections. Curr Opin Microbiol Elsevier Current Trends 39:48–56. https://doi.org/10.1016/J.MIB.2017.09.004

Pompilio A et al (2011) Antibacterial and anti-biofilm effects of cathelicidin peptides against pathogens isolated from cystic fibrosis patients. Peptides 32(9):1807–1814. https://doi.org/10.1016/j.peptides.2011.08.002

Popova C, Dosseva-Panova V, Panov V (2013) Microbiology of periodontal diseases. A review. Biotechnol Biotechnol Equip 27(3):3754–3759. https://doi.org/10.5504/BBEQ.2013.0027

Qayyum S, Khan AU (2014) From chip-in-a-lab to lab-on-a-chip: towards a single handheld electronic system for multiple application-specific lab-on-a-chip (ASLOC). Med Chem Commun 7:1479–1498. https://doi.org/10.1039/c6md00124f

Rabin N et al (2015a) Agents that inhibit bacterial biofilm formation. Future Med Chem 7(5):647–671. https://doi.org/10.4155/fmc.15.7

Rabin N et al (2015b) Biofilm formation mechanisms and targets for developing antibiofilm agents. Future Med Chem 7(4):493–512. https://doi.org/10.4155/fmc.15.6

Ramamurthy T et al (2014) Current perspectives on viable but non-Culturable (VBNC) pathogenic Bacteria. Front Pub Health Frontiers Media SA 2:103. https://doi.org/10.3389/fpubh.2014.00103

Reen FJ et al (2018) Coumarin: a novel player in microbial quorum sensing and biofilm formation inhibition. Appl Microbiol Biotechnol Springer Berlin Heidelberg 102(5):2063–2073. https://doi.org/10.1007/s00253-018-8787-x

Rijnaarts HHM et al (1995) Reversibility and mechanism of bacterial adhesion. Coll Surf B Biointerfaces Elsevier 4(1):5–22. https://doi.org/10.1016/0927-7765(94)01146-V

Römling U et al (1998) Curli fibers are highly conserved between Salmonella typhimurium and *Escherichia coli* with respect to operon structure and regulation. J Bacteriol 180(3):722–731. Available at: http://www.ncbi.nlm.nih.gov/pubmed/9457880. Accessed 9 Sept 2018

Rodríguez-Rubio L, Martínez B, Donovan DM, Rodríguez A, García P (2013) Bacteriophage virion-associated peptidoglycan hydrolases: potential new enzybiotics. Crit Rev Microbiol 39(4):427–434

Roy V et al (2013) AI-2 analogs and antibiotics: a synergistic approach to reduce bacterial biofilms. Appl Microbiol Biotechnol 97(6):2627–2638. https://doi.org/10.1007/s00253-012-4404-6

Rutherford ST et al (2011) AphA and LuxR/HapR reciprocally control quorum sensing in vibrios. Genes & Develop Cold Spring Harbor Laboratory Press 25(4):397–408. https://doi.org/10.1101/gad.2015011

Saginur R et al (2006) Multiple combination bactericidal testing of staphylococcal biofilms from implant-associated infections. Antimicrob Agents Chemother 50(1):55–61. https://doi.org/10.1128/AAC.50.1.55-61.2006

Saini H, Chhibber S, Harjai K (2015) Azithromycin and ciprofloxacin: a possible synergistic combination against *Pseudomonas aeruginosa* biofilm-associated urinary tract infections. Intl J Antimicrob Agents Elsevier 45(4):359–367. https://doi.org/10.1016/j.ijantimicag.2014.11.008

Samad T et al (2017) Swimming bacteria promote dispersal of non-motile staphylococcal species. ISME J 11(8):1933–1937. https://doi.org/10.1038/ismej.2017.23

Sauer K (2003) The genomics and proteomics of biofilm formation. Genome Biol BioMed Central 4(6):219. https://doi.org/10.1186/gb-2003-4-6-219

Sauer K et al (2004) Characterization of nutrient-induced dispersion in *Pseudomonas aeruginosa* PAO1 biofilm. J Bacteriol 186(21):7312–7326. https://doi.org/10.1128/JB.186.21.7312-7326.2004

Sharma A et al (2017) Fosfomycin resistance in *Acinetobacter baumannii* is mediated by efflux through a major facilitator superfamily (MFS) transporter—AbaF. J Antimicrob Chemother Oxford University Press 72(1):68–74. https://doi.org/10.1093/jac/dkw382

Silva IN et al (2018) The OmpR regulator of Burkholderia multivorans controls mucoid-to-nonmucoid transition and other cell envelope properties associated with persistence in the cystic fibrosis lung. J Bacteriol American Society for Microbiology Journals 200(17):e00216–e00218. https://doi.org/10.1128/JB.00216-18.

Singh PK et al (2002) A component of innate immunity prevents bacterial biofilm development. Nature 417(6888):552–555. https://doi.org/10.1038/417552a

Singh V et al (2015) Enzymatic degradation of bacterial biofilms using Aspergillus clavatus MTCC 1323. Microbiology Pleiades Publishing 84(1):59–64. https://doi.org/10.1134/S0026261715010130

Singh S et al (2017) Understanding the mechanism of bacterial biofilms resistance to antimicrobial agents. Open Microbiol J Bentham Science Publishers 11:53–62. https://doi.org/10.2174/1874285801711010053

Spaulding CN et al (2017) Selective depletion of uropathogenic E. coli from the gut by a FimH antagonist. Nature Nature Publishing Group 546(7659):528. https://doi.org/10.1038/nature22972

Stevens WW et al (2015) Chronic rhinosinusitis pathogenesis. J Allergy Clin Immunol NIH Public Access 136(6):1442–1453. https://doi.org/10.1016/j.jaci.2015.10.009

Stoodley P et al (2002) Biofilms as complex differentiated communities. Annu Rev Microbiol 56(1):187–209. https://doi.org/10.1146/annurev.micro.56.012302.160705

Su H-L et al (2009) The disruption of bacterial membrane integrity through ROS generation induced by nanohybrids of silver and clay. Biomaterials Elsevier 30(30):5979–5987. https://doi.org/10.1016/J.BIOMATERIALS.2009.07.030

Sutherland IW (1999) Biofilm exopolysaccharides. In: Microbial extracellular polymeric substances. Springer, Berlin/Heidelberg, pp 73–92. https://doi.org/10.1007/978-3-642-60147-7_4

Sutherland IW et al (2004) The interaction of phage and biofilms. FEMS Microbiol Lett 232(1):1–6. Available at.: http://www.ncbi.nlm.nih.gov/pubmed/15061140. Accessed 10 Sept 2018

Thomason MK, Storz G (2010) Bacterial antisense RNAs: how many are there, and what are they doing? Annu Rev Genet 44(1):167–188. https://doi.org/10.1146/annurev-genet-102209-163523

Tielker D et al (2005) *Pseudomonas aeruginosa* lectin LecB is located in the outer membrane and is involved in biofilm formation. Microbiology 151(5):1313–1323. https://doi.org/10.1099/mic.0.27701-0

Toledo-Arana A et al. (2001) The enterococcal surface protein, Esp, is involved in *Enterococcus faecalis* biofilm formation. Appl Environ Microbiol American Society for Microbiology (ASM) 67(10):4538–4545. https://doi.org/10.1128/AEM.67.10.4538-4545.2001

Tribedi P, Sil AK (2014) Cell surface hydrophobicity: a key component in the degradation of polyethylene succinate by *Pseudomonas* sp. AKS2. J Appl Microbiol 116(2):295–303. https://doi.org/10.1111/jam.12375

Uroz S, Dessaux Y, Oger P (2009) Quorum sensing and quorum quenching: the yin and Yang of bacterial communication. Chem Bio Chem Wiley-Blackwell 10(2):205–216. https://doi.org/10.1002/cbic.200800521

Valle J et al (2006) Broad-spectrum biofilm inhibition by a secreted bacterial polysaccharide. Proc Nat Acad Sci USA National Academy of Sciences 103(33):12558–12563. https://doi.org/10.1073/pnas.0605399103

Vijayakumar SRV et al (2004) RpoS-regulated genes of Escherichia coli identified by random lacZ fusion mutagenesis. J Bacteriol American Society for Microbiology (ASM) 186(24):8499–8507. https://doi.org/10.1128/JB.186.24.8499-8507.2004

Vorkapic D, Pressler K, Schild S (2016) Multifaceted roles of extracellular DNA in bacterial physiology. Current genetics Springer 62(1):71–79. https://doi.org/10.1007/s00294-015-0514-x

Wakimoto N et al (2004) Quantitative biofilm assay using a microtiter plate to screen for enteroaggregative *Escherichia coli*. Am J Trop Med Hyg 71(5):687–90. Available at.: http://www.ncbi.nlm.nih.gov/pubmed/15569806. Accessed: 26 Aug 2018

Walters MC, Roe F, Bugnicourt A, Franklin MJ, Stewart PS (2003) Contributions of antibiotic penetration, oxygen limitation, and low metabolic activity to tolerance of *Pseudomonas aeruginosa* biofilms to ciprofloxacin and tobramycin. Antimicrob Agents Chemother 47(1):317–323

Wang BY, Chi B, Kuramitsu HK (2002) Genetic exchange between *Treponema denticola* and *Streptococcus gordonii* in biofilms. Oral Microbiol Immunol 17(2):108–112. Available at: http://www.ncbi.nlm.nih.gov/pubmed/11929558. Accessed 26 Aug 2018

Wang L, Hu C, Shao L (2017) The antimicrobial activity of nanoparticles: present situation and prospects for the future. Intl J Nanomed Dove Press 12:1227–1249. https://doi.org/10.2147/IJN.S121956

Whitchurch CB et al (2002) Extracellular DNA required for bacterial biofilm formation. Science 295(5559):1487–1487. https://doi.org/10.1126/science.295.5559.1487

Yap PSX et al (2014) Essential oils, a new horizon in combating bacterial antibiotic resistance. Open Microbiol J Bentham Science Publishers 8:6–14. https://doi.org/10.2174/1874285801408010006

Zeng X et al (2011) Synergistic effect of 14-alpha-Lipoyl Andrographolide and various antibiotics on the formation of biofilms and production of exopolysaccharide and Pyocyanin by *Pseudomonas aeruginosa*. Antimicrob Agents Chemother 55(6):3015–3017. https://doi.org/10.1128/AAC.00575-10

Zhao G et al (2013) Biofilms and inflammation in chronic wounds. Adv Wound Care Mary Ann Liebert, Inc 2(7):389–399. https://doi.org/10.1089/wound.2012.0381

Zobell CE, Allen EC (1935) The significance of marine Bacteria in the fouling of submerged surfaces. J Bacteriol American Society for Microbiology (ASM) 29(3):239–251. Available at: http://www.ncbi.nlm.nih.gov/pubmed/16559784. Accessed 12 Aug 2018

7 Antimicrobial Resistance in Microbes: Mode of Action of TolC Like Protein and Their Mechanism of Regulating Stress Resistance and Physiology

Vijaya Bharathi Srinivasan and Govindan Rajamohan

Abstract

Although *Acinetobacter baumannii* is considered the most common nosocomial species that cause severe infections, only few studies have demonstrated the physiological role of outer membrane proteins (OMPs) and its pathogenic potential. Here, we discuss the functions of putative outer membrane protein in physiology and multidrug resistance. The *abuO* like genes were functionally characterized in genetically distinct *A. baumannii* obtained from a tertiary medical center in India. This sequel study provides evidence for the strain non-specific unanimous function of *abuO* in stress response and antimicrobial resistance in *A. baumannii*.

Keywords

Nosocomial pathogen · Biocide resistance · Molecular epidemiology · Active efflux · Outer membrane proteins

7.1 Introduction

Acinetobacter baumannii is a pleomorphic, aerobic, rod-shaped Gram-negative bacterium that belongs to family Moraxellaceae in class Gammaproteobacteria (Dijkshoorn et al. 2007; Boucher et al. 2009). *A. baumannii* affects severely ill or immunocompromised patients (Villegas and Hartstein 2003) and causes severe infections like ventilator-associated pneumonia, urinary tract infections, BSI and

V. B. Srinivasan (✉) · G. Rajamohan
Bacterial Signaling and Drug Resistance Laboratory, Council of Scientific Industrial Research, Institute of Microbial Technology, Chandigarh, India
e-mail: vijirmohan@gmail.com

S. M. Mandal, D. Paul (eds.), *Bacterial Adaptation to Co-resistance*,
https://doi.org/10.1007/978-981-13-8503-2_7

surgical wound infections (Blossom and Srinivasan 2008); associated with high morbidity and mortality (Dijkshoorn et al. 2007; Boucher et al. 2009). Carbapenems are used to treat illnesses caused by *A. baumannii*; however, continued emergence of resistant strains has led to therapeutic failure (Pogue et al. 2013). *A. baumannii* has emerged as a successful nosocomial pathogen due to their capacity to sustain on desiccated and abiotic surfaces for long periods of time due to their ability to form biofilms (Tomaras et al. 2003), their ability to resist disinfectants being used in the hospitals (Rajamohan et al. 2010), and their ability to resist wide variety of antimicrobials and propensity to acquire multiple resistance determinants (Srinivasan et al. 2009). The worldwide prevalence of multi-drug resistant (MDR) *A. baumannii* was recently reported (Lob et al. 2016) in a surveillance study carried from 2011–2014 declaring that MDR rates were of 47% in North America and 93% in the Europe and the Middle East. In developing country like India, the incidence of MDR *A. baumannii* infections have gradually increased mainly in ICU (Tiwari et al. 2012), and it is responsible for around 10% of the HAI in India (Rynga et al. 2015; Nachimuthu et al. 2015). In *A. baumannii*, overproduction of enzymes, inactivation/modification/alteration of drug targets, loss of porins, most significantly over expression of efflux pumps are known to confer high-level multi-drug resistance (Coyne et al. 2011; Vila et al. 2007). *The multi-drug efflux pumps are categorized into five supefamilies:* major facilitator super family (MFS), small multi-drug resistance super family (SMR), the multi-drug and toxic compound extrusion super family (MATE), resistance-nodulation cell division family (RND) and ATP- binding cassette transporter (ABC) (Poole 2005; Piddock 2006). Of these, the tripartite RND pump pumps [consisting of membrane fusion protein, membrane transporters and outer membrane protein (OMP)] remain the most predominant pump conferring in drug resistance in this nosocomial pathogen and the well characterized ones include AdeABC, AdeIJK, AdeFGH (Magnet et al. 2001; Damier-Piolle et al. 2008; Coyne et al. 2010).

The promiscuous OMPs are known to have a pivotal role in bacterial physiology and pathogenesis (Delcour 2009). TolC is a classical OMP which has been well documented with regard to drug resistance. Several reports suggest that AcrAB-TolC forms a functional RND efflux system and assists in effluxing out of substrates directly out of the cell through the OMP TolC (Lomovskaya et al. 2002; Nikaido et al. 1998; Nikaido and Zgurskaya 2001). In *E. coli* it has been further reported that around eight drug transporters co-operate with TolC to extrude various substrates including antibiotics. The RND drug transport systems namely AcrAB, AcrD, AcrEF, MdtEF and MdtABC (Elkins and Nikaido 2002; Hirakawa et al. 2003; Nagakubo et al. 2002; Nishino and Yamaguchi 2001), two MFS systems namely EmrAB and EmrKY (Lomovskaya and Lewis 1992) and one ABC drug transport system MacAB (Kobayashi et al. 2001). Recently, the role of TolC homolog designated as AbuO was elucidated by generating its mutant in a highly resistant strain *A. baumannii* AYE (Srinivasan et al. 2015). It was shown to be playing a role in providing resistance to a variety of broad spectrum antibiotics and also to hospital based disinfectants. Our initial study began with the characterization of Indian *Acinetobacter* isolates obtained from clinical settings to determine the prevalence of MDR pattern in Indian scenario. With such an aim, we focused on characterizing the current trends of antibiotic resistome in *A. baumannii* clinical isolates (n = 116) obtained from different Medical centers in India. During this preface work, we observed the undisputed presence (confirmed by PCR and Southern) and increased expression (confirmed by RT-PCR) of *abuO* across clonally distinct, biocide tolerant, and MDR *A. baumannii* clinical isolates. Hence it prompted us to understand the general biological functions of these AbuO homologs (from Indian clinical isolates AB1, AB2 and AB3) with a comprehensive analysis of its involvement in maintaining *A. baumannii* physiology and conferring broad-spectrum antimicrobial resistance. In this chapter, we describe the results of the study based on various genotypic and phenotypic assays using the WT strains (AB1, AB2 and AB3), generated mutants ($\Delta abuO^1$, $\Delta abuO^2$, $\Delta abuO^3$) and complemented strains ($\Delta abuO\Omega abuO^1$, $\Delta abuO\Omega abuO^2$ and $\Delta abuO\Omega abuO^3$).

7.2 Materials and Methods

7.2.1 Bacterial Strains and Media

The *A. baumannii* isolates AB1, AB2 and AB3 were collected during 2010–2011 as a kind gift from medical centre in India. The cultures were grown at 37 °C in Luria-Bertani (LB) broth with constant shaking (220 rpm) or on LB agar (Difco), supplemented with 400 μg/ml hygromycin for mutant and 50 μg/ml Zeocin for complemented strains. Primers for the present study customized from Eurofins MWG operons, Germany (Srinivasan et al. 2015).

7.2.2 Generation of Mutants in *A. baumannii*

The *A. baumannii* strains WT strains (AB1, AB2 and AB3), (confirmed by *gyrB* and 16S rRNA sequencing) and recombinant plasmid pUC-*abuO* was used to generate isogenic mutants of *abuO* namely $\Delta abuO^1$, $\Delta abuO^2$, $\Delta abuO^3$ respectively. Following the similar strategy as described before (Srinivasan et al. 2015), the complemented constructs obtained were denoted $\Delta abuO\Omega abuO^1$, $\Delta abuO\Omega abuO^2$, and $\Delta abuO\Omega abuO^3$.

7.2.3 Bacterial Growth Curves

The growth kinetics of strains (AB1, AB2, and AB3), ($\Delta abuO^1$, $\Delta abuO^2$, $\Delta abuO^3$, $\Delta abuO\Omega abuO^1$, $\Delta abuO\Omega abuO^2$, and $\Delta abuO\Omega abuO^3$) was monitored in LB at different pH. Optical densities were measured at OD_{600nm} using Bioscreen C automated growth analysis system (Labsystems, Helsinki, Finland) for 18 h at 37 °C shaking and automatically recorded for each well after every 30 mins. Independent experiments were performed with the freshly autoclaved medium in triplicates, at least, three independent times. The growth inactivation assays with slight modifications were done to assess the impact on drug efflux capacity (Srinivasan et al. 2015b).

7.2.4 Motility and Biofilm Forming Assays

In motility assay, *A. baumannii* (AB1, AB2, and AB3), ($\Delta abuO^1$, $\Delta abuO^2$, $\Delta abuO^3$), $\Delta abuO\Omega abuO^1$, $\Delta abuO\Omega abuO^2$, and $\Delta abuO\Omega abuO^3$) cultures grown to OD_{600nm} = 1.0 were pierced into LB agar plates of different agar concentrations and incubated for 12 hrs at 37 °C. Upon incubation bacteria would grow in an outward manner, and the diameter of their growth would reflect their extent of mobility (Srinivasan et al. 2015).

The crystal violet binding (classical biofilm) assay evaluated the ability of *A. baumannii* to form biofilm as described before (Srinivasan et al. 2015). The glass tubes had LB medium with 150 μL of each *A. baumannii* strain diluted to OD_{600nm} = 0.01. Further, they were incubated at 37 °C for 24 h. After incubation, cells were washed three times with phosphate buffered saline to remove the planktonic growth. The methanol treated biofilms once stained with 1% crystal violet for five mins were washed with water and dried. Biofilm thickness was measured by adding 33% glacial acetic acid and taking a reading at OD_{570nm} using an automated plate reader.

7.2.5 Oxidative Stress Tolerance Assays

In this test, small Whatman 3MM paper disks (6 mm) were impregnated with the different amount of hydrogen peroxide (H_2O_2) and later air dried. The (AB1, AB2, and AB3), ($\Delta abuO^1$, $\Delta abuO^2$, $\Delta abuO^3$), $\Delta abuO\Omega abuO^1$, $\Delta abuO\Omega abuO^2$, and $\Delta abuO\Omega abuO^3$) strains were spread on LB agar plate (OD_{600nm} = 0.01) uniformly. In the centre of the plate, filter disks soaked with varied concentrations of H_2O_2 were placed. The diameter of inhibition is measured (in mm) following incubation for 16 h. The liquid growth assay was monitored in the presence of different amounts of H_2O_2 respectively and compared with the WT by measuring the absorbance at OD_{600nm} periodically in Bioscreen C automated growth analysis system (Labsystems, Helsinki, Finland). The *A. baumannii* strains were treated with varied H_2O_2 concentrations in killing assay to evaluate its survival ability as described before (Hennequin and Forestier 2009).

7.2.6 Nitrosative Stress Tolerance Assays

Acidified sodium nitrite and sodium nitroprusside (SNP) were used to generate the nitrosative stress to check cell growth against these NO-releasing agents (Stevanin et al. 2000). Growth of cultures against sodium nitrite was determined as described before (Srinivasan et al. 2015). Briefly, strains were grown aerobically in LB medium at pH 6.0 up to OD_{600nm} of 0.01. Later, *A. baumannii* was exposed to varied concentrations of 5 mM, 10 mM, 15 mM, 20 mM and 25 mM of sodium nitrite and monitored for growth at OD_{600nm} at an interval of every 1 h. To check the response of cultures against SNP, growth profile of different strains, were also determined at pH 7.0 in LB medium supplemented with varied concentrations of SNP and compared with the WT by measuring the absorbance at OD_{600nm} periodically in Bioscreen C automated growth analysis system (Labsystems, Helsinki, Finland) (Srinivasan et al. 2015).

7.2.7 Antibiotic Susceptibility Testing

The *A. baumannii* (AB1, AB2, and AB3), (Δ*abuO*^{1}, Δ*abuO*^{2}, Δ*abuO*^{3}), Δ*abuO*Ω*abuO*^{1}, Δ*abuO*Ω*abuO*^{2}, and Δ*abuO*Ω*abuO*^{3}) strains underwent susceptibility testing using different commercial available antibiotic discs (Hi Media, Bombay, India) as described before and data analyzed as per Clinical and Laboratory Standards Institute [CLSI] guidelines (Clinical and Laboratory Standards Institute 2014). The MIC of antibiotics was tested using E-strips. From different *A. baumannii* strains, the OMPs were purified as per the method described elsewhere (Limansky et al. 2002). Fluorimetric accumulation assays with substrate ethidium bromide were done to assess loss in efflux capability in *A. baumannii* strains as described before (Mortimer and Piddock 1991).

7.2.8 Various Stress Challenge Assays

The (AB1, AB2, and AB3), (Δ*abuO*^{1}, Δ*abuO*^{2}, Δ*abuO*^{3}), Δ*abuO*Ω*abuO*^{1}, Δ*abuO*Ω*abuO*^{2}, and Δ*abuO*Ω*abuO*^{3}) cells were exposed to different stress challenges at varied concentrations such as bile salt deoxycholate (16, 64, 256, 1024, 4096 and 16,348 μg/ml), and sodium chloride (NaCl) (0.075, 0.15, 0.25, 0.5, 0.75, 1 and 2 M); efflux pump substrates as acriflavine (0.5, 4, 16, 64, 256 and 512 μg/ml), acridine orange (0.5, 4, 16, 64, 256 and 512 μg/ml), EtBr (0.5, 4, 16, 64, 256 and 512 μg/ml), rhodamine (0.5, 4, 16, 64, 256 and 512 μg/ml) and saffranine (0.5, 4, 16, 64, 256 and 512 μg/ml); antibiotics as ampicillin (0.5, 4, 16, 64, 256 and 512 μg/ml), chloramphenicol (0.5, 4, 16, 64, 256 and 512 μg/ml), ciprofloxacin (0.5, 4, 16, 64, 256 and 512 μg/ml), neomycin (0.5, 4, 16, 64, 256 and 512 μg/ml) and tetracycline (0.5, 4, 16, 64, 256 and 512 μg/ml); biocides as benzalkonium chloride (3.2, 6.4, 12.8, 25.6, 51.2 and 102.4 μg/ml), chlorhexidine (3.2, 6.4, 12.8, 25.6, 51.2 and 102.4 μg/ml) and triclosan (0.0005, 0.001, 0.005, 0.01, 0.05 and 0.1 μg/ml) and number of colonies were scored after incubation at 37 °C for 16 h. The percentage of survival for each culture was calculated as mentioned before (Srinivasan et al. 2015).

7.2.9 Expression Analysis

The RNA from the mid log-phase cultures was extracted using the RNeasy Mini kit following manufacturer's instructions. The DNase I treated total RNA of 500 ng served as the template for cDNA production. The quantitative PCR reactions for different genes were performed using gene specific primers. By real-time RT-PCR, using Maxima SYBR Green qPCR master mix (Fermentas) in an iCycler thermal cycler (Bio-Rad), the expression levels was monitored, and the melting curve analysis was carried out to confirm amplification of a single product. From at least three separately grown replicate cultures, total RNA was isolated. All real-time RT-PCR experiments were performed more than three times with *rpoB* as an internal control as described previously (Srinivasan et al. 2015).

7.2.10 Statistics Analysis

Homology search, similarities, identities, and domain analysis were performed using NCBI Internet server. All data are presented as means ± the standard error. Plotting and calculation of the standard deviation was done in Microsoft Excel. Statistical analysis of raw data done by using paired Student *t-test*. The P value of >0.001 was considered significant.

7.3 Results and Discussion

7.3.1 Strains of Indian Origin Used in This Sequel Study

In our temporal studies on characterizing the antibiotic resistome of *A. baumannii* clinical isolates (n = 116) obtained from a medical centre in India, we identified the undisputed presence (confirmed by PCR and Southern) of AbuO across genetically unrelated *A. baumannii* clinical isolates. To elucidate the unanimous functions of *abuO* homologs, we selected three high-level MDR and biocide tolerant *A. baumannii* strains which exhibited increased expression of *abuO* as confirmed by RT-PCR. For this purpose namely *A. baumannii* strain AB1 harboring resistance genes *rmtC*, $bla_{\text{OXA-58}}$, $bla_{\text{OXA-51}}$ like genes, *A. baumannii* strain AB2 harboring resistance genes *rmtC*, *rmtD*, *qnrA*, *strA*, *strB*, $bla_{\text{OXA-58}}$, $bla_{\text{OXA-51}}$ like genes, *A. baumannii* strain AB3 harboring resistance genes $bla_{\text{OXA-58}}$, $bla_{\text{OXA-51}}$ like genes, were selected to inactivate *abuO* and decipher its function. The generated constructs *i.e. abuO* mutants ($\Delta abuO^1$, $\Delta abuO^2$, $\Delta abuO^3$; inactivation confirmed by southern) and complemented strains ($\Delta abuO\Omega abuO^1$, $\Delta abuO\Omega abuO^2$ and $\Delta abuO\Omega abuO^3$) were characterized by performing various phenotypic assays as discussed below.

7.3.2 Role of AbuO-Like Protein in Cellular Physiology in *Acinetobacter baumannii*

The maintenance of bacterial physiology is of central importance for the survival of Gram-negative bacteria. Therefore, it is not surprising that several mechanisms exist that act in parallel to respond to perturbations of the bacterial environment. Since *A. baumannii* strains are able to grow in diverse conditions in the host, it seems likely that they have a mechanism(s) for survival in adverse environments, like the high and low pH. TolC together with inner membrane proteins forms the outer membrane component of a protein conducting channel, thus it is important to know whether such proteins have a role in growth or survival in such adverse conditions.

7.3.3 Bacterial Growth Curves

The growth characteristics of WT (AB1, AB2, and AB3), *abuO* mutants ($\Delta abuO^1$, $\Delta abuO^2$, $\Delta abuO^3$), ($\Delta abuO\Omega abuO^1$, $\Delta abuO\Omega abuO^2$, and $\Delta abuO\Omega abuO^3$) revealed unique patterns. Like the Δ*abuO* in our previous study (1.34 fold), the Indian mutants also exhibited a stunted growth by 1.32 fold; ±0.0057 {fold differences at 12 h of growth}; ($\Delta abuO^1$), 1.29 fold; ±0.0061 ($\Delta abuO^2$), 1.64 fold; ±0.0087 ($\Delta abuO^3$) when compared with their respective control strains respectively in LB at pH 7.0, while transcomplemented strains restored the ability to grow [$P < 0.001$] (laboratory observations). Overall, results demonstrated the role of TolC-like outer membrane (Fig. 7.1) protein in influencing growth capability in *A. baumannii.*

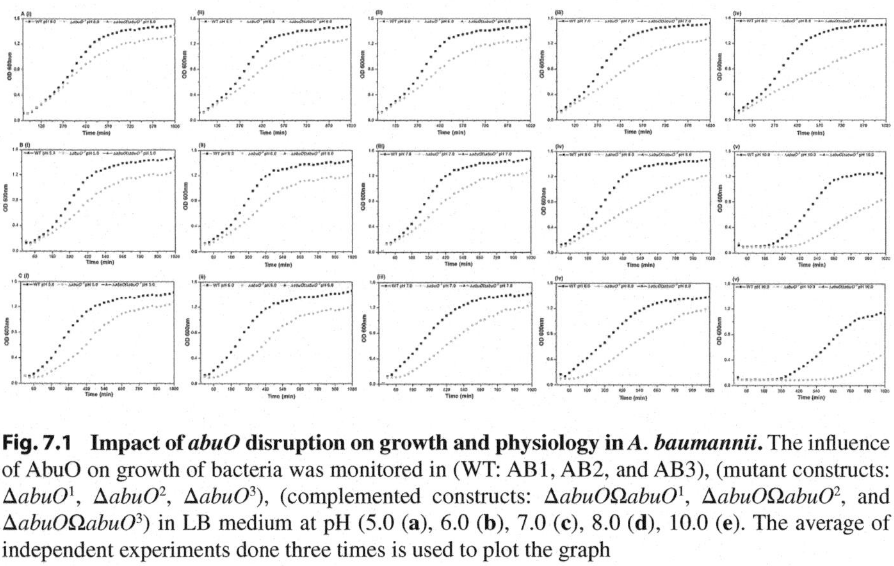

Fig. 7.1 Impact of *abuO* disruption on growth and physiology in *A. baumannii*. The influence of AbuO on growth of bacteria was monitored in (WT: AB1, AB2, and AB3), (mutant constructs: $\Delta abuO^1$, $\Delta abuO^2$, $\Delta abuO^3$), (complemented constructs: $\Delta abuO\Omega abuO^1$, $\Delta abuO\Omega abuO^2$, and $\Delta abuO\Omega abuO^3$) in LB medium at pH (5.0 (**a**), 6.0 (**b**), 7.0 (**c**), 8.0 (**d**), 10.0 (**e**). The average of independent experiments done three times is used to plot the graph

7.3.4 Motility and Biofilm Forming Profiles

On testing the cultures on plates with different agar concentrations, the difference in growth between the parental strains and the mutant were negligible (Fig. 7.2a). The difference in biofilm forming ability was also found negligible (Fig. 7.2b) which tells that role of AbuO in such phenomenon could be indirect.

7.3.5 Functions of AbuO-Like Protein in Stress Response in *Acinetobacter baumannii*

To determine the role of TolC-like protein in intestinal colonization, different strains underwent specific gastrointestinal stress associated with osmotic and bile challenges.

7.3.6 Osmotic and Deoxycholate Challenge Assay

When mid-log cultures were exposed to different concentrations of osmotic challenge it was observed that the total CFU of control strains AB1, AB2 and AB3 (measure of surviving capacity) was higher as compared to mutant. The ability of isogenic mutants to tolerate the osmotic stress at 0.75 M was significantly lower by 1.3 ± 0.12 fold, 3.74 ± 0.24 fold and 3.29 ± 0.15 fold respectively compared to the WT strains (laboratory observations). It is important to note that Δ*abuO* of *A. baumannii* AYE exhibited a 2.15-fold reduced survival compared to its parental strain [$P < 0.01$].

The ability of isogenic mutants to tolerate the deoxycholate stress at 64 μg/ml was significantly lower by 1.05-fold, 1.53-fold and 0.92-fold respectively (Fig. 7.3) compared to the WT strains. It is important to note that Δ*abuO* of *A. baumannii* AYE exhibited a 1.34-fold reduced survival compared to its parental strain [$P < 0.01$]. Collectively, results displayed here reveal the role of TolC-like outer membrane protein in high osmotic and bile tolerance in *A. baumannii* clinical isolates of Indian origin too. High osmolarity and elevated bile level is usually encountered by *A. baumannii* in humans during infection, and thus AbuO might contribute to the persistence of *A. baumannii* during colonization.

7.3.7 Oxidative Stress Tolerance Capability

There is little data to suggest that *A. baumannii* that frequently causes nosocomial infections have evolved specialized mechanisms to protect themselves from oxidative stresses, so to decipher functions of AbuO-like protein of *A. baumannii* in the same was prudent. To decipher functions of AbuO in oxidative stress firstly, disc assay was performed using different concentrations. The mutants exhibited sensitivity at different concentrations tested (Fig. 7.4a). While the *A. baumannii* AYE

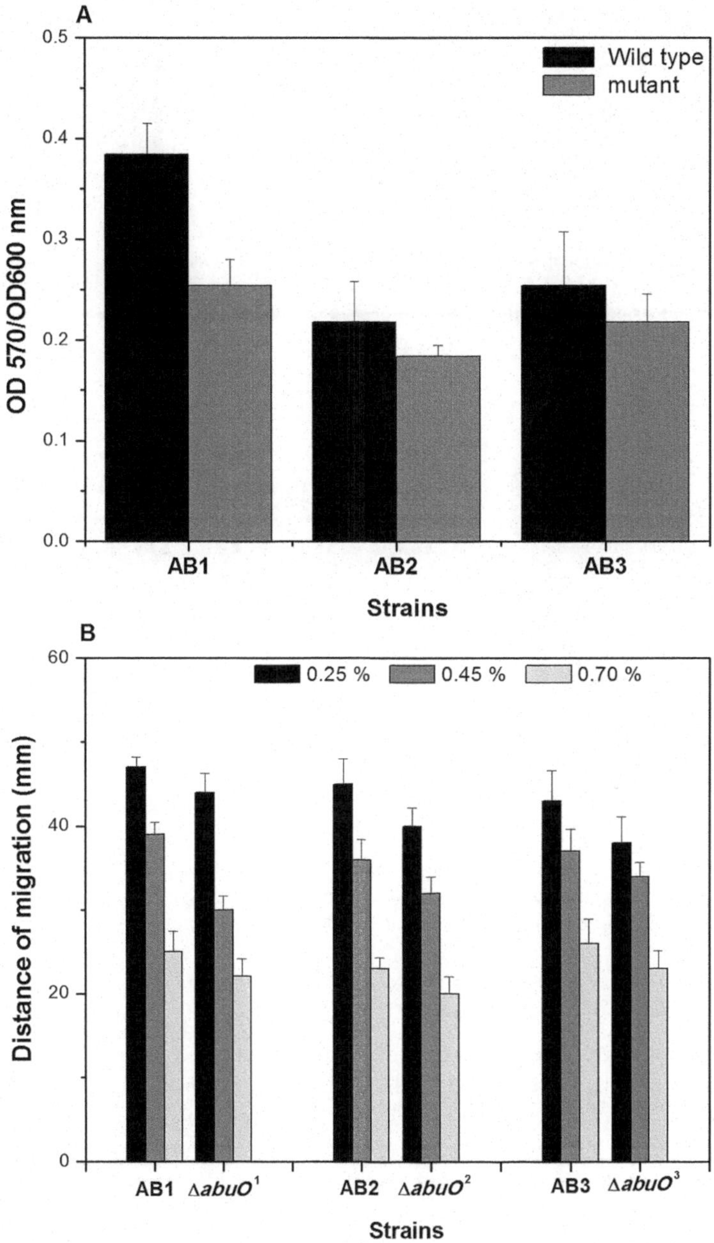

Fig. 7.2 Motility and biofilm forming ability. (a) The mean diameter of halos obtained from independent experiments is plotted with standard deviations. P value for the differences between different strains were <0.01. (**b**) The ability of mutants and WT cells in forming biofilm on glass tubes. The data are the means of measurements performed three times

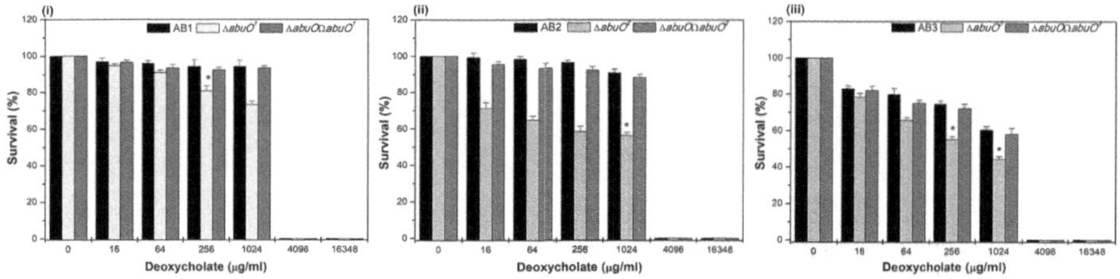

Fig. 7.3 The survival capability of strains (WT: AB1, AB2, and AB3), (mutant constructs: Δ*abuO*[1], Δ*abuO*[2], Δ*abuO*[3]), to grow in the presence of deoxycholate was monitored and the trans-complemented strain restored the ability to tolerate the stress

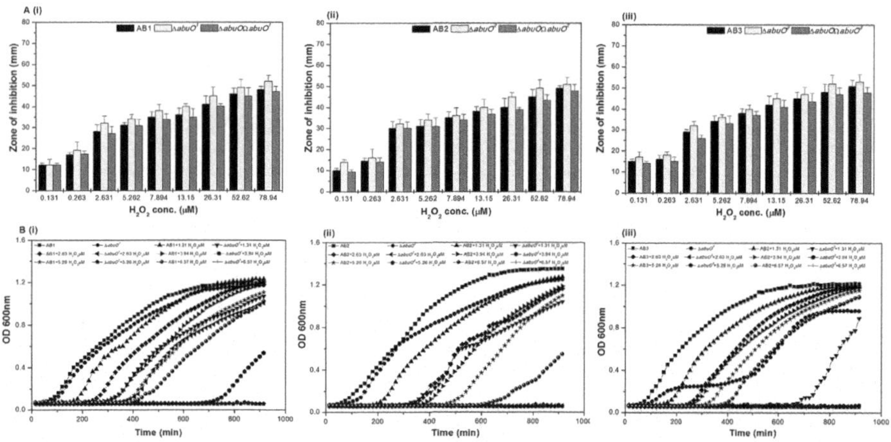

Fig. 7.4 The capability of strains (WT: AB1, AB2, and AB3), (mutant constructs: Δ*abuO*[1], Δ*abuO*[2], Δ*abuO*[3]), (complemented constructs: Δ*abuO*Ω*abuO*[1], Δ*abuO*Ω*abuO*[2], and Δ*abuO*Ω*abuO*[3]) to combat different levels of hydrogen peroxide stress was measured by disc diffusion assay (**a**) and growth (**b**). The data is the means of measurements made in triplicate and performed three times

mutant Δ*abuO* exhibited 12.75-fold impaired growth, the mutants of Indian origin displayed stunted growth of [5.75 fold; ±0.0025 (Δ*abuO*[1]), 8.06 fold; ±0.0025 (Δ*abuO*[2]), 9.26 fold; ±0.0007 (Δ*abuO*[3])] respectively at 3.9465 μM of hydrogen peroxide compared to control strains (Fig. 7.4b) (laboratory observations). Overall, suggesting that AbuO has an equivocal role in oxidative stress tolerance in *A. baumannii*. In *E. coli*, TolC belongs to the *marA*/*soxS*/*rob* regulon and over 40 genes that promote resistance to reactive oxygen species (ROS) (Storz and Imlay 1999).

AbuO may possibly help in efflux of such ROS and help bacterial survival inside host. However studies pertaining to know the exact role of AbuO in response to oxidative stress are highly warranted.

7.3.8 Nitrosative Stress Tolerance Ability

Bacteria encounter a myriad of stresses in the host which includes reactive nitrogen species that generates nitrosative stress, thus role of AbuO-like protein of *A. baumannii* in nitrosative stress was elucidated by comparing the growth profiles of different strains in LB broth with different acidified nitrite concentrations. It was found that mutants did not impart significant growth defect (Fig. 7.5) indicating that AbuO has no direct role in conferring nitrosative stress tolerance in *A. baumannii*.

7.3.9 Role of AbuO-Like Protein in Antimicrobial Resistance in *Acinetobacter baumannii*

Antibiotic resistance is the most common phenomenon in pathogenic bacteria. Multi-drug resistance in microbes is generally confered by intrinsic and/or acquired determinants. The mode of resistance seen in bacteria are of various types such as enzymatic modification of antibiotics, target gene mutation, altered outer membrane permeability and up regulated multi drug efflux pump activity. Efflux involves membrane proteins that transport substances to decrease the drug

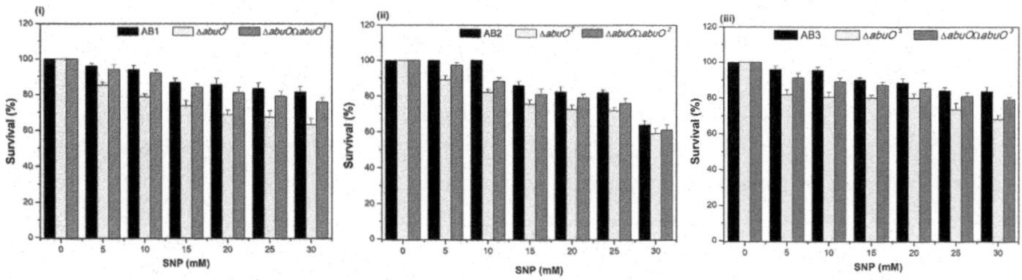

Fig. 7.5 Nitrosative stress tolerance. The survival ability of strains (WT: AB1, AB2, and AB3), (mutant constructs: Δ*abuO*[1], Δ*abuO*[2], Δ*abuO*[3]), (complemented constructs: Δ*abuO*Ω*abuO*[1], Δ*abuO*Ω*abuO*[2], and Δ*abuO*Ω*abuO*[3]) towards varied concentrations of SNP was elucidated as described previously. The differences between WT and Δ*abuO* are statistically significant ($P < 0.05$) for all concentrations used

concentration in the cytoplasm thereby leading to antimicrobial resistance. All efflux pumps utilize periplasm/OMP component to become fully functional (Coyne et al. 2011).

7.3.10 Antibiotic Susceptibilities

Analysis of disc diffusion assay for mutants displayed increased susceptibility to various antibiotics. The MIC values for the WT and the various *abuO* mutants are shown in Table 7.1. While the Δ*abuO* exhibited increased susceptibility to carbenicillin, ceftriaxazone, meropenem, amikacin, streptomycin, tigecycline, the current mutant Δ$abuO^2$ exhibited susceptibilities to amikacin, clindamycin, gentamicin, polymyxin-B, minocycline and tetracycline and Δ$abuO^3$ exhibited susceptibilities to cefepime, ceftazidime, amikacin and clindamycin.

7.3.11 Survival Assay Using Antibiotics

The survival capacity of isogenic mutants to various antibiotics ampicillin, chloramphenicol (Fig. 7.6b), ciprofloxacin (Fig. 7.6c), neomycin and tetracycline shows its direct role in mediating antibiotic resistance. It is worthy to note that the Δ*abuO* mutant exhibited a 1.73-fold reduced survival at 512 μg/ml of ampicillin, whereas the Δ$abuO^1$, Δ$abuO^2$ and Δ$abuO^3$ exhibited 1.28-fold, 1.13-fold and 1.11-fold reduced survival compared to their respective parental strains (Fig. 7.6a). Similarly, the Δ*abuO* mutant exhibited a 1.40-fold reduced survival at 256 μg/ml of neomycin, whereas the Δ$abuO^1$, Δ$abuO^2$ and Δ$abuO^3$ exhibited 2.18-fold, 1.20-fold and 1.18-fold reduced survival compared to their respective parental strains (Fig. 7.6d). Likewise the Δ*abuO* mutant exhibited a 1.25-fold reduced survival at 16 μg/ml of tetracycline, whereas the Δ$abuO^1$ exhibited 1.79-fold reduced survival compared to their respective parental strain and interestingly the Δ$abuO^2$ and Δ$abuO^3$ could grow only till 0.5 μg/ml of tetracycline compared to their respective parental strains (Fig. 7.6e) (laboratory observations).

7.3.12 Survival Assay Using Efflux Pump Based Substrates

To elucidate whether AbuO-like protein confers drug resistance in *A. baumannii* by exploiting active efflux mechanism or not, we tested the ability of mutants for its survival ability when exposed to different efflux based substrates acriflavine (Fig. 7.7a), acridine orange (Fig. 7.7b), ethidium bromide (EtBr) (Fig. 7.7c), rhodamine (Fig. 7.7d) and safranin (Fig. 7.7e) which clearly demonstrated that mutants had dimished ability to survive under the exposed challenges.

Table 7.1 Determination of MIC for clinical *A. baumannii* isolates of Indian origin and *abuO* mutants

Antibiotics	AB1	$\Delta abuO^1$	Fold change	AB2	$\Delta abuO^2$	Fold change	AB3	$\Delta abuO^3$	Fold change
Amoxicillin	240	240	1	240	240	1	240	240	1
Carbenicillin	128	128	1	512	256	2	64	64	1
Cefepime	64	64	1	256	128	2	256	64	4
Ceftazidime	256	128	2	256	128	2	256	64	4
Ceftriaxone	2	2	1	2	2	1	2	2	1
Ticarcillin	60	30	2	60	30	2	60	30	2
Amikacin	256	256	1	256	64	4	256	64	4
Clindamycin	30	30	1	30	10	3	30	5	6
Gentamicin	1024	1024	1	1024	128	8	1024	1024	1
Kanamycin	240	120	2	240	120	2	120	120	1
Streptomycin	240	240	1	240	240	1	60	30	2
Norfloxacin	240	240	1	240	240	1	240	240	1
Sparfloxacin	0.01	0.001	10	1	1	1	2	1	2
Ciprofloxacin	30	30	1	60	30	2	60	30	2
Colistin	0.1	0.1	1	0.1	0.1	1	0.01	0.01	1
Polymyxin B	0.1	0.1	1	0.1	0.01	10	0.1	0.1	1
Minocycline	0.1	0.1	1	1	0.1	10	0.1	0.1	1
Tetracycline	0.01	0.01	1	1	0.1	10	0.1	0.1	1
Trimethoprim	240	120	2	240	240	1	240	120	2
Co-Trimoxazole	240	240	1	240	120	2	240	240	1

MIC determination for *A. baumannii* AB1, AB2, AB3 and $\Delta abuO^1$, $\Delta abuO^2$, $\Delta abuO^3$

The clinical strains and their *abuO* mutants {$\Delta abuO^1$, $\Delta abuO^2$, $\Delta abuO^3$} were tested by E-test to determine the MIC for different drugs. Data represents MIC values in µg/ml. Fold change is the ratio of MIC values for respective WT and their respective $\Delta abuO$

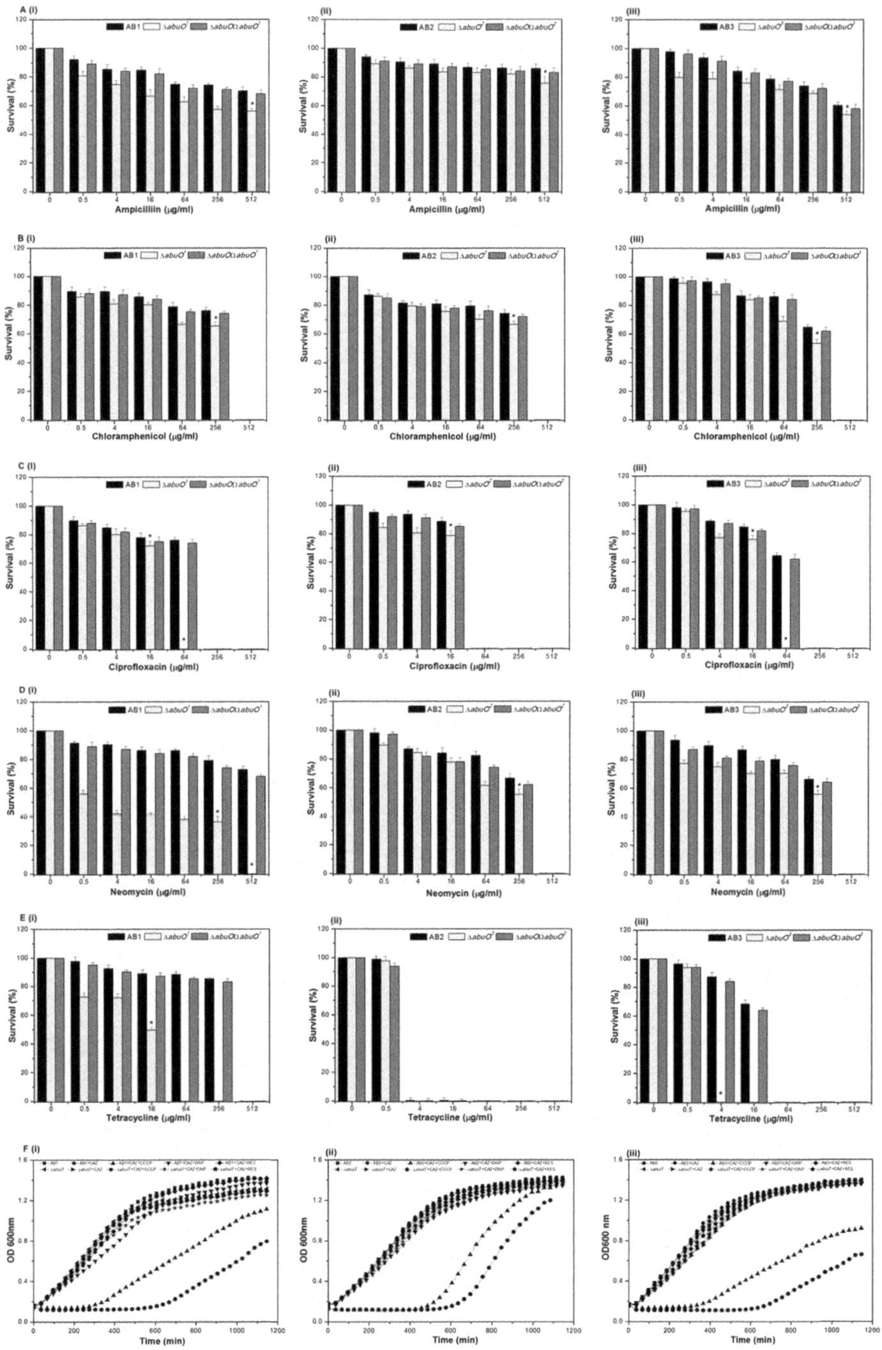

Fig. 7.6 Contributions of AbuO in antibiotic resistance in *A. baumannii*. Survival assays using strains (WT: AB1, AB2, and AB3), (mutant constructs: $\Delta abuO^{1}$, $\Delta abuO^{2}$, $\Delta abuO^{3}$), (complemented constructs: $\Delta abuO\Omega abuO^{1}$, $\Delta abuO\Omega abuO^{2}$, and $\Delta abuO\Omega abuO^{3}$) with ampicillin (**a**), chloramphenicol (**b**), ciprofloxacin (**c**), neomycin (**d**) and tetracycline (**e**) are shown in bar graphs. The growth rate of wild type and mutants (*i*, *ii*, and *iii*) in presence of ceftazidime (**f**) was monitored in the presence of efflux pump inhibitors [CCCP, 2,4 dinitrophenol (DNP) and reserpine]

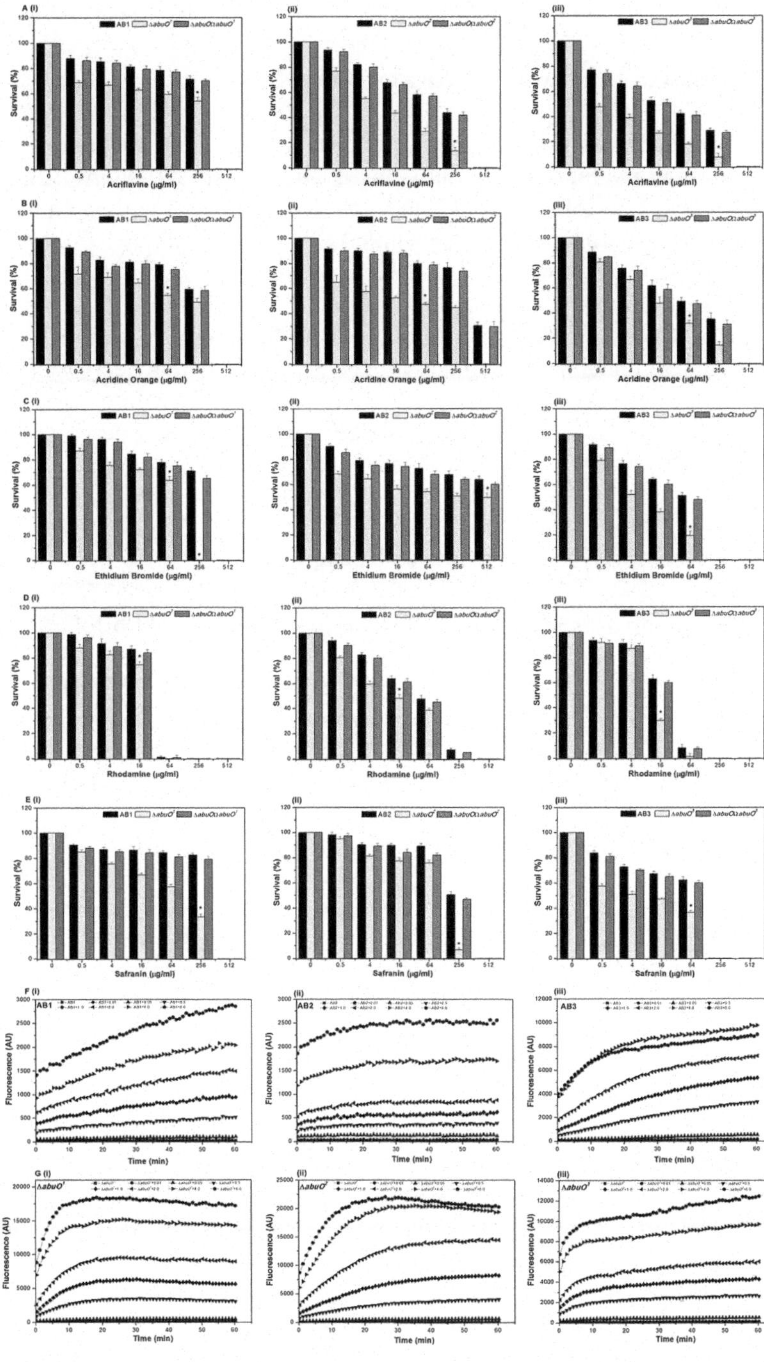

Fig. 7.7 Contributions of AbuO in resistance to efflux based substrates in *A. baumannii*. Survival assays using strains (WT: AB1, AB2, and AB3), (mutant constructs: Δ*abuO*[1], Δ*abuO*[2], Δ*abuO*[3]), (complemented constructs: Δ*abuO*Ω*abuO*[1], Δ*abuO*Ω*abuO*[2], and Δ*abuO*Ω*abuO*[3]) with acriflavine (**a**), acridine orange (**b**), EtBr (**c**), rhodamine (**d**) and safranin (**e**) are shown in bar graphs. EtBr accumulation assays without (**f**) and with CCCP (**g**) at final concentrations that ranged from 0.01 μg/ml to 6.0 μg/ml was tested as described in methods section. The fluorescence was monitored in spectrofluorometer (Hitachi) at 37 °C

7.3.13 Growth Inactivation Assays Using EtBr

The growth inactivation assays were performed using EtBr as fluorescent substrate and found the mutants exhibited lower growth with 2.0 μg/ml EtBr of 1.42 fold (Δ*abuO*[1]), 1.75 fold (Δ*abuO*[2]) and 3.51 fold (Δ*abuO*[3]) compared to control strains respectively [P = 0.001]. The stunted growth profile in the mutant is due to the lack of *abuO* outer membrane efflux component in its functional form. To distinguish whether this is due to the loss of efflux activity in *abuO* mutant, the growth profile was monitored with different efflux pump inhibitors which are known to block/hinder active efflux. As expected a substantial decrease in growth was observed in *abuO* mutant by 1.32 fold (Δ*abuO*[1]), 1.06 fold (Δ*abuO*[2]) and 1.26 fold (Δ*abuO*[3]) in the presence of CCCP as compared to their control strains respectively [P = 0.001]; (laboratory observations, data not shown).

7.3.14 Flourimetric Efflux Assays Using EtBr

On performing flourimetric efflux assay, we found 8.8-fold, 1.08-fold and 10.99-fold higher accumulation of substrate in Δ*abuO*[1], Δ*abuO*[2], Δ*abuO*[3] respectively at 4 μg/ml of EtBr when compared to their respective WT (Fig. 7.7f), and in the presence of CCCP (Fig. 7.7g), clearly indicating that AbuO confers antimicrobial resistance by altering active drug extrusion in *A. baumannii*.

7.3.15 Growth Inactivation Assays Using Ceftazidime

To elucidate the impact on antibiotic efflux capacity, further whole cell ceftazidime accumulations assays were performed to authenticate the observation. As expected a substantial decrease (1.8 fold) in growth was observed in *abuO* mutant in the presence of CCCP as compared to control strains [P = 0.001] (laboratory observations) (Fig. 7.6f).

7.3.16 Survival Assay Using Hospital Based Disinfectants

Bacterial resistance to biocides (antiseptics, disinfectants) can result from enhanced efflux. Therefore, we elucidated the involvement of *abuO* in resistance to different hospital based disinfectants such as benzalkonium chloride, chlorhexidine and triclosan as *A. baumannii* is considered a successful nosocomial pathogen that can stay on abiotic surfaces for long. It is worthy to note that the Δ*abuO* mutant could grow till 6.4 μg/ml of chlorhexidine, the Δ*abuO*[3] could grow only till 3.2 μg/ml of the substrate compared to its parent strain (Fig. 7.8a). Further whole cell chlorhexidine growth inactivation assay authenticated the above mentioned observation [P = 0.001] (Fig. 7.8b).

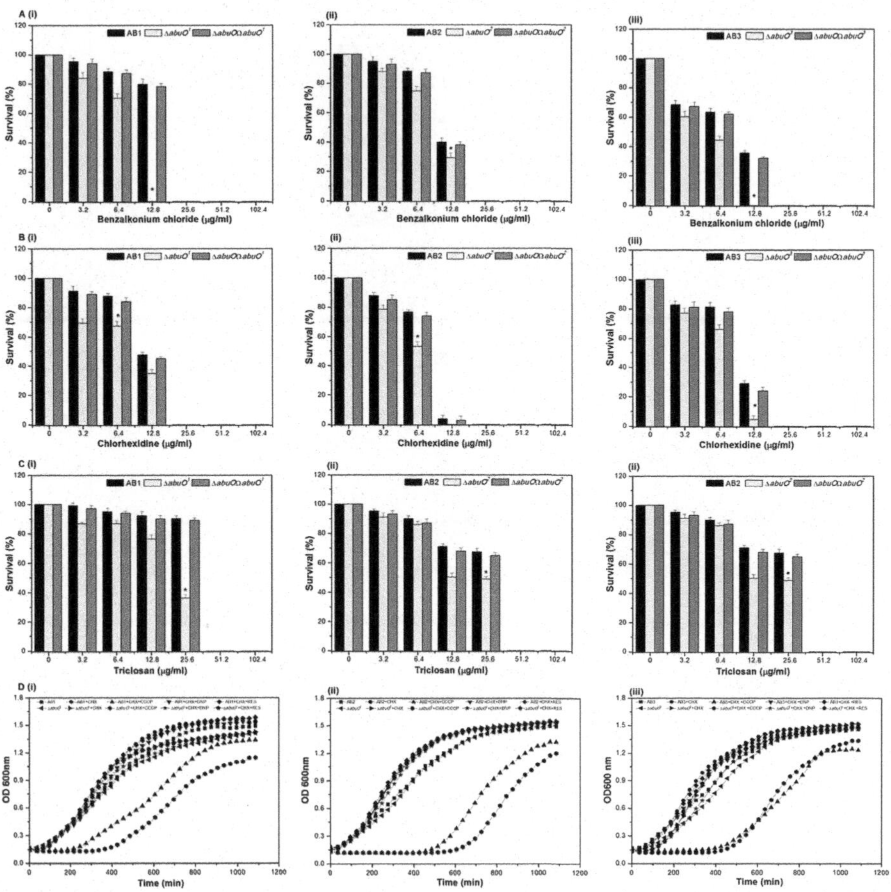

Fig. 7.8 Biocide stress response. Biocide tolerance was tested by performing survival assays using strains (WT: AB1, AB2, and AB3), (mutant constructs: $\Delta abuO^{1}$, $\Delta abuO^{2}$, $\Delta abuO^{3}$), (complemented constructs: $\Delta abuO\Omega abuO^{1}$, $\Delta abuO\Omega abuO^{2}$, and $\Delta abuO\Omega abuO^{3}$) in presence of different concentrations of chlorhexidine [**a**] and growth inactivation assay using chlorhexidine is shown in figure [**b**]

7.3.17 Expression Analysis

The expression analysis of multi-drug resistance genes was found altered as shown in Fig. 7.9. Overall, we decisively concluded that *abuO* has a broad spectrum antimicrobial resistance property, a phenomenon characterized for the first time in genetically distinct *A. baumannii.* The WT and mutants were examined for their abilities to kill *C. elegans* and our findings demonstrate that *abuO* mutant kills *C. elegans* slowly than WT strain ($P < 0.01$; data not shown).

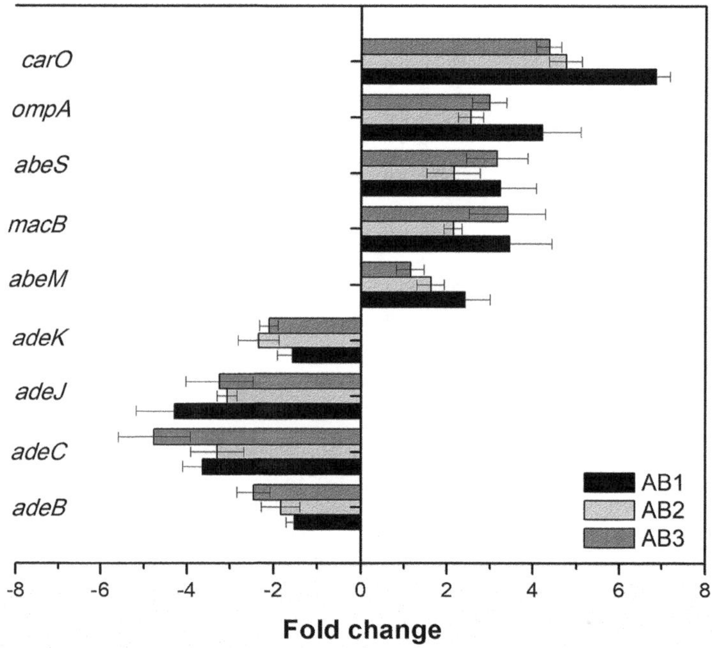

Fig. 7.9 **Expression analysis.** Expression profiles of different antibiotic resistance determinants as observed in (WT: AB1, AB2, and AB3) and (mutant constructs: $\Delta abuO^1$, $\Delta abuO^2$, $\Delta abuO^3$) compared to the internal control strain is shown here

7.4 Conclusions

In the recent era where antibiotic resistance is emerging as a global problem of serious concern, understanding the fundamentals of multi-drug resistance determinants is a great area of interest and potential as inhibitors against them can serve as alternative anti-infectives. In this study we deciphered the conserved functions of AbuO-like protein in clonally unrelated clinical strains of *A. baumannii* with different antibiogram primarily isolated from Indian scenario.

Acknowledgement We are highly thankful to Director, CSIR-Institute of Microbial Technology, Chandigarh, for providing excellent facility to carry out this work. VBS remains thankful to DST (SB/YS/LS-177/2014) for the fellowship. The funding from CSIR, DST (SB/YS/LS-177/2014) and DBT (BT/PR14304/BRB/10/822/2010, BT/HRD/NBA/34/01/2012) are highly acknowledged.

References

Blossom DB, Srinivasan A (2008) Drug-resistant Acinetobacter baumannii-calcoaceticus complex: an emerging nosocomial pathogen with few treatment options. Infect Dis Clin Pract 16:1–3

Boucher HW, Talbot GH, Bradley JS, Edwards JE, Gilbert D, Rice LB, Scheld M, Spellberg B, Bartlett J (2009) Bad bugs, no drugs: no ESKAPE! An update from the Infectious Diseases Society of America. Clin Infect Dis 48:1–12

Clinical and Laboratory Standards Institute (2014) Performance standards for antimicrobial susceptibility testing. Twentieth informational supplement M100-S24U. CLSI, Wayne, PA, USA

Coyne S, Rosenfeld N, Lambert T, Courvalin P, Perichon B (2010) Overexpression of resistance-nodulation-cell division pump AdeFGH confers multidrug resistance in Acinetobacter baumannii. Antimicrob Agents Chemother 54:4389–4393

Coyne S, Courvalin P, Perichon B (2011) Efflux-mediated antibiotic resistance in *Acinetobacter* spp. Antimicrob Agents Chemother 55:947–953

Damier-Piolle L, Magnet S, Brémont S, Lambert T, Courvalin P (2008) AdeIJK, a resistance-nodulation-cell division pump Effluxing multiple antibiotics in Acinetobacter baumannii. Antimicrob Agents Chemother 52:557–562

Delcour AH (2009) Outer membrane permeability and antibiotic resistance. Biochim Biophys Acta 1794:808–816

Dijkshoorn L, Nemec A, Seifert H (2007) An increasing threat in hospitals: multidrug-resistant Acinetobacter baumannii. Nature Rev Microbiol 5:939–951

Elkins CA, Nikaido H (2002) Substrate specificity of the RND-type multidrug efflux pumps AcrB and AcrD of *Escherichia coli* is determined predominantly by two large periplasmic loops. J Bacteriol 184:6490–6498

Hennequin C, Forestier C (2009) *oxyR*, a LysR-type regulator involved in *Klebsiella pneumoniae* mucosal and abiotic colonization. Infect Immun 77:5449–5457

Hirakawa H, Nishino K, Hirata T, Yamaguchi A (2003) Comprehensive studies of drug resistance mediated by overexpression of response regulators of two-component signal transduction systems in *Escherichia coli*. J Bacteriol 185:1851–1856

Kobayashi N, Nishino K, Yamaguchi A (2001) Novel macrolide-specific ABC-type efflux transporter in Escherichia coli. J Bacteriol 183:5639–5644

Limansky AS, Mussi MA, Viale AM (2002) Loss of a 29-kilodalton outer membrane protein in *Acinetobacter baumannii* is associated with imipenem resistance. J Clin Microbiol 40:4776–4778

Lob SH, Hoban DJ, Sahm DF, Badal RE (2016) Regional differences and trends in antimicrobial susceptibility of Acinetobacter baumannii. Int J Antimicro Ag 47:317–323

Lomovskaya O, Lewis K (1992) Emr, an *Escherichia coli* locus for multidrug resistance. Proc Natl Acad Sci U S A 89:8938–8942

Lomovskaya O, Zgurskaya HI, Nikaido H (2002) It takes three to tango. Nat Biotechnol 20:1210–1212

Magnet S, Courvalin P, Lambert T (2001) Resistance-nodulation-cell division-type efflux pump involved in aminoglycoside resistance in Acinetobacter baumannii strain BM4454. Antimicrob Agents Chemother 45:3375–3380

Mortimer PG, Piddock LJ (1991) A comparison of methods used for measuring the accumulation of quinolones by *Enterobacteriaceae*, *Pseudomonas aeruginosa* and *Staphylococcus aureus*. J Antimicrob Chemother 28:639–653

Nachimuthu R, Subramanian R, Maray S, Gothandam KM, Sivamangala K, Manohar P, Bozdogan B (2015) Characterization of carbapenem-resistant Gram-negative bacteria from Tamil Nadu. J Chemother 1973947815Y0000000056

Nagakubo S, Nishino K, Hirata T, Yamaguchi A (2002) The putative response regulator BaeR stimulates multidrug resistance of *Escherichia coli* via a novel multidrug exporter system, MdtABC. J Bacteriol 184:4161–4167

Nikaido H, Zgurskaya HI (2001) AcrAB and related multidrug efflux pumps of *Escherichia coli*. J Mol Microbiol Biotechnol 3:215–218

Nikaido H, Basina M, Nguyen V, Rosenberg EY (1998) Multidrug efflux pump AcrAB of *Salmonella* Typhimurium excretes only those beta-lactam antibiotics containing lipophilic side chains. J Bacteriol 180:4686–4692

Nishino K, Yamaguchi A (2001) Analysis of a complete library of putative drug transporter genes in *Escherichia coli*. J Bacteriol 183:5803–5812

Piddock LJ (2006) Multidrug-resistance efflux pumps – not just for resistance. Nature Rev Microbiol 4:629–636

Pogue JM, Mann T, Barber KE, Kaye KS (2013) Carbapenem-resistant *Acinetobacter baumannii*: epidemiology, surveillance and management. Expert Rev Anti-Infect Ther 1:83–393

Poole K (2005) Efflux-mediated antimicrobial resistance. J Antimicrob Chemother 56:20–51

Rajamohan G, Srinivasan VB, Gebreyes WA (2010) Molecular and functional characterization of a novel efflux pump, AmvA, mediating antimicrobial and disinfectant resistance in *Acinetobacter baumannii*. J Antimicrob Chemother 65:1919–1925

Rynga D, Shariff M, Deb M (2015) Phenotypic and molecular characterization of clinical isolates of *Acinetobacter baumannii* isolated from Delhi, India. Ann Clin Microbiol Antimicrob 14:40

Srinivasan VB, Rajamohan G, Gebreyes WA (2009) Role of AbeS, a novel efflux pump of the SMR family of transporters, in resistance to antimicrobial agents in Acinetobacter baumannii. Antimicrob Agents Chemother 53:5312–5316

Srinivasan VB, Vaidyanathan V, Rajamohan G (2015) AbuO, a TolC-like outer membrane protein of Acinetobacter baumannii, is involved in antimicrobial and oxidative stress resistance. Antimicrob Agents Chemother 59:1236–1245

Srinivasan VB, Venkataramaiah M, Mondal A, Rajamohan G (2015b) Functional characterization of AbeD, an RND-type membrane transporter in antimicrobial resistance in *Acinetobacter baumannii*. PLoS One 10:e0141314

Stevanin TM, Ioannidis N, Mills CE, Kim SO, Hughes MN, Poole RK (2000) Flavohemoglobin Hmp affords inducible protection for Escherichia coli respiration, catalyzed by cytochromes bo' or bd, from nitric oxide. J Biol Chem 275:35868–35875

Storz G, Imlay JA (1999) Oxidative stress. Curr Opin Microbiol 2:188–184

Tiwari V, Vashistt J, Kapil A, Moganty RR (2012) Comparative proteomics of inner membrane fraction from carbapenem-resistant Acinetobacter baumannii with a reference strain. PLoS One 7:e39451

Tomaras AP, Dorsey CW, Edelmann RE, Actis LA (2003) Attachment to and biofilm formation on abiotic surfaces by Acinetobacter baumannii: involvement of a novel chaperone-usher pili assembly system. Microbiology 149:3473–3484

Vila J, Marti S, Sanchez-Cespedes J (2007) Porins, efflux pumps and multidrug resistance in Acinetobacter baumannii. Antimicrob Agents Chemother 59:1210–1215

Villegas MV, Hartstein AI (2003) Acinetobacter outbreaks, 1977–2000. Infect Control Hosp Epidemiol 24:284–295

Efflux Mediated Co-resistance

8

Amit Gaurav, Atin Sharma, and Ranjana Pathania

Abstract

Antimicrobial resistance is one of the major threats to the global health care facilities. The drug resistant microbes take a heavy toll on the human life resulting in huge losses in terms of economy and human resource. Moreover, with current celerity of mobility, the dissemination of these microbes is relatively easy and swift, making the situation even worse. The rising rate of incidence of 'superbugs' that are resistant to all known drugs and dwindling supply of newer antimicrobials, the post-antibiotic era seems an inevitable future. Due to its widespread reach and rapidity, the evolution of antimicrobial resistance (AMR) in bacteria is of particular interest. These microbes resist the action of antimicrobials by various mechanisms, one of which is actively pumping out the antimicrobials from the cellular milieu. This is achieved by specialized proteins, called the efflux pumps, which avoid the effective build-up of antimicrobials and assist survival in otherwise inhibitory concentration of the antimicrobial. These pumps can either be chromosomally encoded or plasmid borne and are generally overexpressed in antimicrobial challenged bacteria. The most striking feature of these efflux pumps is the loss of substrate specificity that enables one pump to efflux out multiple antibiotics. This review focuses on the ability of these pumps to identify multiple substrates and provide selective advantage to the pathogenic bacterial cells.

Keywords

Antibiotic resistance · Efflux pumps · Bacteria · Heavy metals · Metal resistance · Biocides · Transporters

A. Gaurav · A. Sharma · R. Pathania (✉)
Molecular Bacteriology and Chemical Genetics Lab, Department of Biotechnology, Indian Institute of Technology Roorkee, Roorkee, Uttarakhand, India
e-mail: rpathfbs@iitr.ac.in

S. M. Mandal, D. Paul (eds.), *Bacterial Adaptation to Co-resistance*,
https://doi.org/10.1007/978-981-13-8503-2_8

8.1 Introduction

Bacteria are undoubtedly one of the smartest organisms on the planet. From their first appearance about 3 billion years ago, they have successfully colonized the hot springs, cold polar caps, brackish sea waters and the arid hot deserts. Their remarkable ability to thrive under harsh conditions can be attributed to rapid growth and subsequent accelerated rate of evolution. This evolutionary advantage has led them enjoy this lengthy reign and ensures that it will continue for years to come. However, for the past 1 million years, bacteria have faced stiff competition from another species that considers itself to be at the pinnacle the evolutionary tree-humans. In the struggle to ascertain supremacy, both the species have developed strategies to attack and counter-attack each other's advances. Bacteria, especially the pathogenic ones, express multitude of virulence factors that aid in the pathogenesis leading to heavy losses to humans both in terms of material and human resource. As a result, humans had been on the losing side of the battle for majority of the time till the serendipitous discovery of penicillin by Alexander Fleming in 1928 led to the advent of 'the antibiotic era'. Suddenly, the balance of power shifted towards humans and over the period of time a lot of new antibiotics were introduced, especially during the fifth decade of the past century that is fondly remembered as the 'golden era of antibiotics'. Unfortunately, humans undermined the celerity of bacterial evolution which combined with the rampant use and abuse of antibiotics led to appearance of antibiotic resistance in bacteria. Such bacteria resist the antibacterial action of multiple antibiotics and are colloquially referred to as 'superbugs'. Infections by superbugs are difficult to treat and often result in high rates of mortality. The edge that antibiotics gave humans lasted only for a few decades and the human civilization faces the imminent dawn of 'the post-antibiotic era' where antibiotics are no longer a part of human arsenal against bacteria.

Bacteria resist the action of antibiotics by adopting one of the four major mechanisms, namely, altering membrane permeability to avoid antibiotic influx, altering target sites to prevent inhibition, enzymatic modifications of the antibiotic molecule, and expression of efflux pumps to extrude out the antibiotics. The efflux pumps or the transporter proteins have been predicted and verified to carry out multiple processes apart from imparting antibiotic resistance. Interestingly, some efflux pumps can recognize multiple structurally different substrates, expanding the range of resistance that these pumps can confer. This promiscuity in substrate selection leads to resistance to other bioactive compounds like heavy metals and biocides that are not related to antibiotics. But nowadays, disease causing bacteria are becoming smarter by loading themselves with latest classes of arsenal i.e. they are becoming resistant at an alarming rate (Ventola 2015). The reports of deaths due to multiple drug resistant (MDR) bacteria are coming from every corner of the world. One recent study shows that infections due to MDR pathogen are responsible for 58,000 child deaths per year in India alone (Gandra et al. 2018). The increased rate of resistance to existing drugs and lack of discovery of new antibacterial drugs is driving human population to post-antibiotic era ("The world is running out of antibiotics,

WHO report").The present chapter deals with this phenomenon of cross resistance that is brought about by these multi-drug efflux pumps.

8.2 Molecular Mechanisms of Antibiotic Resistance

Some bacteria have intrinsic resistance mechanisms for antibiotics due to composition of the cytoplasmic membrane (Nikaido 2003) while others can acquire or develop resistance to antibiotics (Blair et al. 2015). This resistance is generally mediated by one or more of the several mechanisms (Wright 2011), which can be categorized in to five groups. Diagrammatic representation of all above mentioned mechanisms is depicted in Fig. 8.1.

(a) Reduced permeability
(b) Inactivation of antibiotics by hydrolysis
(c) Modification of antibiotic target site by mutation
(d) Protection of target site by post-translational modification
(e) Increased efflux of antibiotics

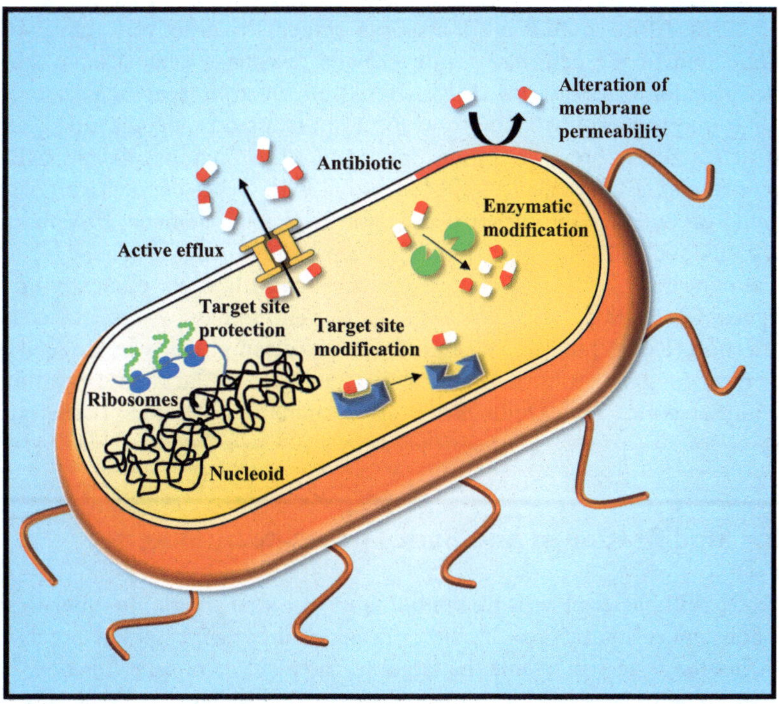

Fig. 8.1 Diagrammatic representation of different mechanisms of antibiotic resistance in a bacterial cell

8.2.1 Reduced Permeability

Due to presence of an extra outer membrane, Gram negative bacteria have an advantage in resisting entry of many antibiotics compared to Gram positive bacteria. Outer membrane proteins, called porins, are trans-membrane proteins that help in diffusion of hydrophilic antibiotics. OmpF and OmpC of *E. coli* are examples of such porins (Kojima and Nikaido 2013). They serve as a nonspecific channel for entry of many essential nutrients (Tamber and Hancock 2003). In order to survive, bacteria reduce permeability of the outer membrane by down regulation of porin biosynthesis. Reduction in porin biosysnthesis significantly increases the resistance to antibiotics of last resort like carbapenems. Such phenomenon is well documented in Enterobacteriaceae, *Pseudomonas* spp. and *Acinetobacter* spp. (Baroud et al. 2013; Lee et al. 2017). Increased antibiotic pressure also selects the population of bacteria with mutation in porin protein as an adaptive evolution (Lavigne et al. 2013).

8.2.2 Inactivation of Antibiotics by Hydrolysis or Inactivation of Antibiotic by Transfer of a Chemical Group

Efflux pump proteins are found in all three domains of life i.e. Archaea, Bacteria and Eukarya. Efflux pumps are transporter proteins that actively transport many antibiotics out of the cell that in turn provide resistance against many clinically important antibiotics (Li et al. 2015). Based on substrate specificity, they can be classified into two groups i.e. narrow spectrum (substrate specific) like Tet efflux pump of *E. coli* or broad spectrum (multidrug efflux pumps) like AcrAB- TolC pump of *E. coli* (Blair et al. 2015; Saier et al. 2006). Genes encoding efflux pumps are found on both on chromosome and mobile genetic elements like plasmids in bacteria (Piddock 2006a). Efflux pump genes present on mobile genetic element are of great concern due to their ability to be transferred to other clinically important pathogens (Poole 2005). One such example is a gene coding for a resistance nodulation division (RND) pump present on an IncH1 plasmid that was isolated from a *Citrobacter freundii* strain that also has a gene for β- lactamase (an antibiotic degrading enzyme) **N**ew **D**elhi **m**etallo-β- lactamase 1 (NDM1) (Dolejska et al. 2013).

8.2.3 Modification of Antibiotic Target Site by Mutation

Antibiotics kill bacterial cells by inhibiting its essential targets. In order to survive under constant antibiotic pressure, bacteria can modify their essential targets in such a way that they can still enable the target to carry out its normal function, but can effectively confer antibiotic resistance. One such example is the point mutation in gene encoding GyrA subunit of DNA gyrase protein that leads to resistance against quinolones class of antibiotics (Lambert 2005; Walsh 2000; Weigel et al. 1998).

Another example is point mutation in *rpoB* gene that encodes β-subunit of RNA polymerase, this gives to high level of resistance to Rifampicin (Kapur et al. 1994; Telenti et al. 1993).

8.2.4 Protection of Target Site by Post-Translational Modification

Bacteria can also protect their target sites by modifying them without mutational changes in the respective genes encoding the target site. Nowadays protection by modification of the target is being reported as clinically important phenomenon; for example 16S rRNA of bacteria is modified by methylation using erythromycin ribosome methylase (erm) enzymes and alter drug binding site and thus prevent binding of macrolides like Erythromycin (Kumar et al. 2014). Another recently identified example is the chloramphenicol–florfenicol resistance (cfr) methyltransferase, which specifically methylates 23S rRNA; this confers resistance to a wide variety of drugs that have targets near the modified site, which includes phenicols (chloramphenicol) and linezolid (Long et al. 2006).

8.2.5 Increased Efflux of Antibiotics

Just after the discovery of penicillin, the enzyme that degrades it (β-lactamase) was reported (Abraham and Chain 1988). Enzymes catalyzing the degradation of antibiotics are major mechanism of antibiotic resistance. Thousands of enzymes have been identified since 1940 that can degrade and modify antibiotics of different classes, including β-lactams, aminoglycosides, phenicols and macrolides (Livermore 2008; Voulgari et al. 2013). Extended-spectrum β-lactamases (ESBLs) and carbapenemases are examples of such antibiotic hydrolyzing enzymes (Pitout and Laupland 2008). The treatment options for the infections caused by pathogens harboring these enzymes have become extremely difficult. The first report of occurrence of ESBLs was on the chromosome, but several recent reports have confirmed the presence of ESBLs genes on the plasmids in Enterobacteriaceae, *P. aeruginosa* and *A. baumannii* (Pitout and Laupland 2008). Antibiotics can also be inactivated by addition of a chemical functional group in such a way that the modified antibiotic is unable to bind to its target site due to steric hindrance. Bacterial enzymes can transfer different chemical groups like acyl, phosphate, nucleotidyl and ribitoyl groups (Wright 2005). Aminoglycoside antibiotics are large molecules with many exposed hydroxyl and amide groups; bacterial enzymes can easily modify these sites and can become resistant (Norris and Serpersu 2013). There are three main classes of aminoglycoside-modifying enzymes i.e. acetyltransferases, phosphotransferases and nucleotidyltransferases (Romanowska et al. 2013). These aminoglycoside-modifying enzymes are not just capable of modifying antibiotic side chain, but they are equally capable of binding them as, the active site of the enzymes mimic the target environment as shown by molecular dynamic studies (Romanowska et al. 2013).

8.3 Efflux Pumps

As mentioned in an earlier section, efflux pumps are transport proteins involved in the expulsion of toxic substrates outside the cell. Genes encoding efflux pumps are present in all microorganisms. With only a few exceptions, they are chromosomally encoded and are conserved, both at the genetic and at the protein levels. Hence, it indicates that the evolution of efflux pumps in bacterial genomes occurred long before the recent use of antibiotics for human and animal therapy. Here, it is worth mentioning that microorganisms isolated from pristine sites like isolated caves and permafrost are known to harbor resistance genes even without any anthropogenic action (Allen et al. 2010; D'Costa et al. 2011). Hence, it is clear that development of resistance to antibiotics is a natural ecological phenomenon and is the product of billions of years of evolution. Both antibiotic-producing microorganisms and antibiotic-susceptible microorganisms live in close association in any given niche. Most natural antibiotics are well exposed to other species in the local environment. The classical example of this phenomenon is isolation of genes that encode resistance to the latest antibiotic, daptomycin (Licensed in 2003, US – FDA), where resistance was present in microorganisms even before clinical use of this antibiotic (Forsberg et al. 2012; Perry and Wright 2013). Nevertheless, rampant use of antibiotics as medicine or misuse of it in horticulture, poultry, apiculture and animal feed etc. by humans has escalated the antibiotic resistance problem (Paterson and van Duin 2017; Tuševljak et al. 2013; Walsh and Wu 2016).

The first efflux pump was reported as P-glycoprotein in mammalian cells (Juliano and Ling 1976). And the first report of efflux pump in bacterial cells was reported as Tet proteins in *E. coli* that made cells resistant to tetracycline (Levy 1992). In 1990s efflux pumps belonging to resistance-nodulation-division (RND) superfamily exporters in *E. coli* and *P. aeruginosa*; capable of exporting variety of substrates (called MDR pumps) were reported (Li et al. 1994; Ma et al. 1993). Since then, identification and characterization of MDR pumps in several bacterial species especially in the ESKAPE pathogens – (the leading cause of antimicrobial resistant nosocomial infections worldwide) [***Enterococcus** faecium*, ***Staphylococcus aureus***, ***Klebsiella** pneumoniae*, ***A**cinetobacter baumannii*, ***P**. aeruginosa*, and ***Enterobacter*** species] is increasingly reported (Li and Nikaido 2004, 2009; Rice 2008). For instance, a recent screening of 25 clinical *A. baumannii* strains isolated from human in Chandigarh, India, revealed a strong correlation of overexpression of the MFS efflux pumps with the high-level of fosfomycin resistance in 23 multi drug resistant strains (Sharma et al. 2017).

Bacterial efflux transporters have been classified into five families based on protein sequences similarity (Blair et al. 2015).

(a) The **r**esistance-**n**odulation-**d**ivision (RND) family

(b) The **m**ajor **f**acilitator **s**uperfamily (MFS)
(c) The **A**TP (adenosine triphosphate)-**b**inding **c**assette (ABC) superfamily
(d) The **s**mall **m**ultidrug **r**esistance (SMR) family
(e) The **m**ultidrug **a**nd **t**oxic compound **e**xtrusion (MATE) family

There are several facts that have been considered while classification of MDR pumps. These include number of components (single protein or multiple proteins), the number of trans membrane regions, the energy source that the pump uses and the types of substrate that the pump exports. Efflux pumps of the RND family, which are only found in Gram negative bacteria, are organized as tripartite systems. These efflux pumps contain the following proteins: a transporter or efflux protein (e.g. AcrB), which is located in the inner (cytoplasmic) membrane of the bacterium; an accessory protein located in the periplasmic space (also known as a membrane-fusion protein) (e.g. AcrA), and an outer-membrane protein (also known as an outer membrane protein channel) (e.g. TolC) (Fig. 8.2) (Du et al. 2014). The biological significance of tripartite system (AcrAB-TolC) can be predicted by the fact that Gram-negative bacteria like *E. coli* have two cell membranes: a cytoplasmic membrane and an outer membrane separated by the periplasmic space. Target of β-lactam antibiotics are located in the periplasmic space. Thus, to confer resistance against a broad spectrum of antibiotics, active transporters must not only pump them out of the cytoplasm but also across the outer membrane (Higgins 2007).

ABC transporters operate by hydrolyzing ATP for the energy source, but all others transporters utilizes proton motive force (PMF) as energy source. The pumps

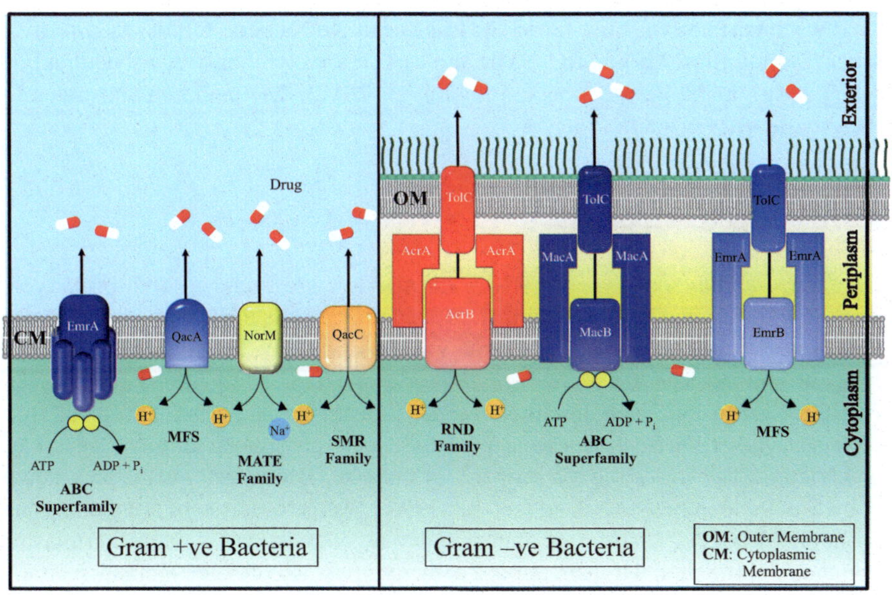

Fig. 8.2 Major categories of efflux pumps in Gram-positive and Gram-negative bacteria

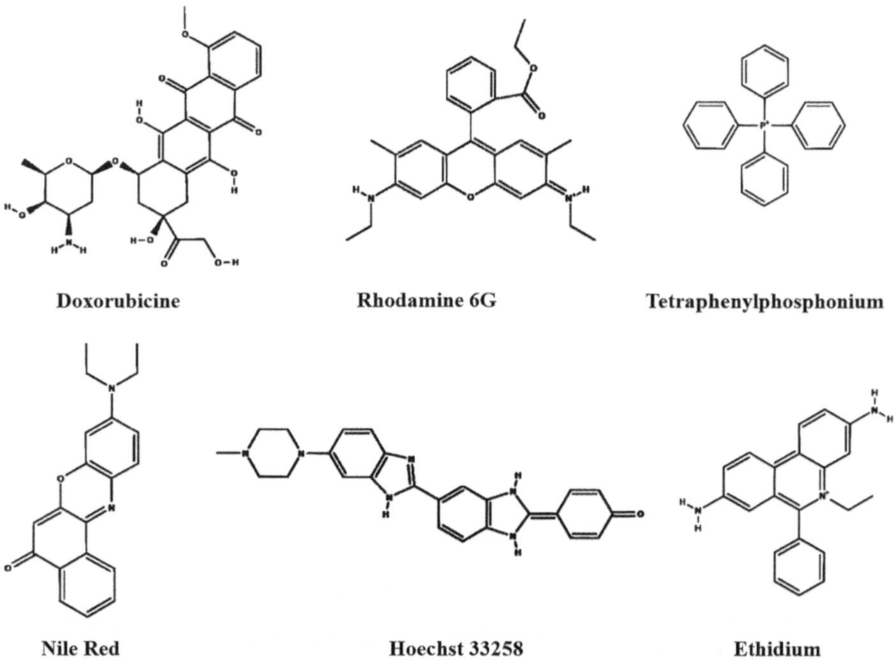

Fig. 8.3 Some common substrates of efflux pumps

which utilizes PMF as energy source are also called secondary transporters or proton/drug antiporters (Blair et al. 2015).

RND superfamily is only found in Gram-negative bacteria, efflux systems of the other four families: MFS, ABC, SMR and MATE are commonly distributed in both Gram-positive and negative bacteria (Li et al. 2015). The common substrates for these pumps are given in Fig. 8.3.

8.3.1 RND Transporters

Resistance-nodulation-division (RND) transporters are found in both prokaryotic and eukaryotic cells. Most extensively studied prokaryotic RND pump is AcrB of *E. coli* (Sun et al. 2014). It has an extremely wide specificity, including practically all types of antibacterial agents (except aminoglycosides), detergents, biocides, dyes, free fatty acids, and even simple solvents (Tsukagoshi and Aono 2000). In an *in vitro* assay, AcrB was shown to also extrude modified phospholipids (Zgurskaya and Nikaido 1999). A common property of these AcrB substrates is the presence of a hydrophobic moiety (Nikaido et al. 1998). AcrB consists of 12 membrane-spanning α-helices that forms transmembrane domain and a large periplasmic domain. The functional transporter is a trimer with a total of 36 membrane-spanning α –helices. Two other proteins, AcrA and TolC, are required for AcrB to pump antibiotics out of the cell. TolC is a pore-like molecule comprising of α -helical pore (diameter ~100 Å) that spans the periplasm and a β-barrel (diameter ~40 Å) that spans the outer membrane (Murakami et al. 2002). These transporters have a large binding site that ensures accommodation of a large range of substrates without any great change in free binding energy. The detailed knowledge about AcrB came after solving the crystal structure in 2002 (Murakami et al. 2002). In 2006, co-crystallization of AcrB with its substrate minocycline or doxorubicin was achieved, this study showed how the large hydrophobic pocket is able to accommodate molecules of different sizes within the pump (Murakami et al. 2006). AcrF and AcrD are homologous to AcrB in *E. coli* and have a wide substrate specificity. AcrD is an aminoglycoside efflux pump that works with AcrA and TolC. MexB is a homologue of AcrB in *P. aeruginosa* (Sennhauser et al. 2009).

8.3.2 MFS Transporters

Major facilitator superfamily (MFS) transporters and ABC transporters cover the two largest and most functionally diverse superfamilies. MFS transporters can be classified into at least 74 families on the basis of sequence homology (Reddy et al. 2012). *E. coli* K-12 alone contains about 70 MFS transporters. The plasmid-encoded TetA pumps were the first bacterial drug efflux pumps identified. TetA is an example of the MFS exporter. In Gram-negative bacteria TetA pumps have 12-TMS (Trans Membrane Segment), whereas 14-TMS Tet pumps are present in Gram-positive bacteria. Most of the MFS pumps are located in the inner membrane and can only transport drugs from the cytosol to the periplasm in Gram negative bacteria (Li et al. 2015). Usually most of the antimicrobial agents reach the cytosol by diffusion across the membrane bilayer, the effluxed drug molecules have a good chance of re-entering the cytosol through diffusion process. However, other constitutively expressed RND pumps, such as AcrAB-TolC and MexAB-OprM, may capture such effluxed drug molecules in the periplasm and further pump them out. Since RND pumps are inefficient in capturing drugs from the cytosol and thus both of these pump works synergistically to enhance the activity of each other in achieving resistance (Tal and Schuldiner 2009). The first X-ray structures of *E. coli* MFS transporter (LacY) was determined in 2003 (Abramson et al. 2003). Subsequently, the structure of a single multidrug transporter from this superfamily, EmrD from *E. coli*, was determined (Yin et al. 2006). EmrD extrudes a range of cytotoxic molecules from the cell. No crystal structure with bound substrate has yet been obtained and there is no direct evidence how drugs bind in, or are transported through it. However, some information is documented for LacY MFS pump. For each molecule of lactose transported, a proton is transferred across the membrane via conserved and essential acidic residues. Proton movement induces a conformational change that exposes the lactose-binding site to the external face of the membrane, reducing the affinity for lactose binding and thus lactose is released. The proton is then released and the transporter returns to its basal state with a high-affinity lactose-binding site exposed to the cytoplasm (Higgins 2007).

8.3.3 ABC Transporters

ABC transporters are most documented efflux pumps in fungi and animal cells responsible for drug resistance (Cannon et al. 2009; Gottesman and Ling 2006). They use the energy of ATP hydrolysis to transport substrates across cell membranes. Typically, they transport inorganic ions, amino acids, sugars, and polypeptides. ABC transporters are generally specific for a given ligand. The minimal functional unit of all ABC transporters consists of four domains. Two cytoplasmic, nucleotide-binding domains (NBDs); hydrolysis of ATP is accomplished by this domain. Two transmembrane domains (TMDs), also called membrane domains, which consist of six membrane-spanning α-helices form the channel through which substrates can cross the membrane. The domains may be arranged with an N-terminal TMD and C-terminal NBD, or vice versa. The driving force for drug transport is mediated by hydrolysis of ATP. ATP binding induces conformational change in NBD dimer which in turn helps in opening and closing of the channel. The best studied bacterial ABC drug exporter is MacB of *E. coli*, which functions together with the periplasmic adaptor MacA and the outer membrane channel TolC (Kobayashi et al. 2001). MacAB-TolC is responsible for macrolide resistance in *E. coli* (Kobayashi et al. 2001).

8.3.4 SMR Transporters

Small multidrug resistance (SMR) proteins are the smallest multidrug transporters. SMR proteins are composed of four transmembrane α-helices of approximately 100–140 amino acids (Schuldiner et al. 1997). They are only restricted to prokaryotic cells (Sun et al. 2014). They function as dimers and thus minimal functional unit is a bundle of eight α -helices with short hydrophilic loops making them very hydrophobic. Unlike other multidrug transporter proteins, the SMR protein family can only transport lipophilic compounds, like quaternary ammonium compounds as well as a variety of antibiotics like ampicillin, penicillin, tetracycline, erythromycin and trimethoprim (Bay and Turner 2012). SMR transporters efflux monocationic molecule like ethidium bromide, tetraphenylphosphonium by H^+ exchange. EmrE is a well characterized SMR transporter found in *E. coli*. A gene (*abeS*) encoding SMR transporter protein is responsible for providing resistance against chloramphenicol, ciprofloxacin, and erythromycin in *A. baumannii* has been reported, which highlights its clinical importance (Srinivasan et al. 2009).

8.3.5 MATE Transporters

The MATE (Multidrug And Toxic Compound Extrusion) family is the most recently categorized among all five transporter families. The first of these transporters was identified as Na^+/cationic agent antiporter NorM from *Vibrio parahaemolyticus* in

1998 (Morita et al. 1998). It contains 12 trans-membrane segment (TMSs). These transporters are prevalent in bacteria, animals and plants. Although the range of substrates for the MATE transporters is narrower than that for the RND transporters, they are able to transport antibiotics like ciprofloxacin, norfloxacin, ofloxacin, gentamicin, kanamycin, erythromycin, chloramphenicol and trimethoprim. They also efflux common dyes like acriflavine, DAPI, ethidium bromide, Hoechst 33342 and rhodamine 6G (Kuroda and Tsuchiya 2009). AbeM is an example of such transporter in *A. baumannii* that provides resistance against many antibiotics (Su et al. 2005).

8.4 Functional Role of Efflux Pumps Other Than Resistance

It is well established that MDR efflux pumps encoded by bacteria can confer clinically relevant resistance to antibiotics. But it seems like their role is not limited to providing resistance to antibiotics they also help in other physiological process essential for bacterial survival (Piddock 2006b). The role of these systems that helps in survival of these pathogens in its ecological niche is described below.

8.4.1 Role of Efflux Pumps in Bacterial Pathogenicity

Role of RND-family efflux pumps in bacterial pathogenicity is well established by several research groups. *S. typhimurium* typically cause gastroenteritis and invasive salmonellosis in mammals. The importance of efflux pump in bacterial pathogenicity can be derived from the fact that the *S. typhimurium* lacking AcrAB–TolC efflux pump is was unable to infect BALB/c mice (Buckley et al. 2006). Another study showed that *P. aeruginosa* lacking the RND-family efflux pump MexAB–OprM could not kill leukocytes deficient mice whereas the parent strain of this mutant *P. aeruginosa* caused a fatal infection (Hirakata et al. 2002). Our recent observations also suggest the role of an MFS transporter in maintaining *A. baumannii* virulence in *Caenorhabditis elegans* infection model (Sharma et al. 2017). Several other examples like importance of RND-family efflux pump MtrCDE in *N. gonorrhoeae* is linked with urinary tract infection (UTI) in female mice (Jerse et al. 2003). Similarly, the importance of efflux pumps in other systems like bacterial pathogenesis in plant is also well established. *Erwinia amylovora* is a plant pathogen which causes the fire-blight disease in apple and pear trees. This bacterium contains homologues of *E. coli* efflux pump gene *acrA* and *acrB*. Disruption of *E. amylovora acrB* gene was found to cause a significant reduction in its ability to cause fire blight symptoms (Burse et al. 2004). Another important aspect of human pathogens is to modulate host factors. The gastrointestinal tract of humans and animals contains bile salts and fatty acids which act as natural antimicrobial agents. *E. coli* that lack AcrAB are hyper susceptible to these agents clearly highlighting the significance of these pumps in microbial pathogenesis (Sulavik et al. 2001).

8.4.2 Role of Efflux Pumps in Oxidative Stress Response

Hydrogen peroxide (H_2O_2) and nitric oxide (NO) are naturally produced by host immune cells during infection. They are also referred to reactive oxygen species (ROS), in case of H_2O_2 and reactive nitrogen species (RNS), in case of NO, these compounds have significant antimicrobial activity, exerting their effects by damaging several types of bacterial macromolecules, like proteins, lipids, and DNA. MexEF-OprN efflux system of *P. aeruginosa* is up-regulated in response to both of these stress elements (Sobel et al. 2005). Ribosome targeting antibiotics like the aminoglycoside streptomycin cause synthesis of abnormal proteins that are prone to oxidative damage. The common recruitment of these efflux pump helps in removing abnormal proteins that accumulate in response to environmental stresses.

8.4.3 Role of Efflux Pumps in Cell-Cell Communication and Biofilm

In order to survive and colonize in a particular niche, a battery of different mechanisms to sense the constantly changing environment has evolved in bacteria. One of these mechanisms consists of cell-to-cell communication systems. These intercellular signaling systems are regulated by the production of one or more low-molecular weight compounds called quorum sensing molecules, which are sensed by molecular receptors of other cells. In *P. aeruginosa*, RND family multidrug pump (MexAB-OprM) is involved in the transport of extracellular signaling molecule, N-acyl homoserine lactones (HSLs) (Alav et al. 2018; Schuster and Greenberg 2006). RND pumps that export HSLs have also been discovered in *Burkholderia pseudomallei* (Chan et al. 2007). Several studies suggest that efflux pumps might play important roles in biofilm formation by influencing cellular aggregation, transport of exopolysaccharides (EPS) or quorum sensing molecules (Sharma et al. 2017). The transport genes *mdtF* of RND family and *lsrA* of ABC superfamily were reported to be expressed at significantly higher levels during biofilm growth compared with exponential- and stationary-phase growth (Schembri et al. 2003).

8.5 Metal, Biocide and Antibiotic Co-Resistance

Antibiotics, antibacterial biocides and metals have long been used by humans in medicine, agriculture, aquaculture, and in consumer products. Biocides generally have a broad spectrum of antimicrobial activity. Many different biocides that are currently in use, have diverse activities and cellular target sites. Commonly used biocides are alcohols, acids and alkalis, aldehydes, anilides and biguanides, diamides, halogenated compounds, oxidising agents, organic acids, peroxygens, phenolics (phenols, bisphenols and halo phenols) and quaternary ammonium compounds (QACs) (McDonnell and Russell 1999). Biocides are widely used in antiseptics, disinfectants, food preservatives, antifouling compounds and anti-infective in healthcare, agriculture and livestock farming, industry, food preparation and in consumer goods (e.g. toothpastes and cosmetics) (Pal et al. 2014). Heavy metals on the other hand also have a broad spectrum of antimicrobial activity. These metals affect multiple cellular targets and thus have a broad spectrum of activity. Prior to extensive use of antibiotics in health settings, antimicrobial metal compounds were the only option as excellent antimicrobial agents. Few examples of such metals are inorganic and organic mercury compounds, silver, copper, gold, tellurium, potassium, magnesium and zinc salts. Some of these compounds are still listed in WHO list of essential medicines. Antimicrobial metal compounds containing copper, zinc, selenium, cadmium and arsenic are used in agriculture, poultry and animal husbandry as growth promoters, fungicides, herbicides and antimicrobials (Butaye et al. 2003). The use of these metal compounds from ancient times have escalated their release in to environment. Although partially proved, the widespread biological availability of metal compounds leads to metal resistance in many microorganisms that are exposed to it in their environments. Sometime metal resistance is a physiological process as well, for example bacteria that have evolved metal detoxification strategies including copper/zinc resistance determinants are able to avoid killing by protozoan predators (Hao et al. 2016).

8.5.1 Mechanisms of Antimicrobial Metal Resistance

Methylation or demethylation of metals by normal cellular metabolism and generalized antimicrobial efflux through multidrug efflux systems are primary mechanism responsible for metal resistance. Bacteria also have metal ion-specific response regulators, which control the expression of structural resistance genes. The expression of these genes produce a metal ion-specific efflux protein or protein complex and/or enzyme(s) that alter the metal ion into a form less toxic to the bacterial cell.

8.5.2 Mechanisms of Biocide Resistance

Bacterial resistance to biocides has been known since the 1950s (Davies and Davies 2010), particularly with the contamination of cationic biocide formulations. In most instances, bacterial resistance emerged following the improper use like use at lower concentrations than prescribed or improper storage of the formulations, resulting in decrease in the effective concentration. Bacterial resistance to all known preservatives has also been reported (Chapman 2003). In the health care settings, bacterial resistance to biocides has long been reported with compounds such as: chlorhexidine, quaternary ammonium compounds, bisphenol, triclosan, iodine compounds, parabens and sterilizants like biocides such as glutaraldehyde. In a recent study, it has been reported that although biocides may be effective against planktonic bacterial cells of methicillin-resistant *Staphylococcus aureus* (MRSA) and *Pseudomonas aeruginosa*, but they are ineffective against nosocomial pathogens growing as surface attached biofilms. Presence of extracellular matrix in the biofilm of *Salmonella enterica* serovar Typhimurium was reported to be responsible for resistance to triclosan (Tabak et al. 2007).

8.5.3 Mechanisms of Metal/Biocide and Antibiotics Cross Resistance

Co-selection of antibiotic and metal resistance in bacteria can promote antibiotic resistance in bacterial populations even without antibiotic selection pressure. There are three major drivers of co-selection.

(a) **Cross –resistance** occurs when a single mechanism like an efflux pump provide resistance to different toxic compounds simultaneously. The multiple-drug resistance (MDR) pump in *Listeria monocytogenes* can export metals like Cd, Zn, Co, Cu, Cr and As in addition to antibiotics like erythromycin, cefotaxime and clindamycin (Mata et al. 2000). Similarly characterization of the MexGHI–OpmD efflux pump in *Pseudomonas aeruginosa* highlighted that the presence of this pump is responsible for increased resistance to vanadium, ticarcillin and clavulanic acid compared with mutants that lack MexGHI–OpmD (Aendekerk et al. 2002).
(b) **Co-resistance** occurs when two or more different resistance genes are physically co-located on the same genetic element, such as a plasmid or a transposon. Or else it can be present in the same bacterial cell where each resistance gene provides resistance to different compounds. A recent report suggests the presence of metal resistance operon (copper and silver) and beta-lactamases (blaCTX-M) gene on the same plasmid that belongs to Enterobacteriaceae family. Surprisingly this plasmid also has a gene for chlorhexidine, quaternary ammonium compounds (QACs) resistance (Fang et al. 2016).
(c) The presence of **Co-regulatory** machinery which can control multiple resistance genes that confer resistance to different toxic compounds. It been reported in *P. aeruginosa* where CzcRS (two component regulator) increases the expression of the czcCBA efflux system conferring resistance to cadmium, zinc and cobalt, and decreases expression of the OprD porin, leading to increased resistance to imipenem one of the antibiotic from carbapenem class (Perron et al. 2004). Similarly, *mdtABC* operon upregulated in response to stress caused by zinc can increase resistance to certain antibiotics, including novobiocin and the bile-salt component deoxycholate in *E. coli* (Lee et al. 2005). These mechanisms of cross-resistance, co-resistance and co-regulation of metal and antibiotic resistance are diagrammatically represented in Fig. 8.4.

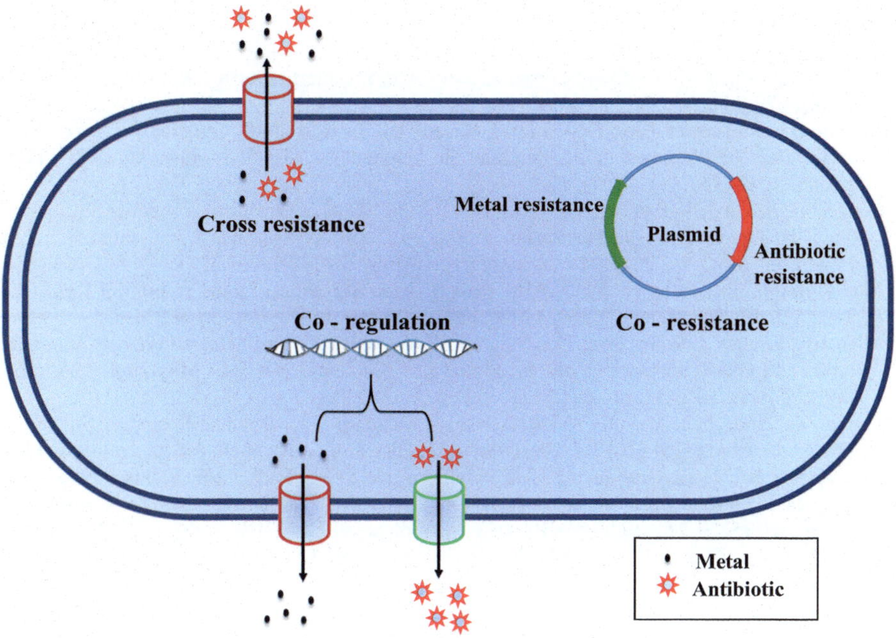

Fig. 8.4 Methods of co-resistance, cross-resistance, and co-regulation of metal- and antibiotic-resistance in bacteria

8.6 Conclusion and Future Perspective

Recent advances in molecular bacteriology has provided in-depth knowledge about MDR efflux pumps. These MDR efflux pump confer resistance not only to clinically relevant antibiotics that are used for infection, but also to a vast array of metals and biocides. There are many studies indicating co-selection of antibiotic resistance by metals and biocide (Li et al. 2017). These findings have direct public health implications since some of the pathogenic strains of several bacterial genera such as *Vibrio*, *Enterobacter* and *Pseudomonas* share common niche with environmental isolates. Resistance genes can be horizontally transferred from environmental organisms to human pathogens. Meanwhile, most of these data correlating the link between co-selection of antibiotic and metal resistance are derived from culture-dependent phenotypic analysis of bacterial isolates, conclusive data for microbial communities are still lacking. Although some studies do indicate co- selection of resistance mechanisms, detailed studies are required for understanding the complex inter-relationships between co-resistance to metals and antibiotics.

References

Abraham EP, Chain E (1988) An Enzyme from Bacteria Able to Destroy Penicillin. Rev Infect Dis 10:677–678

Abramson J, Smirnova I, Kasho V, Verner G, Kaback HR, Iwata S (2003) Structure and Mechanism of the Lactose Permease of Escherichia coli. Science 301:610–615. https://doi.org/10.1126/science.1088196

Aendekerk S, Ghysels B, Cornelis P, Baysse C (2002) Characterization of a new efflux pump, MexGHI-OpmD, from Pseudomonas aeruginosa that confers resistance to vanadium. Microbiol Read Engl 148:2371–2381. https://doi.org/10.1099/00221287-148-8-2371

Alav I, Sutton JM, Rahman KM (2018) Role of bacterial efflux pumps in biofilm formation. J Antimicrob Chemother 73:2003–2020. https://doi.org/10.1093/jac/dky042

Allen HK, Donato J, Wang HH, Cloud-Hansen KA, Davies J, Handelsman J (2010) Call of the wild: antibiotic resistance genes in natural environments. Nat Rev Microbiol 8:251–259. https://doi.org/10.1038/nrmicro2312

Baroud M, Dandache I, Araj GF, Wakim R, Kanj S, Kanafani Z, Khairallah M, Sabra A, Shehab M, Dbaibo G, Matar GM (2013) Underlying mechanisms of carbapenem resistance in extended-spectrum β-lactamase-producing Klebsiella pneumoniae and Escherichia coli isolates at a tertiary care centre in Lebanon: role of OXA-48 and NDM-1 carbapenemases. Int J Antimicrob Agents 41:75–79. https://doi.org/10.1016/j.ijantimicag.2012.08.010

Bay DC, Turner RJ (2012) Small multidrug resistance protein EmrE reduces host pH and osmotic tolerance to metabolic quaternary cation osmoprotectants. J Bacteriol 194:5941–5948. https://doi.org/10.1128/JB.00666-12

Blair JMA, Webber MA, Baylay AJ, Ogbolu DO, Piddock LJV (2015) Molecular mechanisms of antibiotic resistance. Nat Rev Microbiol 13:42–51. https://doi.org/10.1038/nrmicro3380

Buckley AM, Webber MA, Cooles S, Randall LP, La Ragione RM, Woodward MJ, Piddock LJV (2006) The AcrAB-TolC efflux system of Salmonella enterica serovar Typhimurium plays a role in pathogenesis. Cell Microbiol 8:847–856. https://doi.org/10.1111/j.1462-5822.2005.00671.x

Burse A, Weingart H, Ullrich MS (2004) The phytoalexin-inducible multidrug efflux pump AcrAB contributes to virulence in the fire blight pathogen, Erwinia amylovora. Mol Plant-Microbe Interact MPMI 17:43–54. https://doi.org/10.1094/MPMI.2004.17.1.43

Butaye P, Devriese LA, Haesebrouck F (2003) Antimicrobial growth promoters used in animal feed: effects of less well known antibiotics on gram-positive bacteria. Clin Microbiol Rev 16:175–188. https://doi.org/10.1128/CMR.16.2.175-188.2003

Cannon RD, Lamping E, Holmes AR, Niimi K, Baret PV, Keniya MV, Tanabe K, Niimi M, Goffeau A, Monk BC (2009) Efflux-mediated antifungal drug resistance. Clin Microbiol Rev 22:291–321. https://doi.org/10.1128/CMR.00051-08

Chan YY, Bian HS, Tan TMC, Mattmann ME, Geske GD, Igarashi J, Hatano T, Suga H, Blackwell HE, Chua KL (2007) Control of quorum sensing by a Burkholderia pseudomallei multidrug efflux pump. J Bacteriol 189:4320–4324. https://doi.org/10.1128/JB.00003-07

Chapman JS (2003) Disinfectant resistance mechanisms, cross-resistance, and co-resistance. Int Biodeterior Biodegrad, Hyg Disinfect 51:271–276. https://doi.org/10.1016/S0964-8305(03)00044-1

D'Costa VM, King CE, Kalan L, Morar M, Sung WWL, Schwarz C, Froese D, Zazula G, Calmels F, Debruyne R, Golding GB, Poinar HN, Wright GD (2011) Antibiotic resistance is ancient. Nature 477:457–461. https://doi.org/10.1038/nature10388

Davies J, Davies D (2010) Origins and Evolution of Antibiotic Resistance. Microbiol Mol Biol Rev 74:417–433. https://doi.org/10.1128/MMBR.00016-10

Dolejska M, Villa L, Poirel L, Nordmann P, Carattoli A (2013) Complete sequencing of an IncHI1 plasmid encoding the carbapenemase NDM-1, the ArmA 16S RNA methylase and a resistance–nodulation–cell division/multidrug efflux pump. J Antimicrob Chemother 68:34–39. https://doi.org/10.1093/jac/dks357

Du D, Wang Z, James NR, Voss JE, Klimont E, Ohene-Agyei T, Venter H, Chiu W, Luisi BF (2014) Structure of the AcrAB–TolC multidrug efflux pump. Nature 509:512–515. https://doi.org/10.1038/nature13205

Fang L, Li X, Li L, Li S, Liao X, Sun J, Liu Y (2016) Co-spread of metal and antibiotic resistance within ST3-IncHI2 plasmids from E. coli isolates of food-producing animals. Sci Rep 6(25312). https://doi.org/10.1038/srep25312

Forsberg KJ, Reyes A, Wang B, Selleck EM, Sommer MOA, Dantas G (2012) The shared antibiotic resistome of soil bacteria and human pathogens. Science 337:1107–1111. https://doi.org/10.1126/science.1220761

Gandra S, Tseng KK, Arora A, Bhowmik B, Robinson ML, Panigrahi B, Laxminarayan R, Klein EY (2018) The mortality burden of multidrug-resistant pathogens in India: a retrospective observational study. Clin Infect Dis Off Publ Infect Dis Soc Am. https://doi.org/10.1093/cid/ciy955

Gottesman MM, Ling V (2006) The molecular basis of multidrug resistance in cancer: the early years of P-glycoprotein research. FEBS Lett 580:998–1009. https://doi.org/10.1016/j.febslet.2005.12.060

Hao X, Lüthje F, Rønn R, German NA, Li X, Huang F, Kisaka J, Huffman D, Alwathnani HA, Zhu Y-G, Rensing C (2016) A role for copper in protozoan grazing – two billion years selecting for bacterial copper resistance. Mol Microbiol 102:628–641. https://doi.org/10.1111/mmi.13483

Higgins CF (2007) Multiple molecular mechanisms for multidrug resistance transporters. Nature 446:749–757. https://doi.org/10.1038/nature05630

Hirakata Y, Srikumar R, Poole K, Gotoh N, Suematsu T, Kohno S, Kamihira S, Hancock REW, Speert DP (2002) Multidrug efflux systems play an important role in the invasiveness of Pseudomonas aeruginosa. J Exp Med 196:109–118

Jerse AE, Sharma ND, Simms AN, Crow ET, Snyder LA, Shafer WM (2003) A gonococcal efflux pump system enhances bacterial survival in a female mouse model of genital tract infection. Infect Immun 71:5576–5582

Juliano RL, Ling V (1976) A surface glycoprotein modulating drug permeability in Chinese hamster ovary cell mutants. Biochim Biophys Acta 455:152–162

Kapur V, Li LL, Iordanescu S, Hamrick MR, Wanger A, Kreiswirth BN, Musser JM (1994) Characterization by automated DNA sequencing of mutations in the gene (rpoB) encoding the RNA polymerase beta subunit in rifampin-resistant Mycobacterium tuberculosis strains from New York City and Texas. J Clin Microbiol 32:1095–1098

Kobayashi N, Nishino K, Yamaguchi A (2001) Novel macrolide-specific ABC-type efflux transporter inEscherichia coli. J Bacteriol 183:5639–5644. https://doi.org/10.1128/JB.183.19.5639-5644.2001

Kojima S, Nikaido H (2013) Permeation rates of penicillins indicate that Escherichia coli porins function principally as nonspecific channels. Proc Natl Acad Sci U S A 110:E2629–E2634. https://doi.org/10.1073/pnas.1310333110

Kumar N, Radhakrishnan A, Wright CC, Chou T-H, Lei H-T, Bolla JR, Tringides ML, Rajashankar KR, Su C-C, Purdy GE, Yu EW (2014) Crystal structure of the transcriptional regulator Rv1219c of Mycobacterium tuberculosis. Protein Sci Publ Protein Soc 23:423–432. https://doi.org/10.1002/pro.2424

Kuroda T, Tsuchiya T (2009) Multidrug efflux transporters in the MATE family. Biochim Biophys Acta BBA Proteins Proteom Mechan Drug Efflux Strateg Combat Them 1794:763–768. https://doi.org/10.1016/j.bbapap.2008.11.012

Lambert PA (2005) Bacterial resistance to antibiotics: modified target sites. Adv Drug Deliv Rev 57:1471–1485. https://doi.org/10.1016/j.addr.2005.04.003

Lavigne J-P, Sotto A, Nicolas-Chanoine M-H, Bouziges N, Pagès J-M, Davin-Regli A (2013) An adaptive response of Enterobacter aerogenes to imipenem: regulation of porin balance in clinical isolates. Int J Antimicrob Agents 41:130–136. https://doi.org/10.1016/j.ijantimicag.2012.10.010

Lee LJ, Barrett JA, Poole RK (2005) Genome-wide transcriptional response of chemostat-cultured Escherichia coli to zinc. J Bacteriol 187:1124–1134. https://doi.org/10.1128/JB.187.3.1124-1134.2005

Lee C-R, Lee JH, Park M, Park KS, Bae IK, Kim YB, Cha C-J, Jeong BC, Lee SH (2017) Biology of Acinetobacter baumannii: pathogenesis, antibiotic resistance mechanisms, and prospective treatment options. Front Cell Infect Microbiol 7. https://doi.org/10.3389/fcimb.2017.00055

Levy SB (1992) Active efflux mechanisms for antimicrobial resistance. Antimicrob Agents Chemother 36:695–703

Li X-Z, Nikaido H (2004) Efflux-mediated drug resistance in bacteria. Drugs 64:159–204. https://doi.org/10.2165/00003495-200464020-00004

Li X-Z, Nikaido H (2009) Efflux-mediated drug resistance in bacteria: an update. Drugs 69:1555–1623. https://doi.org/10.2165/11317030-000000000-00000

Li XZ, Livermore DM, Nikaido H (1994) Role of efflux pump(s) in intrinsic resistance of Pseudomonas aeruginosa: resistance to tetracycline, chloramphenicol, and norfloxacin. Antimicrob Agents Chemother 38:1732–1741

Li X-Z, Plésiat P, Nikaido H (2015) The challenge of efflux-mediated antibiotic resistance in Gram-negative bacteria. Clin Microbiol Rev 28:337–418. https://doi.org/10.1128/CMR.00117-14

Li L-G, Xia Y, Zhang T (2017) Co-occurrence of antibiotic and metal resistance genes revealed in complete genome collection. ISME J 11:651–662. https://doi.org/10.1038/ismej.2016.155

Livermore DM (2008) Defining an extended-spectrum beta-lactamase. Clin Microbiol Infect 14(Suppl 1):3–10. https://doi.org/10.1111/j.1469-0691.2007.01857.x

Long KS, Poehlsgaard J, Kehrenberg C, Schwarz S, Vester B (2006) The Cfr rRNA methyltransferase confers resistance to Phenicols, Lincosamides, Oxazolidinones, Pleuromutilins, and Streptogramin A antibiotics. Antimicrob Agents Chemother 50:2500–2505. https://doi.org/10.1128/AAC.00131-06

Ma D, Cook DN, Alberti M, Pon NG, Nikaido H, Hearst JE (1993) Molecular cloning and characterization of acrA and acrE genes of Escherichia coli. J Bacteriol 175:6299–6313

Mata MT, Baquero F, Pérez-Díaz JC (2000) A multidrug efflux transporter in Listeria monocytogenes. FEMS Microbiol Lett 187:185–188

McDonnell G, Russell AD (1999) Antiseptics and disinfectants: activity, action, and resistance. Clin Microbiol Rev 12:147–179

Morita Y, Kodama K, Shiota S, Mine T, Kataoka A, Mizushima T, Tsuchiya T (1998) NorM, a putative multidrug efflux protein, of vibrio parahaemolyticus and its homolog in Escherichia coli. Antimicrob Agents Chemother 42:1778–1782

Murakami S, Nakashima R, Yamashita E, Yamaguchi A (2002) Crystal structure of bacterial multidrug efflux transporter AcrB. Nature 419:587–593. https://doi.org/10.1038/nature01050

Murakami S, Nakashima R, Yamashita E, Matsumoto T, Yamaguchi A (2006) Crystal structures of a multidrug transporter reveal a functionally rotating mechanism. Nature 443:173–179. https://doi.org/10.1038/nature05076

Nikaido H (2003) Molecular basis of bacterial outer membrane permeability revisited. Microbiol Mol Biol Rev 67:593–656. https://doi.org/10.1128/MMBR.67.4.593-656.2003

Nikaido H, Basina M, Nguyen V, Rosenberg EY (1998) Multidrug efflux pump AcrAB of Salmonella typhimurium excretes only those beta-lactam antibiotics containing lipophilic side chains. J Bacteriol 180:4686–4692

Norris AL, Serpersu EH (2013) Ligand promiscuity through the eyes of the aminoglycoside N3 acetyltransferase IIa. Protein Sci Publ Protein Soc 22:916–928. https://doi.org/10.1002/pro.2273

Pal C, Bengtsson-Palme J, Rensing C, Kristiansson E, Larsson DGJ (2014) BacMet: antibacterial biocide and metal resistance genes database. Nucleic Acids Res 42:D737–D743. https://doi.org/10.1093/nar/gkt1252

Paterson DL, van Duin D (2017) China's antibiotic resistance problems. Lancet Infect Dis 17:351–352. https://doi.org/10.1016/S1473-3099(17)30053-1

Perron K, Caille O, Rossier C, Van Delden C, Dumas J-L, Köhler T (2004) CzcR-CzcS, a two-component system involved in heavy metal and carbapenem resistance in Pseudomonas aeruginosa. J Biol Chem 279:8761–8768. https://doi.org/10.1074/jbc.M312080200

Perry JA, Wright GD (2013) The antibiotic resistance "mobilome": searching for the link between environment and clinic. Front Microbiol 4(138). https://doi.org/10.3389/fmicb.2013.00138

Piddock LJV (2006a) Clinically relevant chromosomally encoded multidrug resistance efflux pumps in bacteria. Clin Microbiol Rev 19:382–402. https://doi.org/10.1128/CMR.19.2.382-402.2006

Piddock LJV (2006b) Multidrug-resistance efflux pumps – not just for resistance. Nat Rev Microbiol 4:629–636. https://doi.org/10.1038/nrmicro1464

Pitout JD, Laupland KB (2008) Extended-spectrum β-lactamase-producing Enterobacteriaceae: an emerging public-health concern. Lancet Infect Dis 8:159–166. https://doi.org/10.1016/S1473-3099(08)70041-0

Poole K (2005) Efflux-mediated antimicrobial resistance. J Antimicrob Chemother 56:20–51. https://doi.org/10.1093/jac/dki171

Reddy VS, Shlykov MA, Castillo R, Sun EI, Saier MH (2012) The major facilitator superfamily (MFS) revisited. FEBS J 279:2022–2035. https://doi.org/10.1111/j.1742-4658.2012.08588.x

Rice LB (2008) Federal funding for the study of antimicrobial resistance in nosocomial pathogens: No ESKAPE. J Infect Dis 197:1079–1081. https://doi.org/10.1086/533452

Romanowska J, Reuter N, Trylska J (2013) Comparing aminoglycoside binding sites in bacterial ribosomal RNA and aminoglycoside modifying enzymes. Proteins 81:63–80. https://doi.org/10.1002/prot.24163

Saier MH, Tran CV, Barabote RD (2006) TCDB: the transporter classification database for membrane transport protein analyses and information. Nucleic Acids Res 34:D181–D186. https://doi.org/10.1093/nar/gkj001

Schembri MA, Kjaergaard K, Klemm P (2003) Global gene expression in Escherichia coli biofilms. Mol Microbiol 48:253–267

Schuldiner S, Lebendiker M, Yerushalmi H (1997) EmrE, the smallest ion-coupled transporter, provides a unique paradigm for structure-function studies. J Exp Biol 200:335–341

Schuster M, Greenberg EP (2006) A network of networks: quorum-sensing gene regulation in Pseudomonas aeruginosa. Int J Med Microbiol IJMM 296:73–81. https://doi.org/10.1016/j.ijmm.2006.01.036

Sennhauser G, Bukowska MA, Briand C, Grütter MG (2009) Crystal structure of the multidrug exporter MexB from Pseudomonas aeruginosa. J Mol Biol 389:134–145. https://doi.org/10.1016/j.jmb.2009.04.001

Sharma A, Sharma R, Bhattacharyya T, Bhando T, Pathania R (2017) Fosfomycin resistance in Acinetobacter baumannii is mediated by efflux through a major facilitator superfamily (MFS) transporter-AbaF. J Antimicrob Chemother 72:68–74. https://doi.org/10.1093/jac/dkw382

Sobel ML, Neshat S, Poole K (2005) Mutations in PA2491 (mexS) promote MexT-dependent mexEF-oprN expression and multidrug resistance in a clinical strain of Pseudomonas aeruginosa. J Bacteriol 187:1246–1253. https://doi.org/10.1128/JB.187.4.1246-1253.2005

Srinivasan VB, Rajamohan G, Gebreyes WA (2009) Role of AbeS, a novel efflux pump of the SMR family of transporters, in resistance to antimicrobial agents in Acinetobacter baumannii. Antimicrob Agents Chemother 53:5312–5316. https://doi.org/10.1128/AAC.00748-09

Su X-Z, Chen J, Mizushima T, Kuroda T, Tsuchiya T (2005) AbeM, an H+-coupled Acinetobacter baumannii multidrug efflux pump belonging to the MATE family of transporters. Antimicrob Agents Chemother 49:4362–4364. https://doi.org/10.1128/AAC.49.10.4362-4364.2005

Sulavik MC, Houseweart C, Cramer C, Jiwani N, Murgolo N, Greene J, DiDomenico B, Shaw KJ, Miller GH, Hare R, Shimer G (2001) Antibiotic susceptibility profiles of Escherichia coli strains lacking multidrug efflux pump genes. Antimicrob Agents Chemother 45:1126–1136. https://doi.org/10.1128/AAC.45.4.1126-1136.2001

Sun J, Deng Z, Yan A (2014) Bacterial multidrug efflux pumps: Mechanisms, physiology and pharmacological exploitations. Biochem Biophys Res Commun, Integrative Glycobiol Future Perspect 453:254–267. https://doi.org/10.1016/j.bbrc.2014.05.090

Tabak M, Scher K, Hartog E, Romling U, Matthews KR, Chikindas ML, Yaron S (2007) Effect of triclosan on Salmonella typhimurium at different growth stages and in biofilms. FEMS Microbiol Lett 267:200–206. https://doi.org/10.1111/j.1574-6968.2006.00547.x

Tal N, Schuldiner S (2009) A coordinated network of transporters with overlapping specificities provides a robust survival strategy. Proc Natl Acad Sci 106:9051–9056. https://doi.org/10.1073/pnas.0902400106

Tamber S, Hancock REW (2003) On the mechanism of solute uptake in pseudomonas. Front Biosci J Virtual Libr 8:s472–s483

Telenti A, Imboden P, Marchesi F, Matter L, Schopfer K, Bodmer T, Lowrie D, Colston MJ, Cole S (1993) Detection of rifampicin-resistance mutations in Mycobacterium tuberculosis. The Lancet. Originally published as Volume 1, Issue 8846 341: 647–651. https://doi.org/10.1016/0140-6736(93)90417-F

The world is running out of antibiotics, WHO report confirms [WWW Document] (n.d.) World Health Organ. http://www.who.int/news-room/detail/20-09-2017-the-world-is-running-out-of-antibiotics-who-report-confirms. Accessed 25 Sept 2018

Tsukagoshi N, Aono R (2000) Entry into and release of solvents by Escherichia coli in an organic-aqueous two-liquid-phase system and substrate specificity of the AcrAB-TolC solvent-extruding pump. J Bacteriol 182:4803–4810

Tuševljak N, Dutil L, Rajić A, Uhland FC, McClure C, St-Hilaire S, Reid-Smith RJ, McEwen SA (2013) Antimicrobial Use and resistance in aquaculture: findings of a globally administered survey of aquaculture-allied professionals. Zoonoses Public Health 60:426–436. https://doi.org/10.1111/zph.12017

Ventola CL (2015) The antibiotic resistance crisis. Pharm Ther 40:277–283

Voulgari E, Poulou A, Koumaki V, Tsakris A (2013) Carbapenemase-producing Enterobacteriaceae: now that the storm is finally here, how will timely detection help us fight back? Future Microbiol 8:27–39. https://doi.org/10.2217/fmb.12.130

Walsh C (2000) Molecular mechanisms that confer antibacterial drug resistance. Nature 406:775–781. https://doi.org/10.1038/35021219

Walsh TR, Wu Y (2016) China bans colistin as a feed additive for animals. Lancet Infect Dis 16:1102–1103. https://doi.org/10.1016/S1473-3099(16)30329-2

Weigel LM, Steward CD, Tenover FC (1998) gyrA mutations associated with fluoroquinolone resistance in eight species of Enterobacteriaceae. Antimicrob Agents Chemother 42:2661–2667

Wright GD (2005) Bacterial resistance to antibiotics: enzymatic degradation and modification. Adv Drug Deliv Rev 57:1451–1470. https://doi.org/10.1016/j.addr.2005.04.002

Wright GD (2011) Molecular mechanisms of antibiotic resistance. Chem Commun Camb Engl 47:4055–4061. https://doi.org/10.1039/c0cc05111j

Yin Y, He X, Szewczyk P, Nguyen T, Chang G (2006) Structure of the multidrug transporter EmrD from Escherichia coli. Science 312:741–744. https://doi.org/10.1126/science.1125629

Zgurskaya HI, Nikaido H (1999) Bypassing the periplasm: reconstitution of the AcrAB multidrug efflux pump of Escherichia coli. Proc Natl Acad Sci U S A 96:7190–7195

Biofilm and Antibiotic Resistance in *Acinetobacter baumannii* 9

Rajagopalan Saranathan, Sudhakar Pagal, and K. Prashanth

Abstract

Biofilms are the protective armour formed by the bacterial pathogens constituting polysaccharides, proteins and DNA which them inaccessible for antibiotics and other biocides. The strategy employed by nosocomial pathogens such as *Acinetobacter baumannii,* for survival in hospital their exceptional ability to produce highly resistant biofilms. Biofilms assist these bacteria to thrive in adverse conditions such as desiccation, nutrient depletion and physiological stress. *A. baumannii* secretes exopolysaccharides once it has successfully adhered to a surface hydrophilic like glass or hydrophobic such as host cell surfaces. *A. baumannii* being listed by WHO as the leading antibiotic-resistant priority pathogens list for which there is an urgent need to develop new antibiotics, this chapter will summarise the existing knowledge on the stages in biofilm formation, architecture of biofilm matrix, genetic regulation of biofilm formation in *A. baumannii* and recent developments in biofilm inhibiting/dispersing agents.

Keywords

Acinetobacter baumannii · Biofilm · Exopolysaccharides · Antibiotic resistance

R. Saranathan (✉)
Department of Microbiology and Immunology, Albert Einstein College of Medicine, Bronx, NY, USA
e-mail: saraibt@gmail.com

S. Pagal · K. Prashanth
Department of Biotechnology, Pondicherry Central University, Puducherry, India

S. M. Mandal, D. Paul (eds.), *Bacterial Adaptation to Co-resistance*,
https://doi.org/10.1007/978-981-13-8503-2_9

9.1 Introduction

Acinetobacter spp. are strictly aerobic, catalase-positive, oxidase-negative, non-motile, non-fermenting, short, plump coccobacilli, which usually occur in diploid formations or in chains of variable lengths (Peleg et al. 2008; Giamarellou et al. 2008). *Acinetobacter spp.* form smooth, mucoid, grayish-white colonies on solid media that are routinely observed in clinical microbiological laboratories. *A. baumannii* is an important human pathogen and a member of *Acinetobacter spp* which causes a wide range of infections such as ventilator associated pneumonia (VAP), urinary tract infection (UTI), wound, skin and soft tissue infections among ICU patients (Dijkshoorn et al. 2007). Increasing number of studies has been reporting the prevalence of multi-drug resistance (MDR) and extensive drug resistance (XDR) in *A. baumannii* (Poirel and Nordmann 2006; Perez et al. 2007; Zarrilli et al. 2009; Bahador et al. 2013). The most common resistance mechanisms employed by this bacteria to avert the action of antibiotics are the production of β-lactam hydrolyzing enzymes and antibiotic modifying enzymes, target modifications, antibiotic efflux and biofilms (Roca et al. 2012). *A. baumannii* can survive under stressful environmental conditions including hostile hospital environments and persist for extended periods of time on surfaces by forming complex protective structures called biofilms (Gaddy et al. 2009). Biofilms are thin, slimy layers of compact extracellular matrices secreted and developed by bacterial colonies adhering to solid surfaces. *A. baumannii* biofilms may confer resistance to antimicrobial therapies by preventing antibiotic penetration and causing chronic infections by evading phagocytosis (Costerton et al. 2003; Høiby et al. 2010). Gram negative bacteria such as *A. baumannii* live as communities and must interact with and monitor their neighbors, to sense cell density and regulate adaptations to changing environmental conditions. They accomplish this through a highly regulated mechanism called quorum sensing, which involves the synthesis and secretion of auto-inducing signaling molecules such as acyl-homoserine lactones (Bhargava et al. 2010). These signaling molecules interact directly with transcription factors and control gene expression. Biofilm formation, along with multidrug resistance and quorum sensing are the major contributors to the pathogenicity of *A. baumannii.*

9.2 Biofilms

Biofilms are aggregated assemblies of a population of bacterial cells adhered to a surface (biotic or abiotic) which are arranged in a compact tertiary structure and sheathed in polymeric substances matrix composed of exopolysaccharides (EPS), nucleic acids, proteins and other macromolecules (Gaddy et al. 2009). Generally, EPS are a complex mixture of biopolymers constituting the intercellular space of microbial aggregates and forms the structure and architecture of the biofilm matrix (Gaddy et al. 2009). In general, bacteria exist in two life forms during growth and proliferation. Either they are present as single, independent cells (planktonic) or organized into sessile aggregates, commonly referred to as biofilm or biofilm phenotype (Longo et al. 2014). Acute bacterial infections are caused by planktonic bacteria, which are generally treatable with antibiotics (Longo et al. 2014). However, when these bacteria succeed in forming biofilms within human hosts, the infection often turns out to be untreatable and develops into a chronic infection (Longo et al. 2014). *A. baumannii* cells residing inside biofilm matrix are less susceptible to antimicrobial agents compared to planktonic forms due to less penetration of drugs into the biofilm matrix and reduced growth rate and respiration (Høiby et al. 2010). In addition, biofilm forming bacteria are known to overexpress efflux pumps and stress response regulators to survive under severe antibiotic pressure and physiological stress conditions (Høiby et al. 2010). In general, EPS limit the penetration of drugs and prevents them from reaching their target. The conversion of resistant bacterial phenotypes into susceptible ones by biofilm dispersing agents have shown promising results and unlocked strategies for combating drug resistant bacteria (Høiby et al. 2010).

The process of biofilm development as shown in Fig. 9.1, involves five main stages: attachment, proliferation, development and maturation, detachment and dispersion (Stoodley et al. 2001). In vitro biofilms are initiated by planktonic (freely moving) bacteria that explore and interact with surfaces and find niches offering nutrients and then reversibly attach to the surface (Stoodley et al. 2001). Once settled these adherent bacteria subsequently multiply and upregulate genes involved in matrix production. Microcolonies arise in the form of mushrooms or towered structures (Stoodley et al. 2001). These microcolonies are coated with polymeric matrices which adhere to the surface. Chemical signals such as quorum sensing molecules are actively involved in the regulation of biofilm formation and detachment. These signals dictate the precise positioning of cells in the colonies and guide the formation of water channels within the biofilm. Mature biofilms may grow up to 50 μm in thickness and at this stage they may show maximum antibiotic tolerance conferred by the thick polysaccharide matrix surrounding them. Further, focal areas of the biofilm dissolve and disseminate bacterial cells, which can then spread to another location where new biofilms are established (Høiby et al. 2010). This stage is most important in the spread of infections and colonization of pathogens. The process of detachment of cells from bacterial biofilms has been divided into two processes based on the magnitude and frequency of the detachment event: erosion and sloughing (Stoodley et al. 2001).

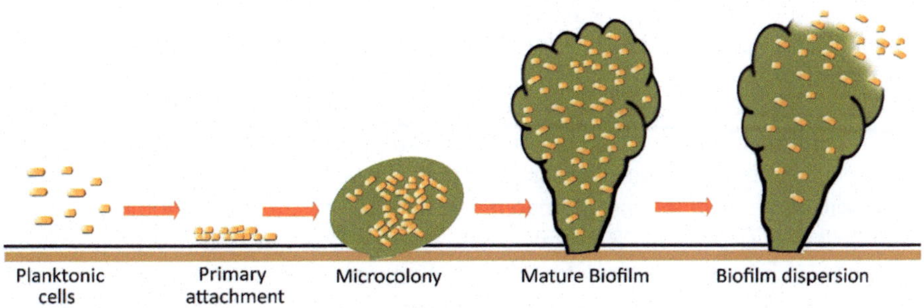

Fig. 9.1 **Stages of development of a biofilm:** (1) Initial attachment of cells to the surface, (2) production of EPS resulting in more firmly adhered attachment (3) early development of biofilm architecture, (4) maturation of biofilm architecture, (5) dispersion of single cells from the biofilm

9.3 Biofilm Architecture and Composition

EPS and proteins have been shown to be key components of the matrix (Branda et al. 2005). Extracellular DNA, lipids and humic substances present in the matrix play important roles in virulence and pathogenicity. Small environmental changes can be responsible for dramatic differences in biofilm architecture. Such changes include variation in oxygen and nitrogen levels, extent of desiccation, temperature, pH, and availability of nutrients (Vu et al. 2009). Structural differences appear to reflect variations in the composition of the extracellular matrix. Biofilm EPS are insoluble in water and often difficult to extract, hence making them challenging to characterize. The proportion of EPS in biofilms can range between approximately 50–90% of total organic matter. Biofilm composition varies with the organism, species or strains involved in its formation. In Gram-negative bacterial biofilms, the presence of uronic acids or ketal-linked pyruvates enhances their anionic properties and allows the association of divalent cations such as calcium and magnesium to increase the binding force in a developed biofilm. The chemical composition of EPS in Gram-positive bacteria could be slightly different as they are cationic in nature (Vu et al. 2009). EPS play a key role in the attachment of cells to substrates/ surfaces during biofilm initiation. EPS bind the cells to the substratum by three types of forces, namely electrostatic interactions, hydrogen bonds and Van der Waal's dispersion forces. These binding forces are likely to contribute to the overall stability of biofilm matrices. It has also been shown that the formation of EPS leads to irreversible attachment with different environmental surfaces which explains the persistence of cells in biofilm structures from days to months in hospital environments (Branda et al. 2005; Sutherland 2001).

Glycosyl composition analysis indicated the presence of high total mannose content across *A. baumannii* strains. *A. baumannii* EPS comprised of 80–90% mannose, whereas the second and third most abundant carbohydrates were galactose and glucose (Bales et al. 2013). Interestingly, the high molecular weight EPS derived from *A. baumannii* were reported to be novel in this study, which was analyzed based on the composition and/or ratio analysis of carbohydrates (Bales et al. 2013). Bales et al. (2013) also determined the glycosydic linkages of *A. baumannii* EPS, which in order of greatest to least, is 2, 6-linked mannopyranosyl, 3-linked mannopyranosyl, 2-linked rhamnopyranosyl, and terminally-linked mannopyranosyl residues.

Early studies concerning *A. baumannii* proposed that its biofilm may be favorable in maintaining the microorganism's growth on solid surfaces in hospital environments and confer protection against antibacterial agents (Vidal et al. 1996); this observation has become more apparent over time as biofilm formation is a common trait in clinical isolates (Rodríguez-Baño et al. 2008). Biofilm forming isolates survive longer on dry surfaces compared to non-biofilm forming counterparts (Espinal et al. 2012). More recently, molecular evidences has related the ability of *A. baumannii* strains with strong biofilm formation and ensuing capacity to cause outbreaks (Gaddy et al. 2009). Established biofilms not only serve as protective defense strategies to prokaryotes during unfavourable environmental conditions, but are also one of the major virulence factors during infection. Although the precise genetic mechanisms underlying pathogenesis and virulence of *A. baumannii* infections remain unclear, biofilm formation however is known to help colonization, resistance and persistence in the host tissues.

9.4 Genetic Regulation of Biofilm Formation

The initiation and formation of biofilms are associated with the occurrence of stress due to nutritional depletion, desiccation, presence of antibiotics and host factors (Doi et al. 2015). At genetic level, however, the interplay of stress, antibiotic resistance and biofilm has remained elusive. The role of quorum sensing is one of the earliest observations pertaining to the regulation of biofilm formation. The biofilm phenotype is expressed at high cell density, as a result of the activation of N-Acyl-Homoserine Lactone (AHL) -mediated quorum sensing signaling system. Therefore, AHL-synthase (AbaI), is a putative target for inhibition of biofilm formation since, *abaI* disruption abrogates biofilm formation by up to 40%. On the other hand, treating *abaI* mutants with AHL reverses this effect (Irie and Parsek 2008). The other essential factor governing biofilm formation is linked to attachment of bacterial cells to surfaces, which appears to be a pre-requisite to biofilm formation. The most pronounced factor mediating this adhesion is the CsuA/B-ABCDE chaperone-usher pili system, when expressed this system, initiates cell adhesion through external appendages, called pili, leading to initiation of biofilm formation on plastic surfaces (Tomaras et al. 2003). Further, *csuC* disruption mutants become pili-deficient as and also fail to produce biofilm. In addition to the pili system, the presence of *A. baumannii* outer membrane protein OmpA has been shown to be essential for adhesion of the organism to host cells *in vitro*. In addition, OmpA also promotes robust biofilm formation on plastic surfaces, in a process that is independent from the CsuA/B-ABCDE chaperone-usher pili system (Gaddy et al. 2009; Choi et al. 2008).

Another factor that is thought to correlate with biofilm formation, and virulence in general is motility. In hyper-motile mutants of *A. baumannii* ATCC 17978, the histone-like nucleoid structuring (H-NS) protein has been found to play an important role in biofilm formation. This protein is a global repressor of virulence factors, the abrogation of which overexpresses genes such as *ata*, genes coding for type VI secretion system and type I pili. Products of all these genes are known for enhancing virulence factors, including biofilm formation (Eijkelkamp et al. 2013). The *ata* gene, which codes for *Acinetobacter* trimeric autotransporter (Ata) is a surface adhesin that plays a pivotal role towards cell adhesion to biological matrices, including collagen and the laminins. This property brought about by Ata is crucial for *A. baumannii* virulence as well as biofilm formation, which enhances the lethality of the organism in a mouse model (Bentancor et al. 2012).

Some of the two-component signaling systems (TCS) also make a significant contribution to biofilm formation in *A. baumannii*. The TCS BfmRS induces biofilm formation, which in turn causes antibiotic resistance. Normally, activation of this system's Response Regulator (RR), BfmR allows transcription of *csu* operon, which directs pili synthesis. However, inactivation of BfmR impairs the *csu* operon, thereby disrupting pili formation and abolishing biofilm development. Pili provide the organism with the ability to adhere to biotic and abiotic surfaces allowing it to initiate biofilm formation. Similarly, GacSA, a global virulence regulator controls pili synthesis, biofilms, and motility (Cerqueira et al. 2014). Recently, a tetR type regulator, adeN was found to be essential for biofilm formation wherein its disruption by *ISAba1* (a transposon exclusively present in *A. baumannii*) led to reduced biofilm production and overexpression of the adeIJK efflux pump (Fig. 9.2) (Saranathan et al. 2017). Though adeN has been identified to be a negative regulator of adeIJK efflux pump, its role in biofilm formation supports the fact that it could be a global regulator in *A. baumannii* which controls multiple physiological processes. The overexpression of adeIJK efflux pumps in the absence of adeN may alter the membrane composition of *A. baumannii* in the sense that biofilm formation may be affected by the inability of the cells to export biofilm components.

Besides the genetic regulation of biofilm production, regulation of the expression of genes, whose products play structural roles also governs biofilm formation. This includes the biofilm associated protein (Bap), PgaABCD, PglC and PglL. Bap is a Staphyloccal homolog of *A. baumannii*, which is similar to bacterial surface adhesins. Bap has an important role in the development and maturation of biofilms and its disruption severely affects the integrity and structure of biofilms (Bentancor et al. 2012; Brossard and Campagnari 2012). The *pgaABCD* locus is responsible for synthesis of cell wall poly-β-(1–6)-N-acetylglucosamine (PNAG), which is a crucial precursor to biofilms (Choi et al. 2009). PglC and PglL mediate a common *O*-glycosylation system, which carries out the glycosylation of proteins. *O*-glycosylated proteins support the building blocks of the capsular polysaccharides

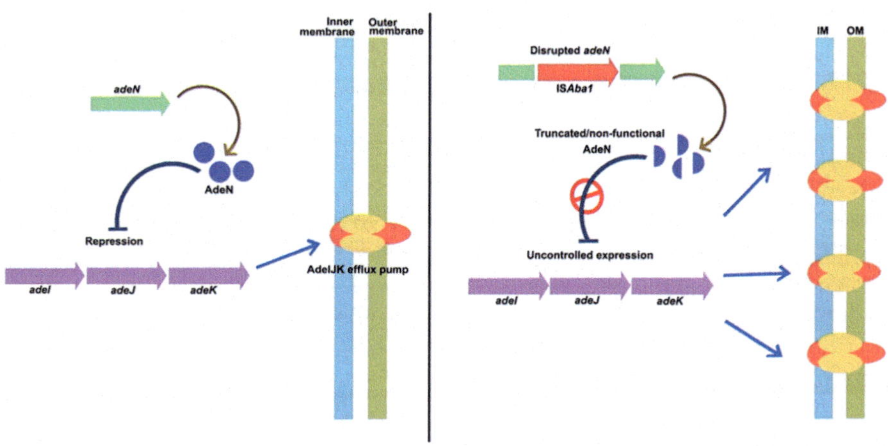

Fig. 9.2 Regulation of adeIJK efflux pump expression by adeN: The disruption of *adeN* by *ISAba1* leads to overexpression of the adeIJK efflux pump and reduced biofilm production due to the alteration in membrane composition

in *A. baumannii*, which are essential components of biofilms (Iwashkiw et al. 2012; Lees-Miller et al. 2013).

9.5 Depolymerases and Phages in Alternative Biofilm Treatment Strategies

Knowing the impact of biofilms in infectious outbreaks, their contribution to virulence and the failure of multiple antibiotics during various aggressive treatment modalities, it has become necessary to conceive alternative remedies. Since microbial biofilm formation is often associated with a decrease in antimicrobial susceptibility, inhibition of biofilm growth as well as dispersion of pre-formed biofilms is very essential (Hughes et al. 1998). EPS can be considered as critical target for biofilm dispersion by devising efficient methods to cleave specific glycosidic bonds in the sugar polymers (Hughes et al. 1998). Nature presents with numerous enzymes which may have plant, animal or microbial origins capable of executing this task efficiently and with greater specificity. These enzymes can break target polysaccharides into smaller units of sugars, hence they are referred to as depolymerases (Hughes et al. 1998). Bacteriophages have also been considered for antimicrobial therapy as they are naturally endowed with the enzymatic capacity to degrade bacterial cell walls as part of their life cycle. They have important properties that render them compelling for alternative therapy (Loc-Carrillo and Abedon 2011); however, concerns regarding their uses in treatment exist for the following reasons: (i) lysogeny, (ii) narrow host ranges, (iii) potential immune rejection by human hosts and (iv) cumbersome culturing processes. Depolymerases of bacterial origin on the other hand, are possibly easier to screen, isolate, characterize and optimize their production in vitro by recombinant DNA technology (Loc-Carrillo and Abedon 2011).

9.6 Biofilm Inhibiting and Dispersing Agents

Strategies of disrupting biofilm formation include inhibition of one or more of the following: quorum sensing signaling, c-di-GMP signaling, (p) ppGpp response, biofilm or EPS synthesis and dispersion/disassembly of biofilms. However, not much of these alternatives have been explored in *A. baumannii*. Most of the studies on combating biofilms of *A. baumannii* relied upon either physically dispersing pre-formed biofilms or inhibiting of biofilm formation without further examination of the mechanism behind such actions. Among the most promising inhibitors is peptide 1018, which can inhibit as well as disperse biofilms of *A. baumannii* at concentrations as low as 20 μg/mL (La Fuente-Núñez et al. 2014). Inhibitors that antagonize c-di-GMP, identified in an *in silico* pharmacophore-based study were recently seen to hinder the development in *A. baumannii* at 200 μM concentration and of them dispersed pre-formed biofilms too (Sambanthamoorthy et al. 2014). More recently, chimeric antimicrobial peptides have been proven to inhibit biofilm formation at minimum biofilm inhibitory concentrations ranging from 6.25 to 50 μM when used in combination with commercially available antibiotics against multidrug resistant isolates of *A. baumannii* (Gopal et al. 2014). Another interesting study showed that the Reverse-Amide 2-Aminoimidazole class of molecules can penetrate through biofilms of *A. baumannii* as well as the bacterial cell membrane to inhibit biofilm development and disperse pre-existing biofilms (Stowe et al. 2015).

In addition to synthetic molecules, the human antimicrobial peptide LL37 has been shown to inhibit biofilm formation in this organism (Feng et al. 2013; Mohamed et al. 2017). Magainin 2, recently isolated from African clawed frog *Xenopus laevis* showed biofilm inhibition against *A. baumannii* at concentration as low as 4 μM, without showing any cytotoxicity in in vivo conditions, while remaining stable at physiological conditions (Kim et al. 2018). Other anti-biofilm agents of biological origin include molecules from Human Milk Oligosaccharides and plant extracts such as *Allium stipitatum* Regel and *Thymus daenensis*, which have showed biofilm inhibition/dispersion (Ackerman et al. 2017; Karunanidhi et al. 2018; Moghimi et al. 2018).

9.7 Conclusion

Since biofilm formation presents one of the major factors in *A. baumannii* infection and antimicrobial resistance, overcoming them can be a advantageous for future discovery of antimicrobials. However, lack of understanding of the interplay between cellular signaling, host factors and biofilm formation pathways prove to be the major bottleneck in progress towards this goal. Nevertheless, the recent encouraging developments have shown that combating biofilm formation can be an excellent strategy to fight multidrug resistant *A. baumannii* infections.

References

Ackerman DL, Craft KM, Doster RS, Weitkamp JH, Aronoff DM, Gaddy JA et al (2017) Antimicrobial and antibiofilm activity of human milk oligosaccharides against Streptococcus agalactiae, Staphylococcus aureus, and *Acinetobacter baumannii*. ACS Infect Dis 4:315–324

Bahador A, Taheri M, Pourakbari B, Hashemizadeh Z, Rostami H, Mansoori N et al (2013) Emergence of rifampicin, tigecycline, and colistin-resistant *Acinetobacter baumannii* in Iran; spreading of MDR strains of novel international clone variants. Microb Drug Resist 19:397–406

Bales PM, Renke EM, May SL, Shen Y, Nelson DC (2013) Purification and characterization of biofilm-associated EPS exopolysaccharides from ESKAPE organisms and other pathogens. PLoS One 8:e67950

Bentancor LV, Routray A, Bozkurt-Guzel C, Camacho-Peiro A, Pier GB, Maira-Litrán T (2012) Evaluation of the trimeric autotransporter Ata as a vaccine candidate against *Acinetobacter baumannii* infections. Infect Immun 80:3381–3388

Bhargava N, Sharma P, Capalash N (2010) Quorum sensing in Acinetobacter: an emerging pathogen. Crit Rev Microbiol 36:349–360

Branda SS, Vik S, Friedman L, Kolter R (2005) Biofilms: the matrix revisited. Trends Microbiol 13:20–26

Brossard KA, Campagnari AA (2012) The *Acinetobacter baumannii* biofilm-associated protein plays a role in adherence to human epithelial cells. Infect Immun 80:228–233

Cerqueira GM, Kostoulias X, Khoo C, Aibinu I, Qu Y, Traven A et al (2014) A global virulence regulator in *Acinetobacter baumannii* and its control of the phenylacetic acid catabolic pathway. J Infect Dis 210:46–55

Choi CH, Lee JS, Lee YC, Park TI, Lee JC (2008) *Acinetobacter baumannii* invades epithelial cells and outer membrane protein A mediates interactions with epithelial cells. BMC Microbiol 8:216

Choi AHK, Slamti L, Avci FY, Pier GB, Maira-Litrán T (2009) The pgaABCD locus of *Acinetobacter baumannii* encodes the production of poly-β-1-6-N-acetylglucosamine, which is critical for biofilm formation. J Bacteriol 191:5953–5963

Costerton W, Veeh R, Shirtliff M, Pasmore M, Post C, Ehrlich G (2003) The application of biofilm science to the study and control of chronic bacterial infections. J Clin Invest 112:1466–1477

Dijkshoorn L, Nemec A, Seifert H (2007) An increasing threat in hospitals: multi-drug resistant *Acinetobacter baumannii*. Nat Rev Microbiol 5:939–951

Doi Y, Murray GL, Peleg AY (2015) *Acinetobacter baumannii*: evolution of antimicrobial resistance - treatment options in seminars in respiratory and critical care medicine, p 85

Eijkelkamp BA, Stroeher UH, Hassan KA, Elbourne LDH, Paulsen IT, Brown MH (2013) H-NS plays a role in expression of *Acinetobacter baumannii* virulence features. Infect Immun 81:2574–2583

Espinal P, Martí S, Vila J (2012) Effect of biofilm formation on the survival of *Acinetobacter baumannii* on dry surfaces. J Hosp Infect 80:56–60

Feng X, Sambanthamoorthy K, Palys T, Paranavitana C (2013) The human antimicrobial peptide LL-37 and its fragments possess both antimicrobial and antibiofilm activities against multidrug-resistant *Acinetobacter baumannii*. Peptides 49:131–137

Gaddy JA, Tomaras AP, Actis LA (2009) The *Acinetobacter baumannii* 19606 OmpA protein plays a role in biofilm formation on abiotic surfaces and in the interaction of this pathogen with eukaryotic cells. Infect Immun 77:3150–3160

Giamarellou H, Antoniadou A, Kanellakopoulou K (2008) *Acinetobacter baumannii*: a universal threat to public health? Int J Antimicrob Agents 32:106–119

Gopal R, Kim YG, Lee JH, Lee SK, Chae JD, Son BK et al (2014) Synergistic effects and antibiofilm properties of chimeric peptides against multidrug-resistant *Acinetobacter baumannii* strains. Antimicrob Agents Chemother 58:1622–1629

Høiby N, Ciofu O, Bjarnsholt T (2010) *Pseudomonas aeruginosa* biofilms in cystic fibrosis. Future Microbiol 5:1663–1674

Hughes KA, Sutherland IW, Clark J, Jones MV (1998) Bacteriophage and associated polysaccharide depolymerases – novel tools for study of bacterial biofilms. J Appl Microbiol 85:583–590

Irie Y, Parsek MR (2008) Quorum sensing and microbial biofilms. In: Bacterial biofilms. Springer, Heidelberg, pp 67–84

Iwashkiw JA, Seper A, Weber BS, Scott NE, Vinogradov E, Stratilo C et al (2012) Identification of a general O-linked protein glycosylation system in *Acinetobacter baumannii* and its role in virulence and biofilm formation. PLoS Pathog 8:e1002758

Karunanidhi A, Ghaznavi-Rad E, Hamat RA, Pichika MR, Lung LTT, Mohd Fauzi F et al (2018) Antibacterial and antibiofilm activities of nonpolar extracts of Allium stipitatum regel against multidrug resistant Bacteria. Biomed Res Int:9845075

Kim M, Kang N, Ko S, Park J, Park E, Shin D et al (2018) Antibacterial and antibiofilm activity and mode of action of Magainin 2 against drug-resistant *Acinetobacter baumannii*. Int J Mol Sci 19:3041

La Fuente-Núñez C, Reffuveille F, Haney EF, Straus SK, Hancock REW (2014) Broad-spectrum anti-biofilm peptide that targets a cellular stress response. PLoS Pathog 10:e1004152

Lees-Miller RG, Iwashkiw JA, Scott NE, Seper A, Vinogradov E, Schild S, Feldman MF (2013) A common pathway for O-linked protein-glycosylation and synthesis of capsule in *Acinetobacter baumannii*. Mol Microbiol 89:816–830

Loc-Carrillo C, Abedon ST (2011) Pros and cons of phage therapy. Bacteriophage 1:111–114

Longo F, Vuotto C, Donelli G (2014) Biofilm formation in *Acinetobacter baumannii*. New Microbiol 37:119–127

Moghimi R, Aliahmadi A, Rafati H, Abtahi HR, Amini S, Feizabadi MM (2018) Antibacterial and anti-biofilm activity of nanoemulsion of Thymus daenensis oil against multi-drug resistant *Acinetobacter baumannii*. J Mol Liq 265:765–770

Mohamed MF, Brezden A, Mohammad H, Chmielewski J, Seleem MN (2017) A short D-enantiomeric antimicrobial peptide with potent immunomodulatory and antibiofilm activity against multidrug-resistant *Pseudomonas aeruginosa* and *Acinetobacter baumannii*. Sci Rep 7:6953

Peleg AY, Seifert H, Paterson DL (2008) *Acinetobacter baumannii*: emergence of a successful pathogen. Clin Microbiol Rev 21:538–582

Perez F, Hujer AM, Hujer KM, Decker BK, Rather PN, Bonomo RA (2007) Global challenge of multidrug-resistant *Acinetobacter baumannii*. Antimicrob Agents Chemother 51:3471–3484

Poirel L, Nordmann P (2006) Carbapenem resistance in *Acinetobacter baumannii*: mechanisms and epidemiology. Clin Microbiol Infect 12:826–836

Roca I, Espinal P, Vila-Farrés X, Vila J (2012) The *Acinetobacter baumannii* oxymoron: commensal hospital dweller turned pan-drug-resistant menace. Front Microbiol 23:148

Rodríguez-Baño J, Martí S, Soto S, Fernández-Cuenca F, Cisneros JM, Pachón J et al (2008) Biofilm formation in *Acinetobacter baumannii*: associated features and clinical implications. Clin Microbiol Infect 14:276–278

Sambanthamoorthy K, Luo C, Pattabiraman N, Feng X, Koestler B, Waters CM et al (2014) Identification of small molecules inhibiting diguanylate cyclases to control bacterial biofilm development. Biofouling 30:17–28

Saranathan R, Pagal S, Sawant AR, Tomar A, Madhangi M, Sah S et al (2017) Disruption of tetR type regulator adeN by mobile genetic element confers elevated virulence in *Acinetobacter baumannii*. Virulence 8:1316–1334

Stoodley P, Hall-Stoodley L, Lappin-Scott HM (2001) Detachment, surface migration, and other dynamic behavior in bacterial biofilms revealed by digital time-lapse imaging. Methods Enzymol 337:306–319

Stowe SD, Thompson RJ, Peng L, Su Z, Blackledge MS, Draughn GL et al (2015) Membrane-permeabilizing activity of reverse-amide 2-aminoimidazole antibiofilm agents against *Acinetobacter baumannii*. Curr Drug Deliv 12:223–230

Sutherland I (2001) Biofilm exopolysaccharides: a strong and sticky framework. Microbiology 147:3–9

Tomaras AP, Dorsey CW, Edelmann RE, Actis LA (2003) Attachment to and biofilm formation on abiotic surfaces by *Acinetobacter baumannii*: involvement of a novel chaperone-usher pili assembly system. Microbiology 149:3473–3484

Tomaras AP, Flagler MJ, Dorsey CW, Gaddy JA, Actis LA (2008) Characterization of a two-component regulatory system from *Acinetobacter baumannii* that controls biofilm formation and cellular morphology. Microbiology 154:3398–3409

Vidal R, Dominguez M, Urrutia H, Bello H, Gonzalez G, Garcia A et al (1996) Biofilm formation by *Acinetobacter baumannii*. Microbios 86:49–58

Vu B, Chen M, Crawford RJ, Ivanova EP (2009) Bacterial extracellular polysaccharides involved in biofilm formation. Molecules 14:2535–2554

Zarrilli R, Giannouli M, Tomasone F, Triassi M, Tsakris A (2009) Carbapenem resistance in *Acinetobacter baumannii*: the molecular epidemic features of an emerging problem in health care facilities. J Infect Dev Ctries 3:335–341

Mechanism of Bacterial Co-resistance

10

Piyush Baindara

Abstract

Co-resistance characterized by the transfer of numerous resistant genes into the same bacteria and by the occurrence of mutations in different genetic loci which affects different antimicrobials. Isolates with the co-resistant gene are more dormant in nature as offsetting events avoid fitness cost of co-resistance. Co-resistant isolates might be selected by different antimicrobial however, a single agent might also select co-resistant isolates which known as co-selection. Both co-resistance and co-selection play a vital role in increasing the chances of bacterial persistence and resistant genes. Therefore, it is worth to consider both, co-resistance and co-selection when scheming strategies to fight against antimicrobial resistance. However, in the recent past, rapid increase has been observed in antimicrobial resistance and we have already attained a crisis where several antibiotics are no longer effective against bacterial infections. In this alarming scenario, the emergence of co-resistance could make the situation worse while a steady decline has been observed in the discovery of new and effective antibiotics during the last few decades. Therefore, there is an urgent need for efficient alternatives to conventional antibiotics to overcome drug resistance and co-resistance problem.

Keywords

Co-resistance · Cross-resistance · Co-selection · Selection pressure · Antimicrobial

P. Baindara (✉)
Department of Microbiology and Immunology, University of Arkansas for Medical Sciences, Little Rock, AR, USA
e-mail: piyush.baindara@gmail.com; PBaindara@uams.edu

S. M. Mandal, D. Paul (eds.), *Bacterial Adaptation to Co-resistance*,
https://doi.org/10.1007/978-981-13-8503-2_10

10.1 Introduction

The emergence of resistant in pathogenic bacterial against multiple drugs or antimicrobial agents rapidly increased in recent past (World Health Organisation (WHO) 2014; Marston et al. 2016). Drug resistant becomes a major public health hazard as no potential antimicrobial agents available to treat the infections caused by drug resistant bacteria (Wright 2017; Michael et al. 2014). Antimicrobial agents are naturally produced by a different organism and capable to kill or inhibit the bacterial growth. Antimicrobial action started as soon as antimicrobial agents administered into the patients infected with pathogenic bacteria. Notably, the patient's normal microflora is adversely affected by antimicrobial agents which leads to a selective pressure over the bacterial evolution (Goldstein 2011; Langdon et al. 2016). Under the antibiotic pressure, bacterial variants with diminished antimicrobial susceptibility get selected and later by the acquisition of mutations, resistance genes evolved (Sykes 2010). Slowly with continuous uncontrolled uses of antibiotics, these mutations or resistance genes aggregated in some pathogenic bacterial species. Accumulation of resistance genes appeared as a complex resistant phenotype in pathogenic bacteria which is rapidly documented throughout the world (Ritchie et al. 2015; Yelin and Kishony 2018; Durão et al. 2018). By having resistant phenotypes against multiple antibiotics these bacteria termed as multidrug resistant bacteria while the capacity of survival under pressure of various antimicrobial avail the chances for dissemination (Hernando-Amado et al. 2017; Martínez and Baquero 2014). Further, variation in the usage pattern in different clinical or non-clinical settings might play a role in the increased frequency of emergence and persistence of multidrug resistant isolates, following a co-selection process which results in the development of co-resistance (Perry et al. 2014; Bal et al. 2010).

10.2 Different Forms of Resistance

Several different definitions had been used for drug resistant bacteria in the recent era of drug resistant. Recently, the European Center for Disease Prevention and Control (ECDC) and the Centre for Disease Control and Prevention (CDC) redefined drug resistant bacteria as multidrug resistant (MDR), extensively drug resistant (XDR) and pandrug resistant (PDR) (Magiorakos et al. 2012). These definitions are defined by taking the commonly found resistant microorganisms in the reference which found in health care related infections caused by *Staphylococcus aureus*, *Pseudomonas aeruginosa*, *Acinetobacter baumannii*, *Enterococcus* spp. and *Enterobacteriaceae* (Marasini et al. 2015; Boucher et al. 2009). These definitions for different resistant bacteria is especially based on their phenotype but resistant is transmitted genetically and the specific resistant mechanism is involved during the development of a specific type of resistance.

Co-resistance, cross-resistance and pleiotropic resistance are major definitions comes into consideration during a discussion of multidrug-resistant phenotypes (Table 10.1). Co-resistant defined as the process when several resistant genes

Table 10.1 Definitions of different forms of drug resistance

Term	Definition
Co-resistance	Different resistance mechanisms encoded by mutated or acquired genes
Cross-resistance	Mutated or acquired resistance genes affecting antimicrobials agents from the same class
Pleiotropic resistance	A resistance mechanism affecting several antimicrobial classes owing to the same genetic event such as mutation or acquisition of a resistance gene
Multidrug resistant (MDR)	Non-susceptibility to at least one agent in three or more antimicrobial categories
Extremely drug resistant (XDR)	Non-susceptibility to at least one agent in all but two or fewer antimicrobial categories
Pan-drug resistance (PDR)	Non-susceptibility to all agents in all antimicrobial categories

transferred and physically co-located on the same genetic element (plasmid or transposon) of same bacterial isolate and each resistant gene conferring resistance for different antibacterial agents (Yelin and Kishony 2018; Pal et al. 2015). Sometimes mutations in different genetic loci which involved in resistance against different antibacterial agents also developed as co-resistant (Cantón and Ruiz-Garbajosa 2011).

Cross-resistance involved acquisition/mutations of different resistant genes affecting the antibacterial agents of the same class or a single mechanism confer resistance to different antibacterial agents simultaneously (Chapman 2003). Methicillin resistant *S. aureus* (MRSA) is a good example to understand these terms. MRSA considered as multidrug resistant as it shows co-resistance against different antibacterial classes including beta-lactams, macrolides, aminoglycosides or fluoroquinolones. MRSA also represents an example of cross-resistance as it showed resistance within specific antibiotic classes which includes all fluoroquinolones owing to the mutations in topoisomerases genes or against macrolides by the presence of several erm, mef or mrs genes (Wendlandt et al. 2015).

When the mutations or acquisition of a single resistance gene results in resistance against several antimicrobial classes, the process defined as pleiotropic resistance. *P. aeruginosa* represents an excellent example of pleiotropic resistance when mutations in efflux pump regulatory genes result in its up-regulation which shows pleiotropic resistance phenotype against some beta-lactams, fluoroquinolones, and aninoglycosides (Livermore 2001). Another good example is cfr gene mutation in *S. aureus*, which is liable for production of methylase that affects ribosomes, results in resistance against linezolid, chloramphenicol, lincosamides, pleuromutilins and streptogramin A (Long et al. 2006). These definitions may further be amended by considering the evolutionary point of view which is important for prediction of antimicrobial resistance and their clinical impact in the population. Genetic aspect of resistance should also be considered along with the antimicrobial uses to fight against the emergence of multidrug resistant pathogens.

10.3 Selection and Co-selection Pressure Leads to Drug Resistance

Co-selection of antibiotics is the process by which antibiotic resistance is maintained in the bacterial population, even in the absence of antibiotics. Co-resistance and cross-resistance mechanisms are the major driving force behind co-selection of antibiotics (Fig. 10.1). Other than co-resistance and cross-resistance mechanisms, co-regulatory mechanisms also involved in the co-selection process. These mechanisms include the transfer of multiple resistant genes, which are regulated by a single regulatory gene and argue resistance against different antibacterial agents (Seiler and Berendonk 2012; Baker-Austin et al. 2006). A regulatory protein CzcR, co-regulates resistance against carbapenems (a class of last resort antibiotics) by suppressing the expression of OprD porin which facilitates entry of these antibiotics to the bacterial cell (Perron et al. 2004; Dieppois et al. 2012).

Although the frequency is variable, a resistant subpopulation is always present within the susceptible bacterial population under the selected antibiotic exposure. These resistant subpopulations can be produced between the lower and upper limits of drug concentrations where susceptible or resistant subpopulation is killed or inhibited, respectively. Further, fluctuations in the concentrations of antimicrobial agents during therapeutic administration leads to dis-balance of the natural pharmacokinetic process which is one of the major reason behind the development of

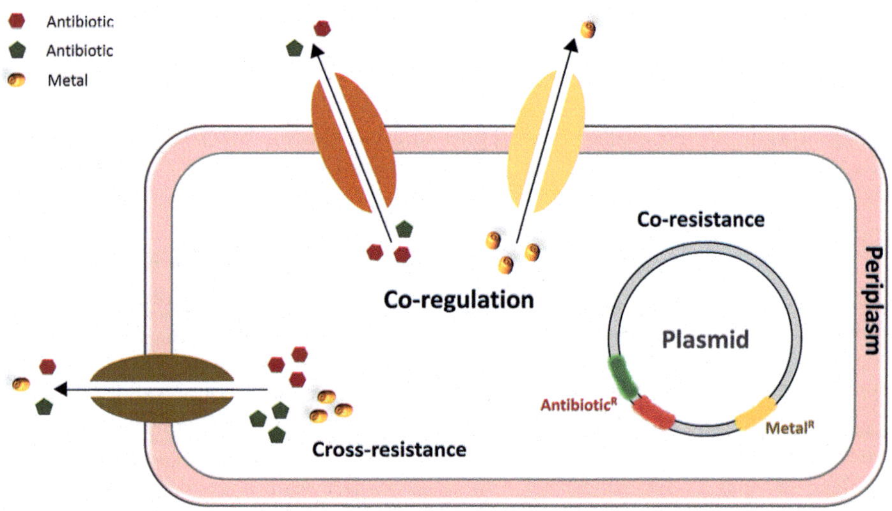

Fig. 10.1 Mechanisms of co-resistance, cross-resistance, and co-regulation or co-expression of antibiotic/antibiotic and metal resistance. Co-resistance occurs when resistance genes for different antibiotics and metals are physically co-located on the same genetic element as a plasmid. When common resistance machinery devoted to resistance to both antibiotics and metal, a phenomenon termed as cross-resistance. Co-regulation or co-expression happened when expression of resistance machinery antibiotics and metals are controlled by a common regulator at the same time. (Figure modified from Baker-Austin et al. 2006)

selection pressure (Andersson and Hughes 2014; Knöppel et al. 2017). Therefore, pharmacokinetics and pharmacodynamics of antibiotics should be optimized to achieve the adequate concentrations for treatment or to reduce the exposure of antimicrobial to the bacteria, thus minimize the chances of resistant development (Muller et al. 2015).

In the presence of different antibiotics or where the antimicrobial drug uses is more frequent like hospital settings, the bacteria with more resistant gene acquisition have higher opportunities of being selected. This process is termed as 'genetic capitalism', the resistant tend to become more resistant (Oz et al. 2014). This is how different antimicrobials might select a single multidrug resistant isolate and a single antimicrobial might select different multidrug resistant isolates. This process defined as co-selection and illustrated in Fig. 10.1.

10.4 Role of Co-resistance in Co-selection of Antibiotics

Understanding the role co-resistance in the co-selection of antibiotics is worth important including genetic linkage of resistant genes and their association with different mobile genetic elements such as plasmids, transposons or integrons. Horizontal gene transfer plays an important role in the transfer of resistant genes between bacteria and this could be explained by understanding the co-resistant mechanism (Fig. 10.1) (Bengtsson-Palme et al. 2018; Bengtsson-Palme and Larsson 2015; Martínez et al. 2015). Biocides or metal resistant and consequences of their interaction with antibiotic resistant is recently known (Scientific Committe on Emerging and Newly Identified Health Risks 2012), however phenomenon of antibiotic resistant was already identified even before the first antibiotic penicillin became widely available (Ventola 2015).

Co-existence of antibiotic and metal resistance genes are known for decades particularly on plasmids (Novick and Roth 1968; Foster 1983). Mercury and antibiotic resistance caused by integron containing mercury transposon is well known and documented since long in a diverse range of bacterial habitats including oral and fecal micro flora of primates (Wireman et al. 1997). Genetic linkage of mercury and antibiotic resistance also found on plasmids as co-transferred during mating between *Enterobacteriaceae* and recipients (Summers et al. 1993). Plasmids from *Salmonella abortus* equi strains found to have co-resistance for ampicillin along with multiple metals including arsenic, chromium, cadmium, and mercury (Ghosh et al. 2000). Recent, studies on mercury exposure to the antibiotic resistant bacteria that colonize in the gastrointestinal tract of the mummichog (*Fundulus heteroclitus*), a small, estuarine fish also revealed the co-selection of mercury of multiple antibiotic resistant genes which further analyzed and confirmed for the position of resistance genes on plasmids (Lloyd et al. 2016). Copper resistance gene *tcrB* also explored to have genetic linkage with macrolide and glycopeptide resistant genes involving horizontal gene transfer (Amachawadi et al. 2011, 2013; Hasman and Aarestrup 2002). In a different study, *oqxAB*, a resistance gene for quaternary ammonium compounds, chlorhexidine, fluoroquinolones etc. reported to have co-existence on the same plasmid with $bla_{CTX\text{-}M}$, *pco* and *sil* resistance genes which confers resistance for β-lactamases, copper and silver respectively (Fang et al. 2016; Mourão et al. 2015). Methicillin resistant gene mec co-existed on the same plasmid with czr gene, responsible for zinc and cadmium can be horizontally transferred and co-selected for methicillin resistance in *S. aureus* (Cavaco et al. 2010, 2011).

In addition, resistant genes can be present within large elements like integrons that exist in transposons, thus transfer of large elements also leads to transfer of resistant genes. Tn21 and Tn21 like transposons are a good example ofe a co-resistance mechanism where mercury resistance operon is co-existed with and integron which have multiple antibiotic resistant genes (Wireman et al. 1997; Bass et al. 1999; Liebert et al. 1999). Tn21 reported for the acquisition of the class 1 integron as crucial for antibiotic resistance genes such as aadA1 which confers resistance for streptomycin (Essa et al. 2003). Class 1 integrons are reported for the presence of trimethoprim resistance genes also (Integrons 2006; Fluit and Schmitz 2004). especially, class 1 integrons involved in human infection also harbor gene cassettes encoding β-lactamase resistance genes (Integrons 2006), while new genes resistant for aminoglycosides have been discovered in recent past (Partridge et al. 2002). Class 2 integrons have an array of gene cassettes containing dihydrofolate reductase (*dfrA1*), streptothricin acetyltransferase (*sat1*), and aminoglycoside adenyltransferase (*aadA1*), which are responsible for resistance to trimethoprim, streptothricin and streptomycin/spectinomycin, respectively (Xu et al. 2009; Hansson et al. 2002). Class 4 integrons also reported to have genetic elements confers resistance to chloramphenicol and fosfomycin (Fluit and Schmitz 2004). Thus, transposons have a straight forward role in co-resistance as a mechanism for the co-selection of the antibiotics. Notably, integrons have been documented for their presence in both pathogenic and non-pathogenic Gram-positive and Gram-negative bacteria (Deng et al. 2015).

Newly discovered colistin resistance gene mcr-1, has also been present in copper resistant isolates and suggested restriction for uses of copper as a growth promoter in animals (Campos et al. 2016). According to recent reports most of the sequenced bacterial strains carry resistance genes involved either biocide, metal, and antibiotic resistance mechanism, however only one out of six bacterial strains carry the resistance genes for both antibiotics and metals. Copper, mercury, silver, arsenic, antimony, cobalt, nickel, cadmium, iron and zinc resistant genes are most commonly co-occur with resistant genes for different classes of antibiotics like sulphonamides, β-lactams, amphenicols, tetracyclines and aminoglycosides (Pal et al. 2015).

Especially, mercury and arsenic resistance genes are frequently presented with integron associated integrases but only mercury resistant gene is highly co-existed with antibiotic resistance genes (Pal et al. 2015). Notably, this common linkage between mercury and antibiotic resistant genes has been reported in bacterial from a wide range of complex environments, including fish gastrointestinal tracts (Akinbowale et al. 2007), fresh water micrococsms (Stepanauskas et al. 2006), oral and fecal microflora of primates (Wireman et al. 1997), oral microflora of patients with amalgam fillings (Summers et al. 1993) and mine sediments (Nemergut et al. 2004). This diverse co-existence of antibiotic and metal resistant genes suggested a diverse and wide range of bacterial adaption to co-resistance. Further, cadmium and zinc resistance genes revealed occasional co-occurrence with antibiotics like macrolides and aminoglycosides (Pal et al. 2015). Thus, not only copper and mercury but other metals like cadmium and zinc could also potentially co-select for antibiotic resistance in bacteria. Accordingly, uncontrolled use of metals should be restricted along with antibiotics to combat the bacterial adaption to co-resistance.

Other than metal or biocide resistance gene which are frequently existed in bacterial isolates from diverse environments, antibiotic resistant gene is more common bacteria isolated from human and domestic animals (Pal et al. 2016; Li et al. 2016). It is a well known fact that humans and domestic animals are two microbial niches which are frequently and purposefully exposed to a strong selection pressure of different type of antibiotic which is the major reason behind the co-occurrence of resistant genes. These facts do not directly provide evidence in favor of metal selection as the driving force behind the evolution of co-resistant bacteria, exposure to metals. However, exposure to metals seemingly has the capacity to deliver the selective advantages to such bacteria via co-resistance mechanism.

10.5 Co-regulation: A Driving Force behind the Development of Resistance

Several transcriptional and translational responses during the co-exposure of antibiotics and metals linked and result into a coordinated response against antibiotics or metals (Baker-Austin et al. 2006; Pal et al. 2017). Multiple recent post-genomic studies confirmed co-regulation as a driving force behind the development of antibiotic resistance determinants. These co-regulatory process involved the regulation of virulence factors, antimicrobial peptides, alginate production, and quorum sensing along with regulation of antibiotic determinants (Navarre et al. 2005; Kong 2005; Balasubramanian et al. 2011).

Chemostat-cultured E. coli strain MG1655 showed up-regulation of mdtABC operon as a response to stress caused by zinc exposure as revealed by microarray analysis (Lee et al. 2005). This up-regulation of mdtABC operon results in overexpression of nodulation-cell division type efflux system which associated with conferring resistance to novobiocin and bile-salt component deoxycholate (Lee et al. 2005; Baranova and Nikaido 2002). Plasmid derived *E. coli* gene robA, (member of XylS-AraC subfamily of DNA-binding proteins) shows an enhanced spectrum of resistance against different antibiotic including tetracycline, chloramphenicol, novobiocin, and metals like silver, cadmium and mercury, upon overexpression (Nakajima et al. 1995). OprD2, a membrane porin of *Pseudomonas aeruginosa*, responsible for carbapenem resistance showed a decrease in expression upon exposure to the zinc eluted from silicone latex urinary catheters and successively results in enhanced resistance to carbapenem class of antibiotics (Conejo et al. 2003). Next, in a recent study, it has been confirmed that presence of arsenate, copper, and zinc associated in stimulating the resistance towards tetracycline in strain LSJC7, a Gram-negative member of the family *Enterobacteriaceae* (Chen et al. 2015*). Heavy* *metal derived co-selection of antibiotics resistance also confirmed in bacterial isolated derived from soil and water bodies which impact mainly aquaculture and agriculture practices* (Seiler and Berendonk 2012). *Further, co-selection for the antibiotics resistance is also observed* among bacteria exposed to biocides used as disinfectants, antiseptics, and preservatives, which results in co-resistance in zoonotic agents that commonly occur in livestock and could also transmit to human populations via food chain (Wales and Davies 2015).

10.6 Resistance Genes as Units for Selection

Co-resistance, the presence of multiples resistance genes, certainly beneficial for bacteria under selective pressure of different antibiotics. It is a well known fact that the emergence of drug resistance is a direct consequence of uncontrolled or unrestricted use of antibiotics while this resistance is developed by mutational events in a Darwinian manner (Sykes 2010). However, selection or co-selection mechanism does not involve mutations while transmitted by lateral transfer (Cantón and Morosini 2011). Antibiotic resistant genes involved in conventional functions of bacterial found in the diverse complex environment (Cantón 2009). Resistant gene also present in antimicrobial producing bacteria, however, resistant traits appeared only when this gene present in different microorganisms other than which naturally produced these resistant genes. Naturally occurring resistance genes acquired and expressing resistant phenotypes in pathogens. Resistant genes along with mutated genes are investigated recently and have increasing prominence in clinical settings (Cantón 2009; Crofts et al. 2017).

In recent past, studies revealed human gut microbiome as a large repertoire of antibiotic resistant genes (van Schaik 2015; Carlet 2012). Further, microarray, mutant library screen, and mutation frequency analysis revealed the presence of resistant genes in animal and insect vectors also (Lamendella et al. 2011; Allen et al. 2009). Different units of antimicrobial resistant employed during the development of co-resistant or resistant shown in (Table 10.2). As the importance of resistance units for selection process unveiled in the last few decades, the mechanism of the selection process should be carefully applied and understand while designing strategies to fight against antimicrobial resistance.

Additionally, expression and maintenance of resistant gene determinants are also important that is regulated by their genetic linkage with other resistant units. Certain insertion sequences like IS26 or ISCR1 has been confirmed to have an important role in the mobilization of resistant genes and also in their phenotypic expression (Shahid 2010). Integrons also played important role in maintenance and regulation of resistant genes while as an inherited element, integrons are able to arrest and assimilate numerous resistant genes (Deng et al. 2015; Hall 2013). Integrons are found linked with transposons or plasmids but also spotted in the chromosomes with large arrangements of gene cassettes. Chromosomal integrons can also exchange genetic material with pathogenic organisms. Fresh water biofilm studies showed a vibrant exchange of qac gene cassettes from class 1 integrons to other

Table 10.2 Different resistance units for selection

Resistance unit	Example	Description
Gene	*ermA*	Erythromycin resistance methylase gene in *Streptococcus pneumoniae*
	mecA	β-lactams resistance gene in Staphylococci encoding PBP2a
	oqxAB	A resistance gene for quaternary ammonium compounds, chlorhexidine, and fluoroquinolones
Insertion sequence	IS26, ISCR1	An insertion sequence involved in the mobilization and expression of different resistance genes, including bla_{ESBLs}
Integrons	Class-1 integrons	Genetic structures able to recruit different resistance genes such as carbapanemases and aminoglycoside modifying enzymes.
	Class 2 integrons	Genetic structures having an array of gene cassettes responsible for resistance to trimethoprim, streptothricin, and streptomycin.
	Class 4 integrons	Genetic structures having an array of gene cassettes responsible for resistance to chloramphenicol and fosfomycin.
Transposons	Tn*1546*	A transposon having vancomycin resistance genes in enterococci.
	Tn*21*	Confers resistance for streptomycin by the acquisition of the class 1 integron.
Sequence type	ST*116*, *ST6*	Shows high resistance against aminoglycosides in animal and human isolates respectively.
	ST*131*	A high-risk clone of Escherichia coli involved in the spread of the CTX-M-15, an extended-spectrum β-lactamase
	ST*258*	A high-risk clone of *Klebsiella pneumoniae* involved in the spread of KPC (Klebsiella pneumoniae carbapenemase) carbapenemases
	ST*147*	Reported to produce, $bla_{NDM\text{-}1}$ and $bla_{CTX\text{-}M\text{-}15}$ genes in *Klebsiella pneumoniae* and showed resistant to almost all antibiotics
Plasmids	IncFII	Associated with the dissemination of $bla_{CTX\text{-}M\text{-}15}$, shows resistance to almost all antibiotics
Clonal complex	CC17	Associated with the dissemination of ampicillin and vancomycin in *Enterococcus faecium*
	CC1, CC2, CC3	Associated with multidrug resistance in *A. baumannii*

integrons of bacteria belongs to diverse habitats and develops resistance to ammonium quaternary compounds (Farkas et al. 2013; Gillings et al. 2009).

Integrons and insertion sequences are not only restricted to environmental bacteria, but they are also reported more frequently from pathogenic and commensal organisms of human and animal origins (Bailey et al. 2011; Wang et al. 2018; Deng et al. 2011). An Australian study explored that class 1 integrons present in 10% of E. coli isolate from animals while in 23% from humans (Dawes et al. 2010). In a recent study co-located resistant genes and integrons along with bla SHV-12 plasmid were analyzed by PCR and sequencing. Results suggested the distribution of bla SHV-12 genes occurred by vertical and horizontal gene transfer along with IncI1 plasmid and exacerbate resistance to β-lactams, chloramphenicol, and streptomycin (Alonso et al. 2017). All these evidence suggest facilitating selection under pressure of antimicrobial uses.

10.7 High Risk Clones and Clonal Complexes as Units for Selection

In the present scenario of multidrug resistance emergence thorough various complex mechanism, research is focused on supra-levels of selection including high risk clones and clonal complexes (Table 10.2) (Willems et al. 2011; Woodford et al. 2011). Although the phenomenon of 'high risk clones' is limited to the defined bacterial population within the species, high risk clones show co-resistance in a wide range (García-Castillo et al. 2011; Abdouchakour et al. 2018; Treepong et al. 2018; Oliver et al. 2015).

Conservation of high risk clones in nature enhances the opportunity of aggregation different type of resistant gene which results in the development of co-resistance. Uropathogenic clinic isolate *E. coli* ST131 is reported to produce NDM-7, a novel NDM variant (Metallo-β-lactamase) (Cuzon et al. 2013). At the same time this isolate carrying other resistant genes like $bla_{\text{CTX-M-15}}$, $bla_{\text{SHV-11}}$, $bla_{\text{TEM-1}}$ and qnr_{S1}. The $bla_{\text{NDM-7}}$ gene was located on a conjugative IncX3-type plasmid bearing $bla_{\text{TEM-1}}$ and *qnr*S1 genes confirmed this isolate as highly resistant and virulent (hui et al. 2016). The ST131 *E. coli* strain is already known to carry $\text{bla}_{\text{CTX-M-15}}$, which encrypts the most efficacious extended-spectrum β-lactamase gene while also confirmed to produce carbapenemases and $bla_{\text{NDM-1}}$ along with other resistant genes (Price et al. 2013; Peirano et al. 2011; Mantengoli et al. 2011).

Furthermore, ST147 clonal complex in *Klebsiella pneumoniae* is reported to produce $bla_{\text{NDM-1}}$ and $\text{bla}_{\text{CTX-M-15}}$ genes confer resistant to almost all antibiotics tested with a high level of resistance to the ertapenem. Results suggested the worldwide distribution of this carbapenemase involving the emergence of IncR plasmid and success of ST147 clonal complex (Abderrahim et al. 2017). In a different study, ST105 clonal complex is also reported in *Klebsiella pneumoniae* which harbor $bla_{\text{NDM-1}}$ gene along with multiple resistant genes including $bla_{\text{IMP-4}}$, $bla_{\text{SHV-1}}$, $bla_{\text{TEM-1}}$, $bla_{\text{CTX-M-15}}$, qnr_{S1}, qnr_{B4}, and aac_{A4} (Zheng et al. 2016). One other clonal complex ST258 also recognized as a *Klebsiella pneumoniae*, carbapenemase producer (Samuelsen et al. 2009). Above explained examples may explain, how the co-existence of multiple resistance genes confers the co-selection process and thus the emergence of bacterial co-resistance to antimicrobials.

Next, three international clonal lineage found dominating the population structures of *A. baumannii* which are CC1 (European clone I comprising ST1, ST7, ST8, ST19 and ST20), CC2 (European clone II comprising ST2, ST45 and ST47) and CC3 (European clone III comprising ST3 and ST14) and thus displaying a multidrug resistance phenotype (Woodford et al. 2011).

Carbapenem resistant gene also reported in *Pseudomonas aeruginosa*, producing metallo-β-lactamases and belongs to a particularly high risk clonal complex, for example, STs 111, 233, 235, 357, 654 and 773 (Wright et al. 2015). In addition, ST111, ST175, and ST235 clonal complexes are also investigated and well represented multidrug resistance phenotypes (Cholley et al. 2011).

Enterococcus faecium genogroup CC17 and *E. faecalis* genogroup CC2 and CC9 represents the best example of high risk clones among the Gram-positive bacteria. Both groups showed multidrug resistance to several antibiotics, including vancomycin and confirmed well adapted to hospital and farm animal settings (Freitas et al. 2011; Ruiz-Garbajosa et al. 2006). High level resistance towards aminoglycosides is recently investigated in E. faecalis isolates from human, animals and environmental specimens. Results suggested, aac(6′)-Ie-aph(2″)-Ia and aph(3″)- IIIa genes responsible for the development of high level resistance against aminoglycosides while animal and human isolates found associated to ST116 and ST6, respectively (Ghebremedhin et al. 2014). Conflicting to the Enterococci, high risk clones of *S. aureus* are also evolving in the community, different from those conventionally come across in the hospital settings (David and Daum 2010; Otter and French 2010). Especially, *S. aureus* USA300 clone (ST9-IV) predominant in North America while community associated MRSA in Europe is considered to be evolved by clonal diversity (Vindel et al. 2014).

10.8 The Fitness Cost: An Essential Factor for Resistance Development

Fitness cost defined as the enhanced rate of survival and competitive ability in the presence of antimicrobials (Andersson and Hughes 2010). Evolution of resistance or co-resistance against antibiotics always carries a fitness cost, which is directly proportional to the persistence of the resistance phenotype in an antibiotic selective environment. Fitness cost also plays an important role in the development of resistance dynamics by creating selective pressure for respective antimicrobials when bacteria encounter an antibiotic free environment. Cost of resistance is highly variable while the responsible factors are not fully understood. Recent reports suggested that chromosomal resistance mutations have a higher cost of fitness than the resistance genes acquired via plasmids (Vogwill and Maclean 2015). This is the reason behind, why plasmids as a major source for resistant gene and resistance evolutions. It has been also reported that the cost of plasmid acquisition significantly increases with the range of resistance. This suggested a regulatory role of fitness cost on the evolution of multidrug resistance via plasmids (Vogwill and Maclean 2015; Melnyk et al. 2015).

Bacteria bears a fitness cost during the emergence of resistat mutants, either by mutation or lateral gene transfer. This fitness cost directly or indirectly affects the selection process and thus results in the persistence of resistance phenotypes. In theory, acquisition of multiple resistance genes increases the cost of fitness, however bacteria experience compensatory mechanisms to reduce the fitness cost and permitting the persistence of resistant phenotype (Melnyk et al. 2015; Roux et al. 2015).

10.9 Antimicrobial Uses and Co-resistance

The uncontrolled use of different antibiotics to treat the same bacterial population in a complex environment rapidly results in a resistance mechanism against the antibiotics. The upsurge in the occurrence of antibiotic resistance features is a defined matter of time which is evidently associated with the antimicrobial use. This is supported by well characterized examples like *S. aureus* which is known to acquire new resistance genes along with the exposure to new antimicrobials in clinical settings (Vestergaard et al. 2016). The rigorous use of some antibiotics over a long time results in universal resistant features in *S. aureus* like MRSA and fluoroquinolone resistance which appears after the introduction of ciprofloxacin in therapeutics (Stryjewski and Corey 2014). During the long time course, recent studies revealed the development of new resistant clones which shows a high degree of resistance in hospital originated MRSA isolates (Vindel et al. 2009; Liu et al. 2016). A new clonal complex, ST398 MRSA is also evolved with multiple resistance genes like cfr, vga(A) and vga(C) which shows resistance against chloramphenicol, lincosamides, pleuromutilins and streptogramin A (Argudín et al. 2011). Importantly, ST398 is also found associated in animals with antimicrobial uses (Conceição et al. 2017; Moreno-Grúa et al. 2018). Enterobacteriaceae also represents the other example for the evolution of high risk clones, resistance to cephalosporins and fluoroquinolones along with extended use of an antimicrobial in clinical settings (Church 2015). In a different study carbapenemase producing isolates of Enterobacteriaceae revealed to have resistance against multiple antibiotics like fluoroquinolones and aminoglycosides (Livermore et al. 2011).

10.10 Discussion

In the recent past, the rapid emergence of drug resistance comes in to picture with the increasing prevalence of bacterial isolates with multidrug resistance phenotype which leads to the development of co-resistance. Multidrug resistant bacterial isolates when exposed to several different selection processes like antibiotic selection pressure along with mutations, they acquired various resistance genes. The resistance gene selection mechanism can happen under a combination of different antimicrobial agents employing co-selection process, results in the development of co-resistance phenotypes. Maintenance cost of multiple resistance genes in a single isolate by the same genetic machinery, reduce fitness cost and increase the persistence chances of co-resistance phenotypes. This is how extended uses of antimicrobials under different clinical settings supports the opportunity for transmission of resistance genes and co-resistance as well. Therefore, strategies for the use of antimicrobial uses should consider and understand the co-resistance and co-selection mechanism before application into the clinical settings to circumvent the chances of co-resistance development.

Acknowledgement I thank Julee for stimulating discussions during the manuscript preparation.

Conflict of Interest There is no conflict of interest.

References

Abderrahim A, Djahmi N, Pujol C, Nedjai S, Bentakouk MC et al (2017) First case of NDM-1-producing *Klebsiella pneumoniae* in Annaba university hospital, Algeria. Microb Drug Resist Epub ahead of print 2017. https://doi.org/10.1089/mdr.2016.0213

Abdouchakour F, Aujoulat F, Licznar-Fajardo P, Marchandin H, Toubiana M et al (2018) Intraclonal variations of resistance and phenotype in Pseudomonas aeruginosa epidemic high-risk clone ST308: a key to success within a hospital? Int J Med Microbiol Epub ahead of print 2018. https://doi.org/10.1016/j.ijmm.2017.11.008

Akinbowale OL, Peng H, Grant P, Barton MD (2007) Antibiotic and heavy metal resistance in motile aeromonads and pseudomonads from rainbow trout (Oncorhynchus mykiss) farms in Australia. Int J Antimicrob Agents Epub ahead of print 2007. https://doi.org/10.1016/j.ijantimicag.2007.03.012

Allen HK, Cloud-Hansen KA, Wolinski JM, Guan C, Greene S et al (2009) Resident microbiota of the gypsy moth midgut harbors antibiotic resistance determinants. DNA Cell Biol Epub ahead of print 2009. https://doi.org/10.1089/dna.2008.0812

Alonso CA, Michael GB, Li J, Somalo S, Simón C et al (2017) Analysis of blaSHV-12-carrying Escherichia coli clones and plasmids from human, animal and food sources. J Antimicrob Chemother. Epub ahead of print 2017. https://doi.org/10.1093/jac/dkx024

Amachawadi RG, Shelton NW, Shi X, Vinasco J, Dritz SS et al (2011) Selection of fecal enterococci exhibiting tcrB-mediated copper resistance in pigs fed diets supplemented with copper. Appl Environ Microbiol. Epub ahead of print 2011. https://doi.org/10.1128/AEM.00364-11

Amachawadi RG, Scott HM, Alvarado CA, Mainini TR, Vinasco J et al (2013) Occurrence of the transferable copper resistance gene tcrB among fecal Enterococci of U.S. feedlot cattle fed copper-supplemented diets. Appl Environ Microbiol. Epub ahead of print 2013. https://doi.org/10.1128/AEM.00503-13

Andersson DI, Hughes D (2010) Antibiotic resistance and its cost: is it possible to reverse resistance? Nat Rev Microbiol 8:260–271

Andersson DI, Hughes D (2014) Microbiological effects of sublethal levels of antibiotics. Nat Rev Microbiol Epub ahead of print 2014. https://doi.org/10.1038/nrmicro3270

Argudín MA, Tenhagen BA, Fetsch A, Sachsenröder J, Käsbohrer A et al (2011) Virulence and resistance determinants of German Staphylococcus aureus ST398 isolates from nonhuman sources. Appl Environ Microbiol. Epub ahead of print 2011. https://doi.org/10.1128/AEM.02260-10

Bailey JK, Pinyon JL, Anantham S, Hall RM (2011) Distribution of the blaTEMgene and blaTEM-containing transposons in commensal Escherichia coli. J Antimicrob Chemother. Epub ahead of print 2011. https://doi.org/10.1093/jac/dkq529

Baker-Austin C, Wright MS, Stepanauskas R, McArthur JV (2006) Co-selection of antibiotic and metal resistance. Trends Microbiol Epub ahead of print 2006. https://doi.org/10.1016/j.tim.2006.02.006

Bal AM, Kumar A, Gould IM (2010) Antibiotic heterogeneity: from concept to practice. Ann N Y Acad Sci Epub ahead of print 2010. https://doi.org/10.1111/j.1749-6632.2010.05867.x

Balasubramanian D, Kong KF, Jayawardena SR, Leal SM, Sautter RT et al (2011) Co-regulation of β-lactam resistance, alginate production and quorum sensing in Pseudomonas aeruginosa. J Med Microbiol Epub ahead of print 2011. https://doi.org/10.1099/jmm.0.021600-0

Baranova N, Nikaido H (2002) The BaeSR two-component regulatory system activates transcription of the yegMNOB (mdtABCD) transporter gene cluster in Escherichia coli and increases its resistance to novobiocin and deoxycholate. J Bacteriol Epub ahead of print 2002. https://doi.org/10.1128/JB.184.15.4168-4176.2002

Bass L, Liebert CA, Lee MD, Summers AO, White DG et al (1999) Incidence and characterization of integrons, genetic elements mediating multiple-drug resistance, in avian Escherichia coli. Antimicrob Agents Chemother 43:2925–2929

Bengtsson-Palme J, Larsson DGJ (2015) Antibiotic resistance genes in the environment: prioritizing risks. Nat Rev Microbiol Epub ahead of print 2015. https://doi.org/10.1038/nrmicro3399-c1

Bengtsson-Palme J, Kristiansson E, Larsson DGJ (2018) Environmental factors influencing the development and spread of antibiotic resistance. FEMS Microbiol Rev. Epub ahead of print 2018. https://doi.org/10.1093/femsre/fux053

Boucher HW, Talbot GH, Bradley JS, Edwards JE, Gilbert D et al (2009) Bad bugs, no drugs: no ESKAPE! An update from the Infectious Diseases Society of America. Clin Infect Dis Epub ahead of print 2009. https://doi.org/10.1086/595011

Campos J, Cristino L, Peixe L, Antunes P (2016) MCR-1 in multidrug-resistant and copper-tolerant clinically relevant Salmonella 1,4,[5],12:i:- and S. Rissen clones in Portugal, 2011 to 2015. Eurosurveillance. Epub ahead of print 2016. https://doi.org/10.2807/1560-7917.ES.2016.21.26.30270

Cantón R (2009) Antibiotic resistance genes from the environment: a perspective through newly identified antibiotic resistance mechanisms in the clinical setting. Clin Microbiol Infect. Epub ahead of print 2009. https://doi.org/10.1111/j.1469-0691.2008.02679.x

Cantón R, Morosini MI (2011) Emergence and spread of antibiotic resistance following exposure to antibiotics. FEMS Microbiol Rev. Epub ahead of print 2011. https://doi.org/10.1111/j.1574-6976.2011.00295.x

Cantón R, Ruiz-Garbajosa P (2011) Co-resistance: an opportunity for the bacteria and resistance genes. Curr Opin Pharmacol Epub ahead of print 2011. https://doi.org/10.1016/j.coph.2011.07.007

Carlet J (2012) The gut is the epicentre of antibiotic resistance. Antimicrob Resist Infect Control. Epub ahead of print 2012. https://doi.org/10.1186/2047-2994-1-39

Cavaco LM, Hasman H, Stegger M, Andersen PS, Skov R et al (2010) Cloning and occurrence of czrC, a gene conferring cadmium and zinc resistance in methicillin-resistant Staphylococcus aureus CC398 isolates. Antimicrob Agents Chemother. Epub ahead of print 2010. https://doi.org/10.1128/AAC.00058-10

Cavaco LM, Hasman H, Aarestrup FM, Wagenaa JA, Graveland H et al (2011) Zinc resistance of Staphylococcus aureus of animal origin is strongly associated with methicillin resistance. Vet Microbiol. Epub ahead of print 2011. https://doi.org/10.1016/j.vetmic.2011.02.014

Chapman JS (2003) Disinfectant resistance mechanisms, cross-resistance, and co-resistance. In: International Biodeterioration and Biodegradation. 2003. Epub ahead of print 2003. https://doi.org/10.1016/S0964-8305(03)00044-1

Chen S, Li X, Sun G, Zhang Y, Su J et al (2015) Heavy metal induced antibiotic resistance in bacterium LSJC7. Int J Mol Sci Epub ahead of print 2015. https://doi.org/10.3390/ijms161023390

Cholley P, Thouverez M, Hocquet D, Van Der Mee-Marquet N, Talon D et al (2011) Most multidrug-resistant Pseudomonas aeruginosa isolates from hospitals in eastern France belong to a few clonal types. J Clin Microbiol. Epub ahead of print 2011. https://doi.org/10.1128/JCM.00102-11

Church LWP (2015) Enterobacteriaceae. In: Clinical Infectious Disease, Second Edition. Epub ahead of print 2015. https://doi.org/10.1017/CBO9781139855952.153

Conceição T, De Lencastre H, Aires-De-Sousa M (2017) Frequent isolation of methicillin resistant Staphylococcus aureus (MRSA) ST398 among healthy pigs in Portugal. PLoS One. Epub ahead of print 2017. https://doi.org/10.1371/journal.pone.0175340

Conejo CC, García I, Martínez-Martínez L, Picabea L, Pascual A (2003) Zinc eluted from siliconized latex urinary catheters decreases OprD expression, causing carbapenem resistance in

Pseudomonas aeruginosa. Antimicrob Agents Chemother. Epub ahead of print 2003. https://doi.org/10.1128/AAC.47.7.2313-2315.2003

Crofts TS, Gasparrini AJ, Dantas G (2017) Next-generation approaches to understand and combat the antibiotic resistome. Nat Rev Microbiol Epub ahead of print 2017. https://doi.org/10.1038/nrmicro.2017.28

Cuzon G, Bonnin RA, Nordmann P (2013) First identification of novel NDM Carbapenemase, NDM-7, in Escherichia coli in France. PLoS One. Epub ahead of print 2013. https://doi.org/10.1371/journal.pone.0061322

David MZ, Daum RS (2010) Community-associated methicillin-resistant Staphylococcus aureus: epidemiology and clinical consequences of an emerging epidemic. Clinical Microbiology Reviews Epub ahead of print 2010. https://doi.org/10.1128/CMR.00081-09

Dawes FE, Kuzevski A, Bettelheim KA, Hornitzky MA, Djordjevic SP et al (2010) Distribution of class 1 integrons with IS26-mediated deletions in their 3′-conserved segments in Escherichia coli of human and animal origin. PLoS One. Epub ahead of print 2010. https://doi.org/10.1371/journal.pone.0012754

Deng Y, Zeng Z, Chen S, He L, Liu Y et al (2011) Dissemination of IncFII plasmids carrying rmtB and qepA in Escherichia coli from pigs, farm workers and the environment. Clin Microbiol Infect. Epub ahead of print 2011. https://doi.org/10.1111/j.1469-0691.2011.03472.x

Deng Y, Bao X, Ji L, Chen L, Liu J et al (2015) Resistance integrons: class 1, 2 and 3 integrons. Ann Clin Microbiol Antimicrob. Epub ahead of print 2015. https://doi.org/10.1186/s12941-015-0100-6

Dieppois G, Ducret V, Caille O, Perron K (2012) The transcriptional regulator CzcR modulates antibiotic resistance and quorum sensing in pseudomonas aeruginosa. PLoS One. Epub ahead of print 2012. https://doi.org/10.1371/journal.pone.0038148

Durão P, Balbontín R, Gordo I (2018) Evolutionary mechanisms shaping the maintenance of antibiotic resistance. Trends Microbiol. Epub ahead of print 2018. https://doi.org/10.1016/j.tim.2018.01.005

Essa AMM, Julian DJ, Kidd SP, Brown NL, Hobman JL (2003) Mercury resistance determinants related to Tn21, Tn1696, and Tn5053 in enterobacteria from the preantibiotic era. Antimicrob Agents Chemother Epub ahead of print 2003. https://doi.org/10.1128/AAC.47.3.1115-1119.2003

Fang L, Li X, Li L, Li S, Liao X et al (2016) Co-spread of metal and antibiotic resistance within ST3-IncHI2 plasmids from E. coli isolates of food-producing animals. Sci Rep. Epub ahead of print 2016. https://doi.org/10.1038/srep25312

Farkas A, Butiuc-Keul A, Ciatarâş D, Neamţu C, Crăciunaş C et al (2013) Microbiological contamination and resistance genes in biofilms occurring during the drinking water treatment process. Sci Total Environ. Epub ahead of print 2013. https://doi.org/10.1016/j.scitotenv.2012.11.068

Fluit AC, Schmitz FJ (2004) Resistance integrons and super-integrons. Clin Microbiol Infect. Epub ahead of print 2004. https://doi.org/10.1111/j.1198-743X.2004.00858.x

Foster TJ (1983) Plasmid-determined resistance to antimicrobial drugs and toxic metal ions in Bacteria. Microbiol Mol Biol Rev Epub ahead of print 1983. https://doi.org/10.1016/s1473-3099(05)70243-7

Freitas AR, Coque TM, Novais C, Hammerum AM, Lester CH et al (2011) Human and swine hosts share vancomycin-resistant Enterococcus faecium CC17 and CC5 and Enterococcus faecalis CC2 clonal clusters harboring Tn1546 on indistinguishable plasmids. J Clin Microbiol Epub ahead of print 2011. https://doi.org/10.1128/JCM.01750-10

García-Castillo M, Del Campo R, Morosini MI, Riera E, Cabot G et al (2011) Wide dispersion of ST175 clone despite high genetic diversity of carbapenem-nonsusceptible Pseudomonas aeruginosa clinical strains in 16 Spanish hospitals. J Clin Microbiol. Epub ahead of print 2011. https://doi.org/10.1128/JCM.00753-11

Ghebremedhin B, Ibukun A, Odugbemi T, König W (2014) Clonal lineages of resistant (HLAR) & virulent Enterococcus faecalis isolates from diverse sources in Nigeria. Int J Infect Dis Epub ahead of print 2014. https://doi.org/10.1016/j.ijid.2014.03.1272

Ghosh A, Singh A, Ramteke PW, Singh VP (2000) Characterization of large plasmids encoding resistance to toxic heavy metals in Salmonella abortus equi. Biochem Biophys Res Commun Epub ahead of print 2000. https://doi.org/10.1006/bbrc.2000.2727
Gillings MR, Holley MP, Stokes HW (2009) Evidence for dynamic exchange of qac gene cassettes between class 1 integrons and other integrons in freshwater biofilms. FEMS Microbiol Lett Epub ahead of print 2009. https://doi.org/10.1111/j.1574-6968.2009.01646.x
Goldstein EJC (2011) Beyond the target pathogen: ecological effects of the hospital formulary. Curr Opin Infect Dis Epub ahead of print 2011. https://doi.org/10.1097/01.qco.0000393485.17894.4c
Hall RM (2013) Integrons. In: Brenner's Encyclopedia of Genetics: Second Edition. Epub ahead of print 2013. https://doi.org/10.1016/B978-0-12-374984-0.00804-4
Hansson K, Sundström L, Pelletier A, Roy PH (2002) IntI2 integron integrase in Tn7. J Bacteriol. Epub ahead of print 2002. https://doi.org/10.1128/JB.184.6.1712-1721.2002
Hasman H, Aarestrup FM (2002) tcrb, a gene conferring transferable copper resistance in Enterococcus faecium: occurrence, transferability, and linkage to macrolide and glycopeptide resistance. Antimicrob Agents Chemother. Epub ahead of print 2002. https://doi.org/10.1128/AAC.46.5.1410-1416.2002
Hernando-Amado S, Sanz-García F, Blanco P, Martínez JL (2017) Fitness costs associated with the acquisition of antibiotic resistance. Essays Biochem Epub ahead of print 2017. https://doi.org/10.1042/EBC20160057
Wang L hui, Liu P pan, Wei D dan, Liu Y, Wan LG et al (2016) Clinical isolates of uropathogenic Escherichia coli ST131 producing NDM-7 metallo-β-lactamase in China. Int J Antimicrob Agents. Epub ahead of print 2016. https://doi.org/10.1016/j.ijantimicag.2016.03.009
Integrons MD (2006) Agents of bacterial evolution. Nat Rev Microbiol. Epub ahead of print 2006. https://doi.org/10.1038/nrmicro1462
Knöppel A, Näsvall J, Andersson DI (2017) Evolution of antibiotic resistance without antibiotic exposure. Antimicrob Agents Chemother Epub ahead of print 2017. https://doi.org/10.1128/AAC.01495-17
Kong, K-F (2005) Characterization of the amp genes involved in the regulation of β-lactamase expression in *Pseudomonas aeruginosa* (2005). ProQuest ETD Collection for FIU. AAI3206025. https://digitalcommons.fiu.edu/dissertations/AAI3206025
Lamendella R, Santo Domingo JW, Ghosh S, Martinson J, Oerther DB (2011) Comparative fecal metagenomics unveils unique functional capacity of the swine gut. BMC Microbiol. Epub ahead of print 2011. https://doi.org/10.1186/1471-2180-11-103
Langdon A, Crook N, Dantas G (2016) The effects of antibiotics on the microbiome throughout development and alternative approaches for therapeutic modulation. Genome Med. Epub ahead of print 2016. https://doi.org/10.1186/s13073-016-0294-z
Lee LJ, Barrett JA, Poole RK (2005) Genome-wide transcriptional response of chemostat-cultured Escherichia coli to zinc. J Bacteriol. Epub ahead of print 2005. https://doi.org/10.1128/JB.187.3.1124-1134.2005
Li J, Pawitwar SS, Rosen BP (2016) The organoarsenical biocycle and the primordial antibiotic methylarsenite. Metallomics. Epub ahead of print 2016. https://doi.org/10.1039/c6mt00168h
Liebert CA, Hall RM, Summers AO (1999) Transposon Tn21, flagship of the floating genome. Microbiol Mol Biol Rev 63:507–522
Liu J, Chen D, Peters BM, Li L, Li B et al (2016) Staphylococcal chromosomal cassettes mec (SCCmec): a mobile genetic element in methicillin-resistant Staphylococcus aureus. Microbial Pathogenesis. Epub ahead of print 2016. https://doi.org/10.1016/j.micpath.2016.10.028
Livermore DM (2001) Of Pseudomonas, porins, pumps and carbapenems. J Antimicrob Chemother. Epub ahead of print 2001. https://doi.org/10.1093/jac/47.3.247
Livermore DM, Warner M, Mushtaq S, Doumith M, Zhang J et al (2011) What remains against carbapenem-resistant Enterobacteriaceae? Evaluation of chloramphenicol, ciprofloxacin, colistin, fosfomycin, minocycline, nitrofurantoin, temocillin and tigecycline. Int J Antimicrob Agents. Epub ahead of print 2011. https://doi.org/10.1016/j.ijantimicag.2011.01.012

Lloyd NA, Janssen SE, Reinfelder JR, Barkay T (2016) Co-selection of mercury and multiple antibiotic resistances in Bacteria exposed to mercury in the Fundulus heteroclitus gut microbiome. Curr Microbiol. Epub ahead of print 2016. https://doi.org/10.1007/s00284-016-1133-6

Long KS, Poehlsgaard J, Kehrenberg C, Schwarz S, Vester B (2006) The Cfr rRNA methyltransferase confers resistance to phenicols, lincosamides, oxazolidinones, pleuromutilins, and streptogramin A antibiotics. Antimicrob Agents Chemother. Epub ahead of print 2006. https://doi.org/10.1128/AAC.00131-06

Magiorakos AP, Srinivasan A, Carey RB, Carmeli Y, Falagas ME et al (2012) Multidrug-resistant, extensively drug-resistant and pandrug-resistant bacteria: an international expert proposal for interim standard definitions for acquired resistance. Clin Microbiol Infect 18:268–281

Mantengoli E, Luzzaro F, Pecile P, Cecconi D, Cavallo A et al (2011) Escherichia coli ST131 producing extended-spectrum β-lactamases plus VIM-1 carbapenemase: further narrowing of treatment options. Clinical Infectious Diseases. Epub ahead of print 2011. https://doi.org/10.1093/cid/ciq194

Marasini BP, Baral P, Aryal P, Ghimire KR, Neupane S et al (2015) Evaluation of antibacterial activity of some traditionally used medicinal plants against human pathogenic bacteria. Biomed Res Int Epub ahead of print 2015. https://doi.org/10.1155/2015/265425

Marston HD, Dixon DM, Knisely JM, Palmore TN, Fauci AS (2016) Antimicrobial resistance. JAMA – J Am Med AssocEpub ahead of print 2016. https://doi.org/10.1001/jama.2016.11764

Martínez JL, Baquero F (2014) Emergence and spread of antibiotic resistance: setting a parameter space. Ups J Med Sci Epub ahead of print 2014. https://doi.org/10.3109/03009734.2014.901444

Martínez JL, Coque TM, Baquero F (2015) What is a resistance gene? Ranking risk in resistomes. Nat Rev Microbiol Epub ahead of print 2015. https://doi.org/10.1038/nrmicro3399

Melnyk AH, Wong A, Kassen R (2015) The fitness costs of antibiotic resistance mutations. Evol Appl Epub ahead of print 2015. https://doi.org/10.1111/eva.12196

Michael CA, Dominey-Howes D, Labbate M (2014) The antimicrobial resistance crisis: causes, consequences, and management. Front Public Heal Epub ahead of print 2014. https://doi.org/10.3389/fpubh.2014.00145

Moreno-Grúa E, Pérez-Fuentes S, Muñoz-Silvestre A, Viana D, Fernández-Ros AB et al (2018) Characterization of livestock-associated methicillin-resistant staphylococcus aureus isolates obtained from commercial rabbitries located in the Iberian Peninsula. Front Microbiol. Epub ahead of print 2018. https://doi.org/10.3389/fmicb.2018.01812

Joana Mourão, Carla Novais, Jorge Machado, Luísa Peixe, Patrícia Antunes (2015) Metal tolerance in emerging clinically relevant multidrug-resistant Salmonella enterica serotype 4,[5],12:i:- clones circulating in Europe. Int J Antimicrob Agents. Epub ahead of print 2015. https://doi.org/10.1016/j.ijantimicag.2015.01.013

Muller AE, Theuretzbacher U, Mouton JW (2015) Use of old antibiotics now and in the future from a pharmacokinetic/pharmacodynamic perspective. Clin Microbiol Infect. Epub ahead of print 2015. https://doi.org/10.1016/j.cmi.2015.06.007

Nakajima H, Kobayashi K, Kobayashi M, Asako H, Aono R (1995) Overexpression of the robA gene increases organic solvent tolerance and multiple antibiotic and heavy metal ion resistance in Escherichia coli. Appl Environ Microbiol. Epub ahead of print 1995. https://doi.org/10.1007/s00340-006-2412-x

Navarre WW, Halsey TA, Walthers D, Frye J, McClelland M et al (2005) Co-regulation of Salmonella enterica genes required for virulence and resistance to antimicrobial peptides by SlyA and PhoP/PhoQ. Mol Microbiol Epub ahead of print 2005. https://doi.org/10.1111/j.1365-2958.2005.04553.x

Nemergut DR, Martin AP, Schmidt SK (2004) Integron diversity in heavy-metal-contaminated mine tailings and inferences about Integron evolution. Appl Environ Microbiol Epub ahead of print 2004. https://doi.org/10.1128/AEM.70.2.1160-1168.2004

Novick RP, Roth C (1968) Plasmid-linked resistance to inorganic salts in Staphylococcus aureus. J Bacteriol 95(4):1335–1342

Oliver A, Mulet X, López-Causapé C, Juan C (2015) The increasing threat of Pseudomonas aeruginosa high-risk clones. Drug Resist Updat Epub ahead of print 2015. https://doi.org/10.1016/j.drup.2015.08.002
Otter JA, French GL (2010) Molecular epidemiology of community-associated meticillin-resistant Staphylococcus aureus in Europe Lancet Infect Dis Epub ahead of print 2010. https://doi.org/10.1016/S1473-3099(10)70053-0
Oz T, Guvenek A, Yildiz S, Karaboga E, Tamer YT et al (2014) Strength of selection pressure is an important parameter contributing to the complexity of antibiotic resistance evolution. Mol Biol Evol Epub ahead of print 2014. https://doi.org/10.1093/molbev/msu191
Pal C, Bengtsson-Palme J, Kristiansson E, Larsson DGJ (2015) Co-occurrence of resistance genes to antibiotics, biocides and metals reveals novel insights into their co-selection potential. BMC Genomics Epub ahead of print 2015. https://doi.org/10.1186/s12864-015-2153-5
Pal C, Bengtsson-Palme J, Kristiansson E, Larsson DGJ (2016) The structure and diversity of human, animal and environmental resistomes. Microbiome. Epub ahead of print 2016. https://doi.org/10.1186/s40168-016-0199-5
Pal C, Asiani K, Arya S, Rensing C, Stekel DJ et al (2017) Metal resistance and its association with antibiotic resistance. In: Advances in Microbial Physiology. 2017. Epub ahead of print 2017. https://doi.org/10.1016/bs.ampbs.2017.02.001
Partridge SR, Collis CM, Hall RM (2002) Class 1 integron containing a new gene cassette, aadA10, associated with Tn1404 from R151. Antimicrob Agents Chemother. Epub ahead of print 2002. https://doi.org/10.1128/AAC.46.8.2400-2408.2002
Peirano G, Schreckenberger PC, Pitout JDD (2011) Characteristics of NDM-1-producing Escherichia coli isolates that belong to the successful and virulent clone ST131. Antimicrob Agents Chemother. Epub ahead of print 2011. https://doi.org/10.1128/AAC.01763-10
Perron K, Caille O, Rossier C, Van Delden C, Dumas JL et al (2004) CzcR-CzcS, a two-component system involved in heavy metal and Carbapenem resistance in Pseudomonas aeruginosa. J Biol Chem Epub ahead of print 2004. https://doi.org/10.1074/jbc.M312080200
Perry JA, Westman EL, Wright GD (2014) The antibiotic resistome: What's new? Curr Opin Microbiol. Epub ahead of print 2014. https://doi.org/10.1016/j.mib.2014.09.002
Price LB, Johnson JR, Aziz M, Clabots C, Johnston B et al (2013) The epidemic of extended-spectrum-β-lactamase-producing Escherichia coli ST131 is driven by a single highly pathogenic subclone, H30-Rx. MBio. Epub ahead of print 2013. https://doi.org/10.1128/mBio.00377-13
Ritchie MD, Holzinger ER, Li R, Pendergrass SA, Kim D (2015) Methods of integrating data to uncover genotype-phenotype interactions. Nat Rev Genet Epub ahead of print 2015. https://doi.org/10.1038/nrg3868
Roux D, Danilchanka O, Guillard T, Cattoir V, Aschard H et al (2015) Fitness cost of antibiotic susceptibility during bacterial infection. Sci Transl Med Epub ahead of print 2015. https://doi.org/10.1126/scitranslmed.aab1621
Ruiz-Garbajosa P, Bonten MJM, Robinson DA, Top J, Nallapareddy SR et al (2006) Multilocus sequence typing scheme for Enterococcus faecalis reveals hospital-adapted genetic complexes in a background of high rates of recombination. J Clin Microbiol. Epub ahead of print 2006. https://doi.org/10.1128/JCM.02596-05
Samuelsen Ø, Naseer U, Tofteland S, Skutlaberg DH, Onken A et al (2009) Emergence of clonally related Klebsiella pneumoniae isolates of sequence type 258 producing plasmid-mediated KPC carbapenemase in Norway and Sweden. J Antimicrob Chemother. Epub ahead of print 2009. https://doi.org/10.1093/jac/dkp018
Scientific Committe on Emerging and Newly Identified Health Risks (2012) Assessment of the antibiotic resistance effects of biocides. Report. Epub ahead of print 2012. https://doi.org/10.2772/8624
Seiler C, Berendonk TU (2012) Heavy metal driven co-selection of antibiotic resistance in soil and water bodies impacted by agriculture and aquaculture. Front Microbiol. Epub ahead of print 2012. https://doi.org/10.3389/fmicb.2012.00399
Shahid M (2010) Citrobacter spp. simultaneously harboring blaCTX-M, blaTEM, blaSHV, blaampC, and insertion sequences IS26 and orf513: an evolutionary phenomenon of recent

concern for antibiotic resistance. J Clin Microbiol. Epub ahead of print 2010. https://doi.org/10.1128/JCM.01467-09

Stepanauskas R, Glenn TC, Jagoe CH, Tuckfield RC, Lindell AH et al (2006) Coselection for microbial resistance to metals and antibiotics in freshwater microcosms. Environ Microbiol. Epub ahead of print 2006. https://doi.org/10.1111/j.1462-2920.2006.01091.x

Stryjewski ME, Corey GR (2014) Methicillin-resistant staphylococcus aureus: an evolving pathogen. Clin Infect Dis Epub ahead of print 2014. https://doi.org/10.1093/cid/cit613

Summers AO, Wireman J, Vimy MJ, Lorscheider FL, Marshall B et al (1993) Mercury released from dental 'silver' fillings provokes an increase in mercury- and antibiotic-resistant bacteria in oral and intestinal floras of primates. Antimicrob Agents Chemother. Epub ahead of print 1993. https://doi.org/10.1128/AAC.37.4.825

Sykes R (2010) The 2009 Garrod lecture: the evolution of antimicrobial resistance: a Darwinian perspective. J Antimicrob Chemother. Epub ahead of print 2010. https://doi.org/10.1093/jac/dkq217

Treepong P, Kos VN, Guyeux C, Blanc DS, Bertrand X et al (2018) Global emergence of the widespread Pseudomonas aeruginosa ST235 clone. Clin Microbiol Infect Epub ahead of print 2018. https://doi.org/10.1016/j.cmi.2017.06.018

van Schaik W (2015) The human gut resistome. Philos Trans R Soc B Biol Sci Epub ahead of print 2015. https://doi.org/10.1098/rstb.2014.0087

Ventola CL (2015) The antibiotic resistance crisis. Pharm Ther 40:277–283

Vestergaard M, Leng B, Haaber J, Bojer MS, Vegge CS et al (2016) Genome-wide identification of antimicrobial intrinsic resistance determinants in staphylococcus aureus. Front Microbiol. Epub ahead of print 2016. https://doi.org/10.3389/fmicb.2016.02018

Vindel A, Cuevas O, Cercenado E, Marcos C, Bautista V et al (2009) Methicillin-resistant Staphylococcus aureus in Spain: molecular epidemiology and utility of different typing methods. J Clin Microbiol. Epub ahead of print 2009. https://doi.org/10.1128/JCM.01579-08

Vindel A, Trincado P, Cuevas O, Ballesteros C, Bouza E et al (2014) Molecular epidemiology of community-associated methicillin-resistant Staphylococcus aureus in Spain: 2004-12. J Antimicrob Chemother Epub ahead of print 2014. https://doi.org/10.1093/jac/dku232

Vogwill T, Maclean RC (2015) The genetic basis of the fitness costs of antimicrobial resistance: a meta-analysis approach. Evol Appl Epub ahead of print 2015. https://doi.org/10.1111/eva.12202

Wales A, Davies R (2015) Co-selection of resistance to antibiotics, biocides and heavy metals, and its relevance to foodborne pathogens. Antibiotics. Epub ahead of print 2015. https://doi.org/10.3390/antibiotics4040567

Wang X, Zhu Y, Hua X, Chen F, Wang C et al (2018) F14:A-:B- and IncX4 Inc group cfr-positive plasmids circulating in Escherichia coli of animal origin in Northeast China. Vet Microbiol Epub ahead of print 2018. https://doi.org/10.1016/j.vetmic.2018.02.003

Wendlandt S, Shen J, Kadlec K, Wang Y, Li B et al (2015) Multidrug resistance genes in staphylococci from animals that confer resistance to critically and highly important antimicrobial agents in human medicine. Trends Microbiol Epub ahead of print 2015. https://doi.org/10.1016/j.tim.2014.10.002

Willems RJL, Hanage WP, Bessen DE, Feil EJ (2011) Population biology of gram-positive pathogens: high-risk clones for dissemination of antibiotic resistanceFEMS Microbiol Rev. Epub ahead of print 2011. https://doi.org/10.1111/j.1574-6976.2011.00284.x

Wireman J, Liebert CA, Smith T, Summers AO (1997) Association of mercury resistance with antibiotic resistance in the gram-negative fecal bacteria of primates. Appl Environ Microbiol. Epub ahead of print 1997. https://doi.org/10.1177/1087054712468053

Woodford N, Turton JF, Livermore DM (2011) Multiresistant gram-negative bacteria: the role of high-risk clones in the dissemination of antibiotic resistance. FEMS Microbiol Rev. Epub ahead of print 2011. https://doi.org/10.1111/j.1574-6976.2011.00268.x

World Health Organisation (WHO) (2014) Antimicrobial resistance: global report on surveillance. Bull World Health Organ. Epub ahead of print 2014. https://doi.org/10.1007/s13312-014-0374-3

Wright GD (2017) Opportunities for natural products in 21st century antibiotic discovery. Nat Prod Rep Epub ahead of print 2017. https://doi.org/10.1039/c7np00019g

Wright LL, Turton JF, Livermore DM, Hopkins KL, Woodford N (2015) Dominance of international 'high-risk clones' among metallo-β-lactamase-producing Pseudomonas aeruginosa in the UK. J Antimicrob Chemother. Epub ahead of print 2015. https://doi.org/10.1093/jac/dku339

Xu Z, Li N, Shirtliff ME, Alam MJ, Yamasaki S et al (2009) Occurrence and characteristics of class 1 and 2 integrons in pseudomonas aeruginosa isolates from patients in southern China. J Clin Microbiol. Epub ahead of print 2009. https://doi.org/10.1128/JCM.02027-08

Yelin I, Kishony R (2018) Antibiotic resistance. Cell. Epub ahead of print 2018. https://doi.org/10.1016/j.cell.2018.02.018

Zheng R, Zhang Q, Guo Y, Feng Y, Liu L et al (2016) Outbreak of plasmid-mediated NDM-1-producing Klebsiella pneumoniae ST105 among neonatal patients in Yunnan, China. Ann Clin Microbiol Antimicrob Epub ahead of print 2016. https://doi.org/10.1186/s12941-016-0124-6

Antibiotics and Microbial Antibiotic Resistance in Soil

11

Ali-Akbar Safari-Sinegani, Mehdi Rashtbari, Nayereh Younessi, and Babak Mashkoori

Abstract

Induced antibiotic resistance in both clinical and nonclinical strains, caused by selective agents of antibiotic resistance genes, considered as one of the most important challenges of the present century. Evidences support increasing antibiotic resistance in the organic waste- treated soils which might affect soil biological and functional diversity. Manure, toxic compounds like insecticides, herbicides and chemical fertilizers which contain heavy metals are among the most important origins of antibiotic resistance in soil and dissemination of resistance determinants within ecosystem. Heavy metals could confer antibiotic resistance to microorganisms. Most of heavy metal resistance mechanisms are the same as antibiotic resistance. In most soils, heavy metal concentration is also much higher than antibiotic concentration. Therefore, it seems that the first option to control antibiotic resistance is the evaluating of resistance degree in specific habitats like soil, underground waters and manures which could participate in increasing the antibiotic resistance in the environment. Hence, the present paper aims to show the importance of antibiotics in soil and their impact on microbial functions and antibiotic resistance.

Keywords

Antibiotic sorption · Enzyme activity · Antibiotic function · Gene Transfer · Heavy metals · Multiple antibiotic resistance (MAR)

A.-A. Safari-Sinegani (✉) · M. Rashtbari · N. Younessi · B. Mashkoori
Department of Soil Science, College of Agriculture, Bu-Ali Sina University, Hamadan, Iran
e-mail: aa-safari@basu.ac.ir

S. M. Mandal, D. Paul (eds.), *Bacterial Adaptation to Co-resistance*,
https://doi.org/10.1007/978-981-13-8503-2_11

11.1 Introduction

Antibiotics with diverse applications used as efficient drugs enter the environments due to inappropriate waste disposals and their widespread use. Antibiotics not only select for resistance in clinical and environmental microorganisms, but also indirectly affect plants, animals, habitats and people as antibiotic resistance could be easily dispersed (Verlicchi et al. 2010). In recent years, a particular concern raised through increasing use of pharmaceuticals, especially antibiotics, due to their unfavorable potential ecological and human health effects.

The research conducted on trends in universally antibiotic consumption between 2000 and 2015, showed that the antibiotic use increased by 65% (from 21.1 billion DDDs to 34.11 billion DDDs) and the antibiotic consumption rate increased from 11.3 to 15.7 DDDs per 1000 inhabitants per day (39%). These increases were higher in low-middle income countries. They increased 114% and 77% respectively. The highest consumption values were reported in India, China and Pakistan in 2015, related to increase in economic development. In contrast, in high income countries the total antibiotic consumption increased by 6.6% and the rate of consumption increased by 4% with no correlation with economic growth (Klein et al. 2018).

Alighardashi et al. (2014) reviewed environmental risk of selected antibiotics in Iran. According to their evaluation, antibiotics consumption in Iran is several times greater than European countries, for example, beta lactams (e.g. Penicillin) are consumed approximately 10 times greater in Iran compared with Scandinavian region (Table 11.1). The calculated predicted environmental concentrations (PECs) ranged from 0.0071 to 0.8570 and the predicted no-effect concentration (PNEC) values based on eco-toxicity data were estimated for all studied antibiotics (varied from 0.0037 to 177). The risk quotient (RQ) exceeded one for amoxicillin, penicillin G, sulfamethoxazole, and erythromycin and amoxicillin had the highest risk to aquatic organisms. With respect to the emergence of microbial resistance, monitoring the most frequently-used antibiotics is critical to prevent resistance spread among soil bacteria (Alighardashi et al. 2014).

Environment contamination with antibiotics causes bacteria to be resistant against antibiotics and the dispersion of resistance elements among different ecosystem compartments. After their application, antibiotics, are found in livestock manure, and finally reach to the agricultural products, contaminate soil, sediments, and underground waters (Verlicchi et al. 2010).

Table 11.1 Consumption of some antibiotics in Iran and three European countries (No./1000 habitation year)

Pharmaceutical	Denmark	Sweden	Norway	Iran
Penicillin	7.3	8	7.5	76.25
Cephalosporin	<0.05	–	0.366	2.71
Trimethoprim, sulfamethoxazole	0.8	0.9	1.45	8.95
Aminoglycoside	<0.05	<0.1	0.05	60

Ansari (2001))

Antibiotic resistance along with climate change is among the greatest challenges in twenty-first century. Although, there are some considerations for declining climate change like replacing fossil fuels, there is no proper alternative for reducing antibiotic application. It is believe that the antibiotic era is over and diseases will be deadly in the near future due to loss of antibiotics' efficiency (Levy and Marshall 2004). It is estimated that antibiotic application rate in the world is 100,000 to 200,000 to/ha and only in Germany, more than 250 antibiotic types are used for human, livestock and poultry (Wise 2002). About 10% of whole antibiotics are applied in clinics (Kümmerer 2009). Therefore, the first reports for resistance against antibiotics came from hospitals. The first resistance reported was against Sulfa drug in the 1940s. Considering the numerous cautions about increasing antibiotic resistance in the 1970s and 1980s (Levy 2002), during these decades, pathogenic microorganisms with multidrug resistance (MDRs) were identified. Hence, antibiotic resistance in microorganisms is considered to be as a great challenge and must be evaluated.

11.2 Soil Microorganisms and Antibiotic Production

Most of soil bacteria have the ability to produce antibiotics. Antibiotics like beta-lactams, amino glycosides, streptomycin are made by soil bacteria. Gram-negative bacteria like *Pseudomonas*, *Azotobacter*, *Agrobacterium*, *Burkholderia*, *Serratia* and *Erwinia* strains and Gram-positive strains like *Streptomyces*, *Actinobacter*, *Kocuria* and *Bacillus* are great antibiotic producers and the most of antibiotics produced within the soil are effective against most microorganisms. For instance, Pyrrolnitrin, produced by *Pseudomonades* and *Burkholderia*, is lethal for most fungi and/or 2,4 diacetylphloroglucinol, produced by *pseudomonades*, is lethal for fungi (Raaijmakers and Mazzola 2012). Streptomycin production is very intensive in soils and its production by S. *griseus* has been reported (Huddleston et al. 1997). In a study, it was revealed that about 1 percent of soil actinobacters are able to producing of streptomycin (Baltz 2006). Studying the streptomycin resistance mechanisms and DNA extracted from various soils, sediments and animal manure indicated that streptomycin resistance genes caused the inactivation of bactericidal ability to produce this antibiotic (van Overbeek et al. 2002).

Many antibiotic producers have high populations in soil. However, antibiotic concentration in the nature and within the most habitats is less than minimum inhibitory concentration (MIC), but it has been observed that this low concentration can also be a stimulus for translating antibiotic resistance genes (Davies et al. 2006). Some of antibiotic producing bacteria have no ability to keeping resistance genes. In a study on different soils, Ghosh and LaPara (2007) found that some of known species of Streptomyces bacteria which have ability to production of tetracycline, did not have the ability to maintain any of resistance genes against tetracycline. Nevertheless, if the antibiotic concentration increases, the abundance of resistance genes increases as well and also plasmids harboring resistance genes acclimate with large group of bacteria and hence gene reserves increase (Heuer et al. 2002).

Antibiotics in higher concentrations could be used by specific bacteria as carbon and energy sources (Dantas et al. 2008). Low concentrations of antibiotic may also have significant effects on soil microbial communities. It has been proved that a low concentration of vancomycin (0.032 mg/L) induced the expression of *vanA* gene in enterococci (Obst et al. 2006). It has also been documented that low concentrations of antibiotics may change the gene expression in biofilms (Ohlsen et al. 1998). Therefore, antibiotics in soil, regardless of their concentrations, are needed to be evaluated due to their contribution in different aspects concerning microbial functions, environmental pollutants, and human health.

Jafari et al. (2014) studied antibacterial activity of *Pseudonocardia* sp. JB05, as a rare salty soil actinomycete, against *Staphylococcus aureus*. The authors also isolated different indigenous bacterial strains from alkaline soils of Hoze-Soltan, Qom, Iran, and compared their ability to produce antibacterial compounds. Their results indicated that *Pseudonocardia* sp. JB05 was the most effective candidate in terms of antibacterial activity. The antibacterial compounds detected in JB05 indicated a minimum inhibitory concentration (MIC) of 40 $AUmL^{-1}$ against *Staphylococcus aureus.*

Ebadi et al. (2018) identified antibiotic-producing *Streptomyces* species in Iran's soil phenotypically and genotypically. Fifty two actinomycete isolates were separated from 100 soil samples, and then 30 and 3 isolates were selected from primary and secondary screening, respectively. Strain 28 had a peak (RF) similar to gentamicin and isolates 34 and 4 had similar peaks (RF) to streptomycin in HPLC analysis. They sequenced 16S rRNA genes of the isolates, in which isolate 28 and 4 had 99.93% similarity to *Streptomyces youssoufiensis* and *Streptomyces cyaneofuscatus*, respectively. The results demonstrated that new isolates of productive Streptomyces in soil samples of Iran had the ability to produce antibacterial agents which may be used for industrial production of new antibiotics.

Although the diversity and population of antibiotic-producing bacteria in soil are considerably high but antibiotic activity in soil is considerably low may be due to low production and stability of antibiotics in soil and physico-chemical reactions with soil particles.

11.3 Physico-chemical Reactions of Antibiotics in Soil (Sorption, Desorption and Degradation)

Antibiotics could sorb on soil particles with various physical force and chemical bonds. Antibiotic sorption depends on physical properties of antibiotics like water solubility, lipophilicity, volatility and their sorption ability. Also, antibiotic absorption on soil depends on sorption coefficient (Kd) between soil particles and water which affects antibiotics sorption coefficient. Sulfonamide antibiotics have great mobility in soil which causes their transferring and passing through soil to surface and underground waters. While, tetracycline has little mobility, so tetracycline chance to being in run-off is less than sulfonamides (Davies et al. 2006). Many of antibiotics have ionizable structure and their pKa is in soil natural pH range and could have positive and negative charges in soil. Neutral antibiotics show hydrophobic properties and cationic species absorbs on negatively-charged soil particles and the anionic species went away (Ter Laak and Gebbink 2006) and/or absorbed on soil via cation-bridge (Gao and Pedersen 2005). Soil pH, cation exchange capacity (CEC) and soil organic carbon affect antibiotic sorption (Ter Laak and Gebbink 2006). Addition of sewage sludge to the soil decreased antibiotic absorption which depends on changes in soil pH and organic matter (Boxall et al. 2002). It has been stated that application of manures and sewage sludge after antibiotic application in soil, resulted to 10 to 40 percent increase in sulfonamides content in run-off from agriculture lands compared to run-off from lands without application of manures and sewage sludge (Burkhardt et al. 2005).

Antibiotic dissipation in soil could be depends on irreversible bond with soil particles (Heise et al. 2006). According to finding of Heise et al. (2006), only a little percent of Sulfamethoxazole is available for plants and earthworms. Owing to these complex processes, residuals of antibiotics present in manures observed in many of ecosystem parts. Releasing of antibiotics applied for treating animals has been observed in various studies (Han et al. 2009; Ghosh and LaPara 2007). The dependence of tetracycline to repeated application manures in soil was observed by Hamscher et al. (2002), however, stating that whether the concentration of these antibiotics in the habitats provokes resistance genes or not, is difficult. It has been stated that low concentration of tetracycline is sufficient to transferring and provoking resistance genes (Clewell et al. 1995). In a study, a tenfold increase in plasmid transfer in soil receiving animal manure has been reported (Gotz and Smalla 1997). Other factors like heavy metal concentration, high soil CEC and soil organic matter could increase genetic elements responsible for antimicrobial resistance (Berg et al. 2005; Kong et al. 2006).

Kumar et al. (2005) showed the strong tendency of antibiotics to bind with soil clay particles. Kong et al. (2012) also proved that clay particle type strongly affects antibiotic bioavailability and antimicrobial activity. Lv et al. (2013) evaluated the influence of montmorilonite on antimicrobial activity of tetracycline (TC) and ciprofloxacin (CIP) and found that the antimicrobial activity of tetracycline was affected by the presence of montmorilonite, as the non-resistant bacteria continued to grow even in extremely high TC doses. Higher amounts of these antibiotics, up to 342 mg/g of TC or 337 mg/g of CIP, could be adsorbed on both external surfaces and interlayer spaces of montmorilonite. The adsorbed TC or CIP had stability against desorption by DI water. The strong adsorption of TC or CIP on montmorilonite resulted in a significant decrease in antibiotic activity of TC or CIP which was significantly higher than the MIC at input concentrations. Decreasing antibiotic activity was displayed for both TC-sensitive and TC-resistant strains. On the other hand, the slow but persistent desorption of TC or CIP from montmorilonite surfaces at the sub-MIC or EC50 levels may induce bacteria resistance to the antibiotics adsorbed (Lv et al. 2013).

Wegst-Uhrich et al. (2014) reported that it is difficult, if not impossible, to estimate the fate and mobility of antimicrobials in the environment. The variability in mobility is large, as demonstrated by distribution coefficient and the Freundlich

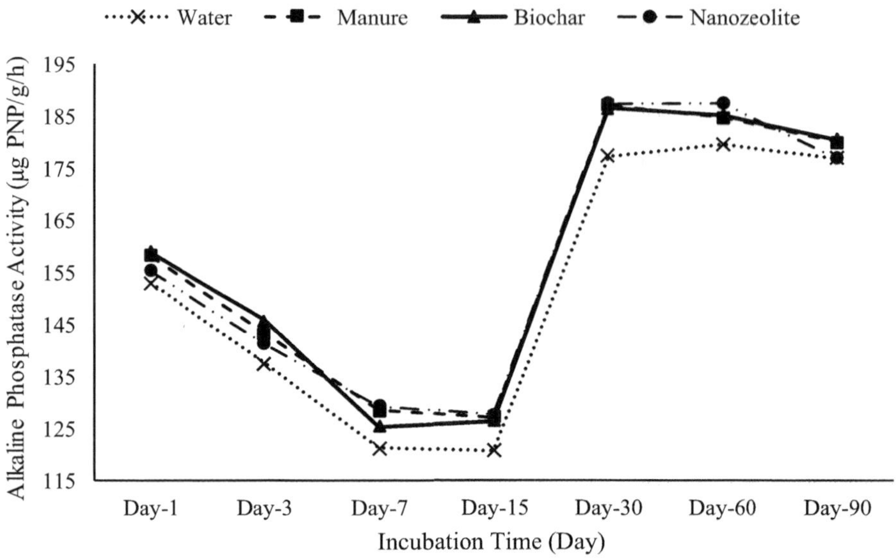

Fig. 11.1 Application of various amendments for reducing negative effects of antibiotics on soil alkaline phosphatase activity during 90-day incubation period

sorption constant values, owing to environmental factors such as pH, ionic strength, and organic strength as well as the multiple chemical functions of the molecule. Microorganisms may be involved in sorption of antibiotics in soil, manure, and biosolids and it may result in degradation or possibly irreversible binding of antibiotics onto manure solids with time.

Soil organic and mineral components and amendments may have different effects on antibiotic activity in soil. Application of biochar and zeolite decreased negative effects of the added antibiotics, even in higher concentrations, on soil microorganisms and their enzyme activity. In our study, during the first incubation times, up to day 7, application of manure and biochar resulted to higher alkaline phosphatase (ALP) activity, while in 7th day up to 30th day, nano-zeolite application maintained higher ALP activity compared to those organic amendments which might be attributed to better antibiotic sorption and stability on nano-zeolite particles and decreasing negative impacts of applied antibiotics. There was no significant difference between days 30 and 60 after antibiotic spiking in nano-zeolite applied treatments, however alkaline phosphatase activity experienced strong decrease up to 90th day. Suring whole study period, antibiotics applied via distilled water had the lowest enzyme activity (Rashtbari 2019, Unpublished data; Fig. 11.1).

The effects of applied antibiotic in soil and manures on population of different groups of bacteria were also studied. They are discussed in the following section.

11.4 The Implications of Antibiotics to Soil Biological Activities

Soil microorganisms by decomposing various organic matters in soil and utilizing them, improve soil health and productivity. Various studies has reported that soil biota and microorganism's processes like biological and functional diversity, dehydrogenase enzyme activity and nitrification efficiency might be used as soil pollution indicators, nutrients cycle and energy metabolism (Ding et al. 2014; Lin et al. 2016). After reaching into the soil, antibiotics could affect soil microbial count and biological and functional diversity. Chen et al. (2013) studied the effect of application of oxy-tetracycline (OTC) on soil enzyme activity and microbial structure diversity and found that low and moderate amounts of OTC, less than 15 ppm, increased microorganisms population and biomass (the abundance of bacteria and fungi) in receiving soils. OTC toxicity for soil fungi was more than toxicity for soil bacteria and the Gram-negative bacteria were prominently more resistant than Gram-positive bacteria. Ding et al. (2014) evaluated the dynamics of soil bacteria in response to repeated manure use (three times) containing sulfadiazine (SDZ; 10 and 100 ppm) in a 133-day period. Presence of SDZ and manure application times significantly affected soil bacteria count and the highest SDZ concentration had the greatest impact. Soil microorganism's response to TC application amended by cow manure studied by Chessa et al. (2016) who evaluated TC bio-availability in a clay and sandy soil and the effect of TC application on soil microbial structure and function. Findings showed that TC had lethal effect on soil microbial structure and efficiency, especially in a short time and the highest concentration. However, effects of TC decreased after 7-days. Cow manure changed soil microbial community in both soils, increased microorganism's efficiency in clay soil and had contribution in recovering soil microbial structure in TC-applied treatments. Lin et al. (2016) studied changes in soil biodiversity via application of TC, Sulfadimethoxine (SMM) and CIP in soil-plant system (100 mg/kg) and identified bacterial respondents. Bacterial respondents were more in proteo-bacteria. CIP- and SMM- resistance increased as soil contamination time increased while for TC, resistance decreased by increasing the contamination time.

The effect of various antibiotics (Oxytetracycline, penicillin and gentamycin) on soil microbial activities was studied at different concentrations at 1, 3, 7, 15, 30, 60 and 90 days by incubation. Gentamycin application had strong negative effect on soil microbial count and enzyme activities until 15-day and then resulted to increase in soil microbial activities, while Oxytetracycline application had moderate negative effects on soil microorganisms. The effects from day 30 were in steady state while afterward soil microbial activity increased significantly. The highest negative effect of penicillin on soil microorganisms was in seventh day of incubation (Rashtbari 2019, Unpublished data). Molaei et al. (2017a) studied the impact of chlortetracycline and sulfapyridine antibiotics on soil enzyme activities at day 1, 4 and 21. Based on results, the chlortetracycline and sulfapyridine antibiotics negatively affected the soil dehydrogenase activity, however the effect of sulfapyridine decreased with time of incubation. In fact, sulfapyridine significantly affected the alkaline phosphatase activity for all time intervals, while chlortetracycline seemed to inhibit its activity within 1 and 4 days of incubation. Molaei et al. (2017a) concluded that chlortetracycline and sulfapyridine antibiotics have harmful effects on soil microbes, with the range of effects varying with the duration of incubation and the type of antibiotics used. Also, Molaei et al. (2017b) assessed some cultural experimental methods to study the influence of sulfamethoxazole and Oxytetracycline on microbial activities in a soil. Although OTC antibiotic positively affected microbial biomass carbon after 1 day of incubation, there was no significant difference in microbial biomass carbon between different treatments of this antibiotic after the first day. Nevertheless, both sulfamethoxazole and Oxytetracycline had significant effects on iron (III) reduction at the concentration of 1 and 10 mg/kg, and iron (III) reduction was completely inhibited at concentrations above 10 mg/kg. Amin et al. (2012) evaluated the impact of Oxytetracycline, tylosin, and amoxicillin on specific methanogenic activity of anaerobic biomass. Based on their findings, the minimum inhibitory concentration of oxytetracycline, amoxicillin, and tylosin were 8000, 9000, and 9000 mg/L, respectively. The authors also observed that increasing the concentration of antibiotics decreased the volume of produced biogas from biomass per unit weight. COD removal was 42–82% probably due to long retention time and adsorption to flocks. Rashtbari (2019, unpublished data) studied the effect of various concentrations of gentamycin, penicillin and Oxytetracycline on soil ALP activity. Results showed that antibiotic application significantly affected ALP activity in soil during 90-day incubation period. Among the studied antibiotics, gentamycin decreased soil ALP activity up to seventh day of incubation and then soil ALP activity increased up to 30th day and then experienced a gradual decrease in activity. Oxytetracycline had moderate effects on soil ALP activity up to 15th day and then increased the enzyme activity. The lowest enzyme activity by 88 μg $PNPg^{-1}$ h^{-1} obtained by gentamycin application on seventh incubation day (Rashtbari 2019, unpublished data; Fig. 11.2). Telesinski et al. (2018) showed the negative impact of penicillin on soil enzyme activities which could disrupt the homeostasis in soil ecosystem. Hammesfahr et al. (2011) and Liu et al. (2015) found no significant effect of OTC on soil alkaline phosphatase activity; attributed to the low OTC concentration in soil in their studies. Wei et al. (2009) indicated increase in soil alkaline phosphatase activity after antibiotic application which might be due to release of ortho-phosphates from lysed cells. Increase in soil phosphatase activity after 14th and 21st day from antibiotic application could be attributed to increase in organic matter and enriching soil microbial community through degradation of antibiotics (Punitha et al. 2012).

Rashtbari (2019, unpublished data) calculated resistance index for soil alkaline phosphatase activity. Alkaline phosphatase had the highest resistance index against OTC antibiotic at 100 and 200 mg/kg concentration and application of 200 mg/kg gentamycin resulted to the lowest resistance index for soil alkaline phosphatase activity (Fig. 11.3). Also, their results showed that gentamycin and penicillin at 50 mg/kg caused the highest resilience index for this enzyme activity in soil (Fig. 11.4). According to the Fig. 11.4, increasing antibiotic concentrations resulted to decrease in resilience index. Telesinski et al. (2018) found similar results when

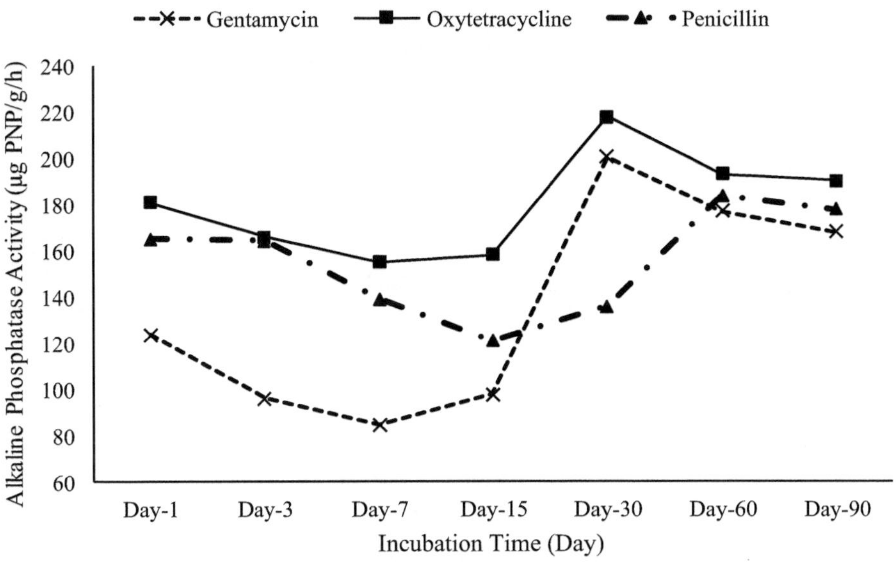

Fig. 11.2 Effect of various antibiotics on soil ALP activity during a 90-day incubation period

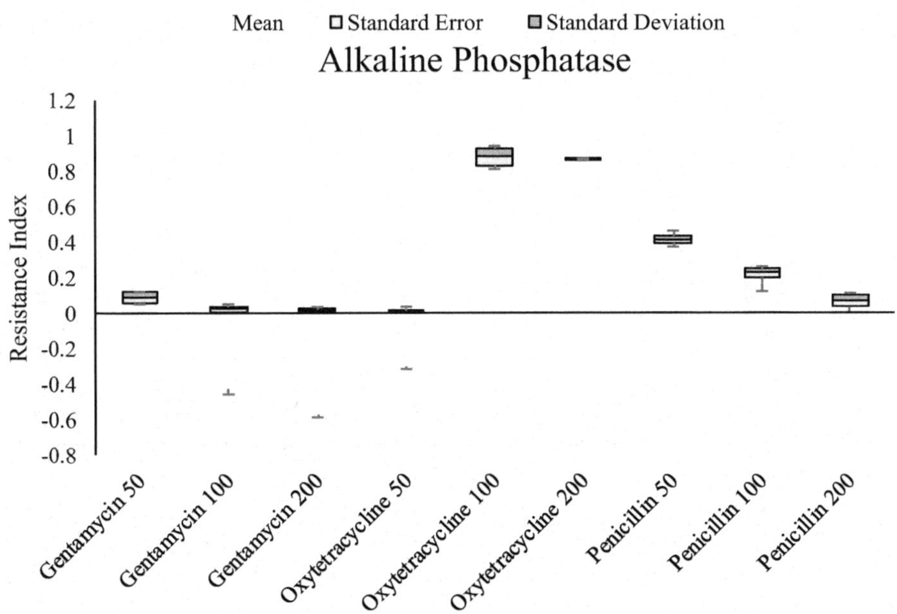

Fig. 11.3 Effect of various antibiotics on resistance index of soil ALP activity at different concentrations

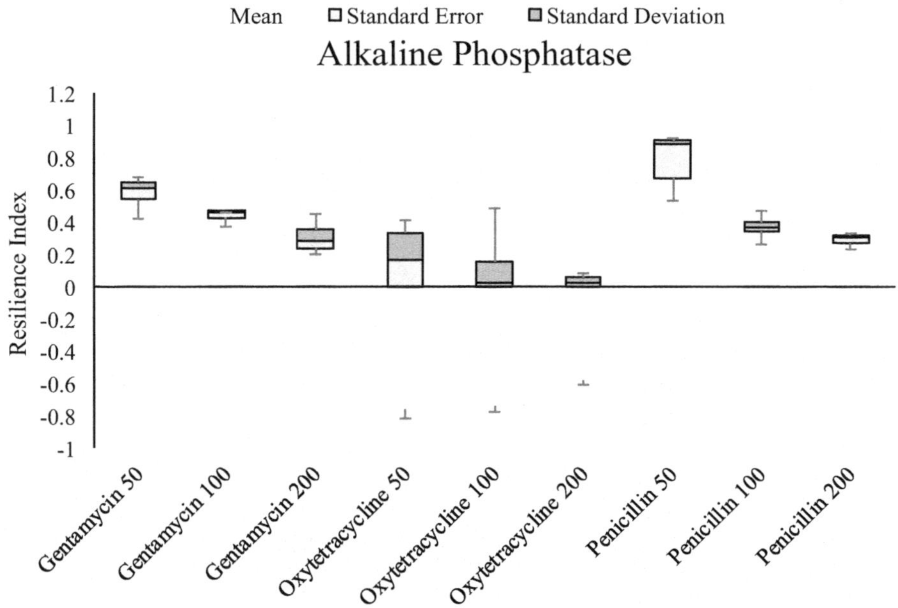

Fig. 11.4 Effect of various antibiotics on resilience index of soil ALP activity at different concentrations

studying soil enzyme activities in a soil containing 200 mg/kg penicillin G which found 50.8, 41, 40.1 and 51.14 percent decrease in enzyme activities for alkaline phosphatase, acid phosphatase, dehydrogenase and urease at different incubation times, respectively.

11.5 Microorganisms and Antibiotic Resistance Gens

In environmental microorganisms, there is a natural reservoir of antibiotic resistance gene whose changes are more than reservoir of microorganisms in clinics in which a few antibiotics were applied. Therefore, antibiotic application causes dissipation of specific resistance genes (Martinez 2009; Baquero et al. 2009; Wright 2007) which in turn causes decrease in intrinsic resistance and increase in induced antibiotic resistance through antibiotic application in the environment (Martinez 2009). According to the reports, soil problematic Gram-positive resistant bacteria are *staphylococcus*, *streptococcus* and *enterococcus* species. Bacteria from enterobacteriaceae family (*Klebsiella pneumonia*, *Escherichia coli*, *salmonella*, and *enterococcus* and so on) and pseudomonas (*P. aerogina*) are stressful (Martinez 2009; Berg et al. 2005; Bradbury 1986). Enterobacteriaceae, which has great clinical implications, is free living, saprophytic or parasitic bacteria, very fast growing and found in every habitats. Emergence and dissipation of antibiotic resistance in enterobacteriaceae caused difficulties in clinical diseases. Younessi et al. (2017)

detected *Beta-lactamase* gene in the culturable bacteria isolated from heavy metal polluted agricultural, pasture, and mining soils around mines in Hamedan, Iran. The highest level of *beta-lactamase* genes was observed in isolates from the pasture soils. In the agricultural and mining soils, a high abundance of bla_{TEM}^{+} isolates was found which also exhibited phenotypically resistance to beta-lactam antibiotics. The identified multi-drug resistant and bla_{TEM}^{+} isolates were from these genera: *Achromobacter, Bacillus, Brevibacillus, Aminobacter* and *Brevundimonas.* The high number of bla_{TEM}^{+} bacteria in all the soils may be attributed to the other unknown functions of *bla* genes which help microorganisms resist contaminated environments. Sensitivity of some bla_{TEM}^{+} bacteria to beta-lactam antibiotics was also interesting. A high abundance of bla_{TEM}^{+} bacteria form the agricultural soil were susceptible to beta-lactams. The tetB and strA genes were not figured out in any of the isolated bacteria from heavy metal polluted soils (Younessi et al. 2019). The vanA gene was detected in 5.71% of all the strains, and only one isolate of Pseudomonas speceies harbored aac(3)-II. The high number of isolated bacteria containing at least one resistance gene indicated the potential resistance and durability of resistance genes in the metal-polluted soils.

11.6 Antibiotic Resistance Gene Transfer

Natural transformation in many bacteria species has been observed and the role of transformation in transferring bacterial antibiotic resistance, extensively reviewed by Lorenz and Wackernagel (1994) who indicated the environmental isolates have the ability to take up exterior DNA from environmental pools. It has been shown that *Bacillus, Streptomyces* and other bacterial species release DNA and plasmids. The released free elements might be absorbed and stabilized by sand and clay particles in soil which protects them against DNAse and became resistant which maintains their transformation ability for weeks or even months (Lorenz and Wackernagel 1994). These DNA fragments and plasmids could enter into soil bacteria and other transformation processes contribute in transferring the newly entered DNA fragments to the other bacterial species (Fig. 11.5) (Lorenz and Wackernagel 1994).

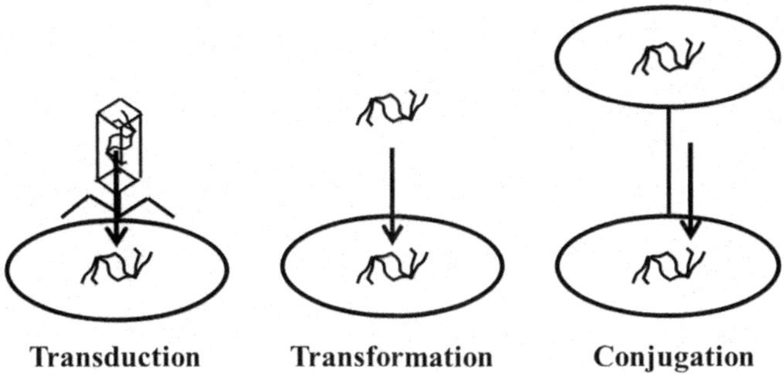

Fig. 11.5 Various processes involving in antibiotic resistance gene transfer

Transformation requires competent cell to receive DNA fragments and some bacteria are naturally competent and some might be competent during their life (Lunsford 1998). Some bacteria, also, might be induced via proper methods like protoplast formation, $CaCl_2$ and EDTA treatments and other methods to take up DNA fragments (Lunsford 1998). Nielsen et al. (1997) found that nutrition induced competence in soil bacteria and transformation rate were higher in soil moisture of 35%.

Gene transfer between bacteria by transduction process is mediated by bacteriophages. Following host cell lysis, transduction genes randomly are transferred in generalized transduction, whereas in specialized transduction, temperate phages transfer only the genes close to their integration sites on the host genome. Transduction is common in nature and in relatively high phage-loaded aquatic environments, it can be considered as a major mechanism of gene transfer (Jiang and Paul 1998).

11.7 Heavy Metals and Antibiotic Resistance

Heavy metals are elements with density more than 5 g/cm^3 (Srivastava and Majumder 2008). These metals with great contamination potential in the environment, are intensively studied. Copper (Cu), Mercury (Hg), Zinc (Zn), Lead (Pb) and Cadmium (Cd) are among the most known metals (Li et al. 2016). Heavy metal toxicity in soil is determined by metal concentration, soil pH, texture, organic carbon and so on. Heavy metals like Pb and Cd could make toxic compounds which affect plant physiological function, however some heavy metals like Zn, Nickel (Ni) and especially Cu are required for oil microorganisms and in high concentrations are toxic (Kachur et al. 1998). It has been stated that metals form strong bonds with thiol groups and oxygen receptors in biomolecules and so entered into protein structure, changes their functions via mutations in DNA structure (Poole and Gadd 1989). One of the most important consequences of metals inhibitory effects on bacteria occurs when the metal bonds with sulfhydryl group in enzyme and causes enzyme inactivation (Kachur et al. 1998). Therefore, expelling heavy metals from cell for stability of cell efficiency is so important (Cobine et al. 1999).

Some of heavy metals resistance mechanisms in bacteria include: (1) decreasing cell wall permeability to metals (Ruiz et al. 2003), (2) decreasing metal ions accumulation in cell through pumping these metals out of the cells (Gupta et al. 2015), (3) placing some cations inside the molecules having sulfhydryl group, and (4) reducing or transformation of metal ions into the less toxic species (Nies 1999). Studies shown that bacteria metals resistance structures are located on their plasmid and easily could be transferred (Ji and Silver 1995). Plasmid coding systems mostly include pumping of toxic metals. Therefore, any mechanism related to the pumping have plasmid origin and would be easily transferred between bacteria (Silver and Walderhaug 1992). Totally, bacteria would be resistant against heavy metals through chromosomal structures, transposons and/or plasmids. In some cases, heavy metal resistance structures which located on chromosomes are more complex than plasmid structures. Therefore resistant bacteria in soil could proliferate their population in a cautious rate (Levy and Marshall 2004). On the other hand, most of pollutant-resistance coding plasmids transferred from environmental bacteria to the pathogenic one (Heuer et al. 2002). Most of heavy metal resistance mechanisms are the same as antibiotic resistance. Also, in most soils heavy metal concentration is so more than antibiotic concentration (Byrne-Bailey et al. 2009; Han et al. 2009). Hence, one of the most important factors which indirectly provokes antibiotic resistance genes is soil heavy metals. Safari Sinegani and Younessi (2017) evaluated antibiotic resistance of bacteria isolated from metal-polluted soils under different land uses. Among all the samples, those from the tailings of mines with the highest heavy metals levels had the lowest number of bacteria, but a relatively higher abundance of heavy metal- and antibiotic-resistant bacteria. A high degree of resistance was observed for ampicillin and amoxicillin in the isolates from all the soils. The agricultural soil isolates had a high prevalence of resistance towards vancomycin, tetracycline, and streptomycin. Among all the tested antibiotics, gentamicin (GM) was the most potent. The most frequent pattern of multiple antibiotic resistance in the isolates from agricultural soils was amoxicillin (AMX), ampicillin (AM), streptomycin (S), vancomycin (VA), tetracycline (TE), and doxycycline (DO). It was concluded that the higher percentage of isolates with multiple antibiotic resistance in the agricultural soils that in the mining waste soils may be related to (1) the level of soil heavy metals, (2) the population and diversity of soil bacteria, (3) the application of manure, and (4) other factors affecting gene transfer between bacteria (Safari Sinegani and Younessi 2017). Among isolated bacteria from heavy metal polluted soils more than 77% of the isolates were resistant to β-lactam antibiotics, and 28.57%, 40%, and 31.43% of them were resistant to streptomycin, vancomycin, and tetracyclines, respectively. The high number of antibiotic resistant bacteria, isolated from the sampled soils indicated the importance of natural environmental reservoirs for resistant microbs (Younessi et al. 2019).

11.8 Manure and Antibiotic Resistance

Manures as an agricultural inputs used for increasing and maintaining soil productivity. However, application of veterinary antibiotics in animal husbandry changed soil microbial structure and increased antibiotic resistance in manure (Schauss et al. 2009). Depending on chemical composition of antibiotic, large portions of antibiotic resistance enter soil through manure application (Heuer and Smalla 2007a, b). Tetracycline antibiotics group including chlortetracycline (CTC), Oxytetracycline (OTC), tetracycline (TC) and Sulfamethoxazole (SMX) and Sulfadiazine (SDZ) are among the most commonly used antibiotics in livestock production. Topp et al. (2013) reported that biodegradation of antibiotics in agricultural soils after prolonged exposure accelerated. Various authors reported a positive correlation between manure application and antibiotic resistance of bacterial isolates in soil and water environments (Bibbal et al. 2007; Peak et al. 2007). There are reports that antibiotic resistance genes (ARGs) seem to be higher in pig manure than cow and sheep manures and it is suggested that the abundance of ARGs may directly correlate with antibiotics applied in husbandry of these animals (Enne et al. 2008; McKinney et al. 2010). The *tet*M and *tet*O genes had higher concentrations in manure of tetracycline-treated animals, in which higher tetracycline residues were detected (Schwaiger et al. 2009). In another study, a higher abundance of *sul1*, *sul2* and bla_{TEM} and resistance plasmids observed in farms with higher antibiotic application and higher animal population (Binh et al. 2008, 2010). In a farm with routinely application of amoxicillin and tetracycline, manure had higher abundance of antibiotic resistance genes (Thompson et al. 2008).

Manure having considerable amounts of antibiotic residuals used for amendment of soils carrying high amount of ARGs (Heuer and Smalla 2007a, b). Ghosh and LaPara (2007) stated that a relationship between manure and the occurrence of antibiotic resistance in soil could not always be established. However, a range of studies confirmed correlation of manure application in soil and occurrence of antibiotic resistance in these soils. For instance, Byrne-Bailey et al. (2009) demonstrated increasing sulfonamide resistant genes after two-year field application of antibiotic containing manure. Jayalakshmi et al. (2017) reviewed antibiotic residues in animal products and its impact on the environments and human health and stated that antibiotic residues may occur in animal products when administration of drug is done? in extralabel fashion and not following withholding period after treatment. Many of the administered drugs are not completely absorbed from gut and excreted through faeces and urine as either parent compound or its toxic metabolites. The application of manure or farm effluents in agricultural lands leads to selection of resistant bacteria, development and transmission of antibiotic resistance genes in the microbes. The antibiotic resistance in animal and human leads to poor response to treatment during illness. The use of antibiotics as feed additives at sub therapeutic dose should be strictly prohibited and for therapeutic purpose, it must be used in proper dose in a certain time.

In a study done by our team in Bu-Ali Sina University, it found that the isolated *E.coli* from industrial and conventional dairy manures had considerable resistance (unlikely respond to high dosage therapy) against ampicillin and vancomycin antibiotics (Mashkoori 2014). However bacterial resistance against some of the other antibiotics (i.e. amoxicilin, doxcyclin, tetracylcline) were intermediate (likely respond to high dosage therapy) in most cases. However the highest ZOI and sensitivity were measured in application of gentamycin for all of the isolates. The *E.coli* sensitivity against this antibiotic in conventional dairy manure compared to those in industrial dairy manure was higher (Figs. 11.6 and 11.7).

In another study it was revealed that the relative abundance of resistant bacteria to amoxicillin, ampicillin and gentamicin in the industrial and conventional poultry manures was not significantly different (Younessi 2017). Relative abundance of resistant coliforms to streptomycin, vancomycin, doxycycline and tetracycline in industrial poultry manures was significantly higher than that in conventional manures (Fig .11.8). The highest resistance in Escherichia coli strains from industrial manure was to tetracycline and doxycycline. In isolates from conventional manure, the highest resistance was to streptomycin. The lowest resistance in isolates

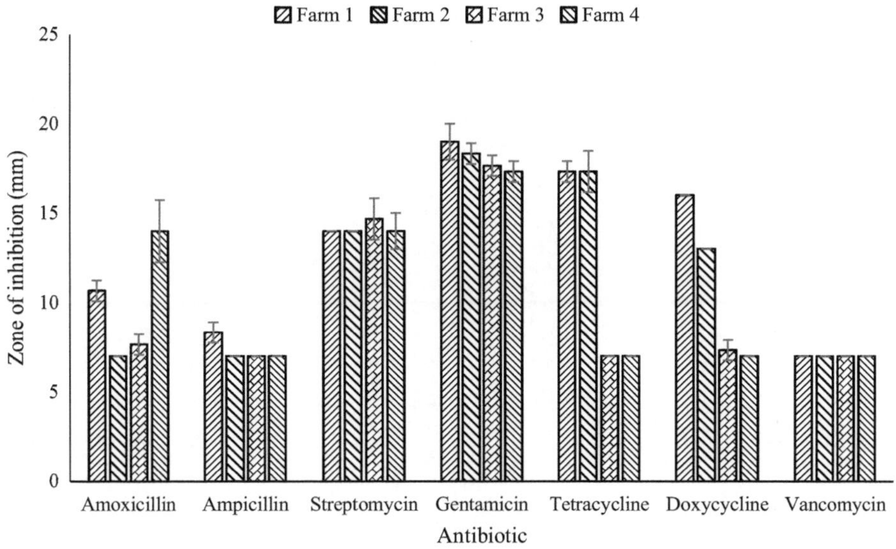

Fig. 11.6 Kirby-Bauer disk susceptibility test on *Esherchia coli* isolated from dairy manure sampled from 4 industrial farms in Hamadan province, Iran in spring 2012

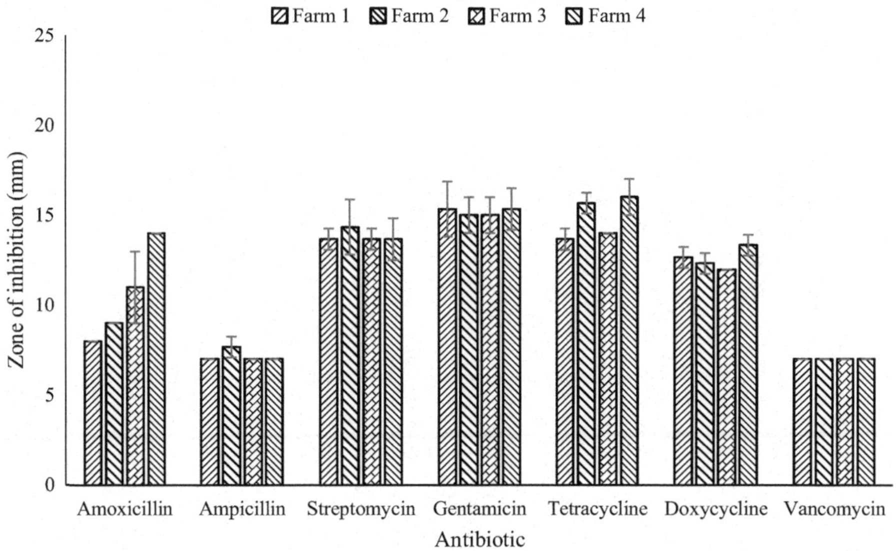

Fig. 11.7 Kirby-Bauer disk susceptibility (Sus.) test on *Esherchia coli* isolated from dairy manure sampled from 4 conventional farms in Hamadan province, Iran in spring 2012

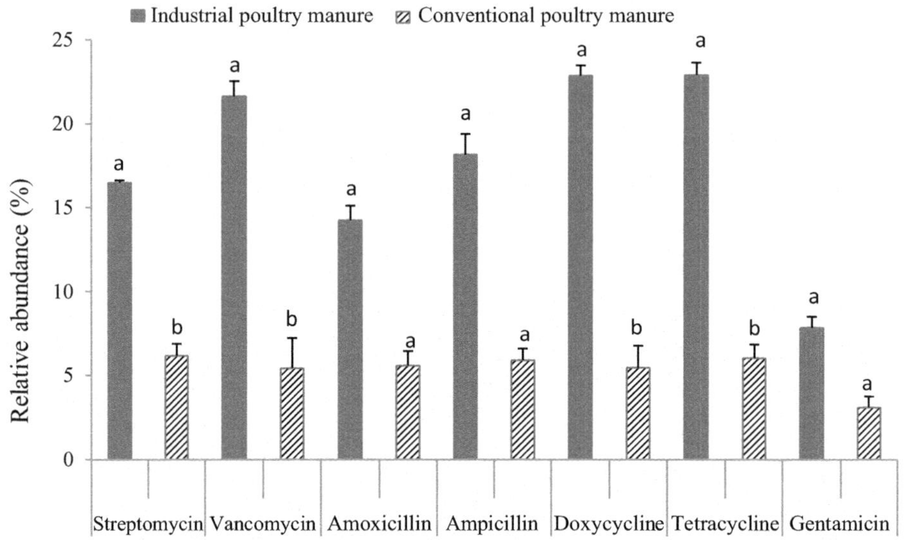

Fig. 11.8 Antibiotic resistance of culturable bacteria of industrial and conventional poultry manure on nutrient agar

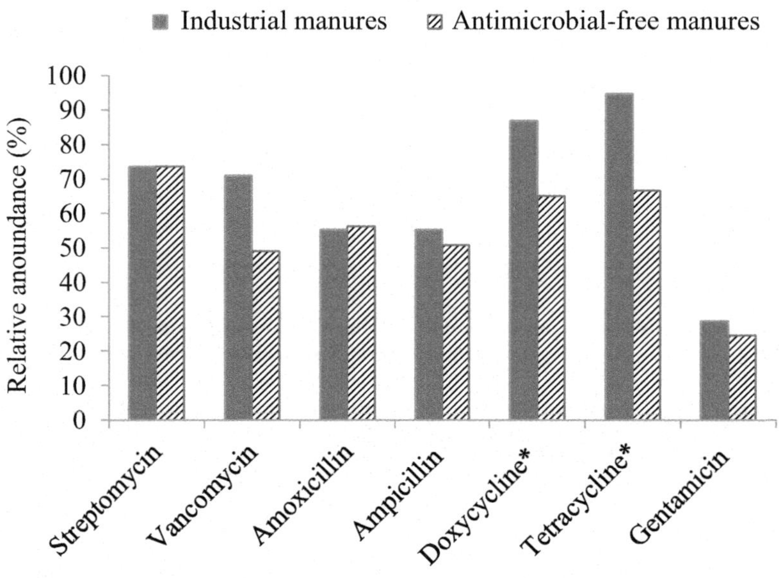

Fig. 11.9 Antibiotic resistance of culturable *Escherichia coli* isolated from industrial and conventional poultry manure on Mueller–Hinton agar

from both manure samples was against gentamicin (Fig. 11.9). Eager et al. (2009) have reported that Anti-inflammatory drugs such as tetracycline are a stimulant for bacterial antibiotic resistance.

11.9 Genetic Manipulation and Antibiotic Resistance

Salyers and Ama'nile-Cuevas (1997) showed that even in the absence of any pressure, resistance genes would be stable in microorganisms. Microorganisms can easily acquire resistance genes and their transferring ways and hardly lose them in ways which are facilitated by cooperation of gene transfer elements and horizontal transfer of antibiotic resistance genes (Salyers and Ama'nile-Cuevas 1997). This cooperation described as "hunting as a pack" by Salyers and Ama'nile-Cuevas (1997). Horizontal gene transfer is also mediated by bacterial genome structure. Bacterial genome is a haploid genome of relatively small size ranging from 500 to 10,000 kbp and within genomes, genes with related functions are clustered into operons to facilitate their transcription and the translation of resultant polycistronic mRNA (Salyers and Ama'nile-Cuevas 1997). To date, sequencing bacterial genomes provided information and new sights about bacterial genes functions and overall organization of the sequenced genomes (Casjens 1998). The genomes of different isolates of the same bacterial species show multiple insertion/deletion differences, each in a few kbp to 200 kbp range. These are thought to be integrated accessory elements such as transposons, integrons, retrons, invertrons, prophages, defective prophages, pathogenicity islands, and plasmids.

Prokaryotic cells naturally acquire exogenous DNA through different pathways including transduction vis bacteriophage infection, conjugation mediated by cell–cell contact, or directly from the environment through a natural physiological state of competence developed by the bacterium (Zeaiter et al. 2018). Researchers utilized these natural mechanisms of horizontal gene transfer (HGT) for cell manipulation. To insert DNA into bacteria, different methods have been developed, including chemical and physical techniques. Because homologous recombination of exogenous DNA in a host cell is limited to sequences with high DNA similarity, shuttle vectors are preferred to compel DNA entry into bacterial cells (Thomas and Nielsen 2005). Plasmids are extra-chromosomal DNA elements that are usually transferred in nature to both closely and distantly related bacterial species according to their replication origin. They are generally used as shuttle vectors in electroporation and conjugation-based procedures owing to their relative ease of manipulation. Certain bacteria remain nevertheless recalcitrant to internalize and express exogenous plasmid DNA (Zeaiter et al. 2018).

11.10 Conclusion

Antibiotics as efficient drugs consumed by human and animals, could enter soil in various ways and affect soil microbial functions and biodiversity and increase antibiotic resistance genes within soil. Antibiotic resistance is one of the greatest challenges in the twenty-first century as it spreads through different gene transfer mechanisms and may develop the population of resistant organisms. Antibiotics can be sorbed to soil particles with various physical forces and chemical bonds. Antibiotic sorption depends on physical properties of antibiotics like water solubility, lipophilicity, volatility and their chemical sorption and binding ability. Heavy metals can increase bacteria resistant to antimicrobial in soil. Most heavy metal resistance mechanisms are the same as antibiotic resistance. Also, in most soils heavy metal concentration is much higher than antibiotic concentration. It seems that the first option to control antibiotic resistance in soil bacteria is the evaluating of the resistance in specific habitats like soil, underground waters and manures which can be used to decreasing the antibiotic resistance in the environment. In spite of considerable understanding of mechanisms of action of antibiotic and the antibiotic resistance in soil, following areas need to be studied, with special emphasize to the soil medium and soil microorganisms:

- Investigating the fate of resistance genes in soil and tracking antibiotic resistance genes within soil microorganisms.
- Finding ways for controlling and combating antibiotic resistance in aquatic and terrestrial environments.
- Recognizing the negative effects of antibiotics on soil microbial functions and diversity.
- Studying the effect of antibiotic producers on soil microbial antibiotic resistance and the effect of soil-produced antibiotics on functional diversity of microorganisms in proximity.

References

Alighardashi A, Rashidi A, Neshat AA et al (2014) Environmental risk of selected antibiotics in Iran. IJHSE 1(3):132–137

Amin MM, Hashemi H, Ebrahimi A et al (2012) Effects of oxytetracycline, tylosin, and amoxicillin antibiotics on specific methanogenic activity of anaerobic biomass. Int J Env Health Eng 1(4):1–4

Ansari F (2001) Use of systematic anti-infectives agent in Iran during 1997-1998. Eur J Clin Pharmacol 57(6–7):547–551

Baltz RH (2006) Marcel Faber roundtable: is our antibiotic pipeline unproductive because of starvation, constipation or lack of inspiration? J Ind Microbiol Biotechnol 33:507–513

Baquero F, Alvarez-Ortega C, Martinez J (2009) Ecology and evolution of antibiotic resistance. Environ Microbiol Rep 1:469–476

Berg J, Tom Petersen A, Nybroe O (2005) Copper amendment of agricultural soil selects for bacterial antibiotic resistance in the field. Lett Appl Microbiol 40:146–151

Bibbal D, Dupouy V, Ferré JP et al (2007) Impact of three ampicillin dosage regimens on selection of ampicillin resistance in Enterobacteriaceae and excretion of blaTEM genes in swine feces. Appl Environ Microbiol 73:4785–4790

Binh CTT, Heuer H, Kaupenjohann M et al (2008) Piggery manure used for soil fertilization is a reservoir for transferable antibiotic resistance plasmids. FEMS Microbiol Ecol 66:25–37

Binh CTT, Heuer H, Gomes NCM et al (2010) Similar bacterial community structure and high abundance of sulfonamide resistance genes in field-scale manures. In Manure: management, uses and environmental impacts. Dellaguardia CS. Hauppauge, NY, New York: Nova Science Publishers, p141–166.

Boxall AB, Blackwell P, Cavallo R et al (2002) The sorption and transport of a sulphonamide antibiotic in soil systems. Toxicol Lett 131:19–28

Bradbury JF (1986) Guide to plant pathogenic bacteria. CAB international, Kew, pp. xviii + 332pp

Burkhardt M, Stamm C, Waul C et al (2005) Surface runoff and transport of sulfonamide antibiotics and tracers on manured grassland. J Environ Qual 34:1363–1371

Byrne-Bailey K, Gaze WH, Kay P, Boxall A, Hawkey PM, Wellington EM (2009) Prevalence of sulfonamide resistance genes in bacterial isolates from manured agricultural soils and pig slurry in the United Kingdom. Antimicrob Agents Chemother 53:696–702

Casjens S (1998) The diverse and dynamic structure of bacterial genomes. Ann Rev Genet 32:339–377

Chen W, Liu W, Pan N et al (2013) Oxytetracycline on functions and structure of soil microbial community. J Soil Sci Plant Nutr 13(4):967–975

Chessa L, Pusino G, Pasqualina Mangia N (2016) Soil microbial response to tetracycline in two different soils amended with cow manure. Environ Sci Pollut Res 23:5807–5817

Clewell DB, Flannagan SE, Jaworski DD (1995) Unconstrained bacterial promiscuity: the Tn916–Tn1545 family of conjugative transposons. Trends Microbiol 3:229–236

Cobine P, Wickramasinghe WA, Harrison MD et al (1999) The Enterococcus hirae copper chaperone CopZ delivers copper (I) to the CopY repressor. FEBS Lett 445:27–30

Dantas G, Sommer MO, Oluwasegun RD, Church GM (2008) Bacteria subsisting on antibiotics. Science 320:100–103

Davies J, Spiegelman GB, Yim G (2006) The world of sub-inhibitory antibiotic concentrations. Curr Opin Microbiol 9:445–453

Ding GC, Radi V, Ter-Hai BS et al (2014) Dynamics of soil bacterial communities in response to repeated application of manure containing sulfadiazine. PLoS One 9(3):1–10

Eager RM, Cunningham CC, Senzer N, Richards DA, Raju RN, Jones B et al (2009) Phase II trial of talabostat and docetaxel in advanced non-small cell lung cancer. Clin Oncol (R Coll Radiol) 21(6):464–472

Ebadi S, Sohrabi H, Peymani A et al (2018) Identification of antibiotic-producing Streptomyces species in Iran's soil by phenotypic and genotypic methods. Biotech Health Sci 5(1):e59854

Enne VI, Cassar C, Sprigings K et al (2008) A high prevalence of antimicrobial resistant Escherichia coli isolated from pigs and a low prevalence of antimicrobial resistant E. coli from cattle and sheep in Great Britain at slaughter. FEMS Microbiol Lett 278:193–199

Gao J, Pedersen JA (2005) Adsorption of sulfonamide antimicrobial agents to clay minerals. Environ Sci Technol 39:9509–9516

Ghosh S, LaPara TM (2007) The effects of subtherapeutic antibiotic use in farm animals on the proliferation and persistence of antibiotic resistance among soil bacteria. ISME J 1:191–203

Gotz A, Smalla K (1997) Manure enhances plasmid mobilization and survival of Pseudomonas putida introduced into field soil. Appl Environ Microbiol 63:1980–1986

Gupta G, Parihar SS, Ahirwar NK et al (2015) Plant growth promoting rhizobacteria (PGPR): current and future prospects for development of sustainable agriculture. J Microb Biochem Technol 7:096–102. https://doi.org/10.4172/1948-5948.1000188

Hammesfahr U, Kotzerke A, Lamshöft M, Wilke BM, Kandeler E, Thiele-Bruhn S (2011) Effects of sulfadiazine-contaminated fresh and stored manure on a soil microbial community. Eur J Soil Biol 47:61–68

Hamscher G, Sczesny S, Höper H et al (2002) Determination of persistent tetracycline residues in soil fertilized with liquid manure by high-performance liquid chromatography with electrospray ionization tandem mass spectrometry. Anal Chem 74:1509–1518
Han I, Congeevaram S, Park J (2009) Improved control of multiple-antibiotic-resistance-related microbial risk in swine manure wastes by autothermal thermophilic aerobic digestion. Water Sci Technol 59(2):267–271
Heise J, Höltge S, Schrader S et al (2006) Chemical and biological characterization of non-extractable sulfonamide residues in soil. Chemosphere 65:2352–2357
Heuer H, Smalla K (2007a) Manure and sulfadiazine synergistically increased bacterial antibiotic resistance in soil over at least two months. Environ Microbiol 9:657–666
Heuer H, Smalla K (2007b) Manure and sulfadiazine synergistically increased bacterial antibiotic resistance in soil over at least two months. Environ Microbiol 9:657–666
Heuer H, Krogerrecklenfort E, Wellington EMH, Egan S, Van Elsas JD, Van Overbeek L et al (2002) Gentamycin resistance genes in environmental bacteria: prevalence and transfer. FEMS Microbiol Ecol 42:289–302
Huddleston AS, Cresswell N, Neves M et al (1997) Molecular detection of streptomycin-producing streptomycetes in Brazilian soils. Appl Environ Microbiol 63:1288–1297
Jafari N, Behroozi R, Farajzadeh D et al (2014) Antibacterial activity of *Pseudonocardia* sp. JB05, a rare salty soil actinomycete against *Staphylococcus aureus*. Biomed Res Int 2014:1–7. https://doi.org/10.1155/2014/182945
Jayalakshmi K, Paramasivan M, Sasikala M et al (2017) Review on antibiotic residues in animal products and its impact on environments and human health. J Entomol Zool Stud 5(3):1446–1451
Ji G, Silver S (1995) Bacterial resistance mechanisms for heavy metals of environmental concern. J Ind Microbiol 14:61–75
Jiang SC, Paul JH (1998) Gene transfer by transduction in the marine environment. Appl Environ Microbiol 64:2780–2787
Kachur AV, Koch CJ, Biaglow JE (1998) Mechanism of copper-catalyzed oxidation of glutathione. Free Radic Res 28:259–269
Klein EY, Van Boeckel TP, Martinez EM et al (2018) Global increase and geographic convergence in antibiotic consumption between 2000 and 2015. Proc Natl Acad Sci USA (PNAS) 115(15):E3463–E3470. https://doi.org/10.1073/pnas.1717295115
Kong WD, Zhu YG, Fu BJ et al (2006) The veterinary antibiotic Oxytetracycline and Cu influence functional diversity of the soil microbial community. Environ Pollut 143:129–137
Kong WD, Li CG, Dolhi JM et al (2012) Characteristics of Oxytetracycline sorption and potential bioavailability in soils with various physical-chemical properties. Chemosphere 87:542–548
Kumar K, Gupta SC, Chander Y et al (2005) Antibiotic use in agriculture and its impact on the terrestrial environment. Adv Agron 87:1–54
Kümmerer K (2009) Antibiotics in the aquatic environment–a review–part I. Chemosphere 75:417–434
Levy SB (2002) The antibiotic paradox: how the misuse of antibiotics destroys their curative powers, 2nd edn. Int Microbiol 5:155–156. https://doi.org/10.1007/s10123-002-0082-z.
Levy SB, Marshall B (2004) Antibacterial resistance worldwide: causes, challenges and responses. Nat Med 10:S122–S129
Li Y, Yu Y, Yang Z et al (2016) A comparison of metal distribution in surface dust and soil among super city, town, and rural area. Environ Sci Pollut Res 23:7849–7860
Lin H, Jin D, Freitag TE et al (2016) A compositional shift in the soil microbiome induced by tetracycline, sulfamonomethoxine and ciprofloxacin entering a plant-soil system. Environ Pollut 212:440–448
Liu B, Li X, Zhang X, Wang J, Gao M (2015) Effects of chlortetracycline on soil microbial communities: comparisons of enzyme activities to the functional diversity via Biolog EcoPlates™. Eur J Soil Biol 68:69–76
Lorenz MG, Wackernagel W (1994) Bacterial gene transfer by natural genetic transformation in the environment. Microbiol Rev 58:563–602

Lunsford RD (1998) Streptococcal transformation: essential features and applications of a natural gene exchange system. Plasmid 39:10–20
Lv G, Pearce CW, Gleason A et al (2013) Influence of montmorilonite on antimicrobial activity of tetracycline (TC) and ciprofloxacin (CIP). J Asian Earth Sci 77:281–286
Martinez JL (2009) Environmental pollution by antibiotics and by antibiotic resistance determinants. Environ Pollut 157(11):2893–2902
Mashkoori B (2014) The study of number of enteric bacteria and their antibiotic resistance in industrial and traditional dairy cow manures in Hamedan. MSc thesis, Bu-Ali Sina University, Hamedan, Iran (in Persian)
McKinney CW, Loftin KA, Meyer MT et al (2010) tet and sul antibiotic resistance genes in livestock lagoons of various operation type, configuration, and antibiotic occurrence. Environ Sci Technol 44:6102–6109
Molaei A, Lakzian A, Datta R et al (2017a) Impact of chlortetracycline and sulfapyridine antibiotics on soil enzyme activities. Int Agrophys 31:209–505
Molaei A, Lakzian A, Gh H et al (2017b) Assessment of some cultural experimental methods to study the effects of antibiotics on microbial activities in a soil: an incubation study. PLoS One 12(7):e0180663
Nielsen KM, van Weerelt MD, Berg TN et al (1997) Natural transformation and availability of transforming DNA to Acinetobacter calcoaceticus in soil microcosms. Appl Environ Microbiol 63:1945–1952
Nies DH (1999) Microbial heavy-metal resistance. Appl Microbiol Biotechnol 51:730–750
Obst U, Schwartz T, Volkmann H (2006) Antibiotic resistant pathogenic bacteria and their resistance genes in bacterial biofilms. J Artif Organs 29:387–394
Ohlsen K, Ziebuhr W, Koller KP, Hell W, Wichelhaus T, Hacker J (1998) Effects of subinhibitory concentrations of antibiotics on alpha-toxin (HLA) gene expression of methicillin-sensitive and methicillin-resistant Staphylococcus aureus isolates. Antimicrob Agents Chemother 42:2817–2823
Peak N, Knapp CW, Yang RK et al (2007) Abundance of six tetracycline resistance genes in wastewater lagoons at cattle feedlots with different antibiotic use strategies. Environ Microbiol 9:143–151
Poole RK, Gadd GM (1989) Metal-microbe interactions. Published for the Society for General Microbiology/Published for the Society for General Microbiology by IRL Press. Oxford/New York
Punitha BC, Hanumantharaju TH, Jayprakash R, Shilpashree VM (2012) Acetamiprid impacton urease and phosphatase activity in selected soils of southern Karnataka. Intl J Basic Appl Chem Sci 2:1–6
Raaijmakers JM, Mazzola M (2012) Diversity and natural functions of antibiotics produced by beneficial and plant pathogenic bacteria. Annu Rev Phytopathol 50:403–424
Rashtbari M (2019) Effect of various agricultural antibiotics and biochar and Nano-zeolite amendments on soil microbial population, biodiversity and biological interactions of chickpea (*Cicer arietinum* L.). PhD thesis in soil biology and biotechnology, Bu Ali Sina University, Hamadan, Iran (unpublished data)
Ruiz N, Montero T, Hernandez-Borrell J et al (2003) The role of Serratia marcescens porins in antibiotic resistance. Microb Drug Resist 9:257–264
Safari Sinegani AA, Younessi N (2017) Antibiotic resistance of bacteria isolated from heavy metal-polluted soils with different land uses. JGAR 10:247–255
Salyers AA, Ama'bile-Cuevas CF (1997) Why are antibiotic resistance genes so resistant to elimination? Antimicrob Agents Chemother 41:2321–2325
Schauss K, Focks A, Heuer H et al (2009) Analysis, fate and effects of antibiotic sulfadiazine in soil ecosystems. Trends Anal Chem 28:612–618
Schwaiger K, Harms K, Ho¨ lzel CS et al (2009) Tetracycline in liquid manure selects for co-occurrence of the resistance genes tet(M) and tet(L) in Enterococcus faecalis. Vet Microbiol 139:386–392

Silver S, Walderhaug M (1992) Gene regulation of plasmid-and chromosome-determined inorganic ion transport in bacteria. Microbiol Rev 56:195–228
Srivastava N, Majumder C (2008) Novel biofiltration methods for the treatment of heavy metals from industrial wastewater. J Hazard Mater 151:1–8
Telesinski A, Platkowski M, Cybulska K, Telesinska N, Wrobel J, Pawlowska B (2018) Response of soil enzymes to two antibiotics: polymyxin B and penicillin G. Fresenius Environ Bull 27(5A):3837–3845
Ter Laak TL, Gebbink WA (2006) Estimation of soil sorption coefficients of veterinary pharmaceuticals from soil properties. Environ Toxicol Chem 25:933–941
Thomas CM, Nielsen KM (2005) Mechanisms of, and barriers to, horizontal gene transfer between bacteria. Nat Rev Microbiol 3(9):711–721
Thompson CL, Wang B, Holmes AJ (2008) The immediate environment during postnatal development has long-term impact on gut community structure in pigs. ISME J 2:739–748
Topp E, Chapman R, Devers-Lamrani M et al (2013) Accelerated biodegradation of veterinary antibiotics in agricultural soil following long-term exposure, and isolation of a sulfamethazine-degrading sp. J Environ Qual 42:173–178
van Overbeek LS, Wellington EM, Egan S et al (2002) Prevalence of streptomycin-resistance genes in bacterial populations in European habitats. FEMS Microbiol Ecol 42:277–288
Verlicchi P, Galletti A, Masotti L (2010) Management of hospital wastewaters: the case of the effluent of a large hospital situated in a small town. Water Sci Technol 61:2507–2519
Wegst-Uhrich SR, Navarro DAG, Zimmerman L et al (2014) Assessing antibiotic sorption in soil: a literature review and new case studies on sulfonamides and macrolides. Chem Cent J 8(5):1–12
Wei X, Wu SC, Nie XP, Yediler A, Wong MH (2009) The effects of residual tetracycline on soil enzymatic activities and plant growth. J Environ Science Health B 44:461–471
Wise R (2002) Antimicrobial resistance: priorities for action. J Antimicrob Chemother 49:585–586
Wright GD (2007) The antibiotic resistome: the nexus of chemical and genetic diversity. Nat Rev Microbiol 5:175–186
Youncssi N (2017) Evaluation of microbial tolerance to some heavy metals and antibiotics in metal-contaminated soils and water resources in Hamadan province. PhD thesis, Bu-Ali Sina University, Hamedan, Iran (in Persian)
Younessi N, Safari Sinegani AA, Khodakaramian GH (2017) Detection Beta-lactamase gene in the culturable bacteria isolated from agricultural, pasture and mining soils around mines in Hamedan, Iran. Biol J Microorg 6(21):36–48. In Persian with English Summary
Younessi N, Safari Sinegani AA, Khodakaramian G (2019) Detection of antibiotic resistance genes in culturable bacteria isolated from soils around mines in Hamedan. Iran Int J Environ Sci Technol. https://doi.org/10.1007/s13762-018-02178-2. (online published)
Zeaiter Z, Mpelli F, Crotti E et al (2018) Methods for the genetic manipulation of marine bacteria. Electron J Biotechn 33:17–28

Microbial Adaptation and Resistance to Pesticides

12

Debarati Paul and Santi M. Mandal

Abstract

Dramatic escalations in fabrication and application of chemicals in the form of pesticides, explosives, dyes, drugs, antimicrobial agents, etc has had pervasive impacts on the microbial population, flora and fauna dwelling in the contaminating habitat, with creation of "mosaic pathways" or "mosaic organisms", leading to evolution of species. It is well understood that gene transfer between microbes found in natural systems, assist them to tolerate and resist the effects of antimicrobials, e.g. in farms where cattle are sheltered and agricultural lands where various pesticides are used time and again. It has been suggested that at low concentrations some antibiotics serve as signaling molecules therefore few of the genes encoding antibiotic resistance were originally selected for metabolic functioning or for signal transductions in their host cells. Higher concentration of antibiotics discharged in specific habitats (e.g., clinics/hospitals) due to irresponsible human activity has the potential to shift these metabolic roles toward development of resistance to antibiotics/drugs. The chapter describes the occurrence of cross or co resistance to biocides and/ antibiotics exhibited under persistent and selective pressure highlights the significance of mobile genes and lateral genetic transfers from one to another microbe. Such phenomenon is now common in particular areas where agricides occur in recalcitrant/ lesser concentrated state in soil and water, and is transmitted directly to human or enter our bodies via food chain thereby facilitating the switch from cross to co-resistance.

D. Paul (✉)
Amity Institute of Biotechnology, Amity University, Noida, Uttar Pradesh, India
e-mail: dpaul@amity.edu

S. M. Mandal
Central Research Facility, Indian Institute of Technology Kharagpur, Kharagpur, India

S. M. Mandal, D. Paul (eds.), *Bacterial Adaptation to Co-resistance*,
https://doi.org/10.1007/978-981-13-8503-2_12

Keywords
Pesticides · Agricides · Mosaic organism/pathways · Cross-resistance and co-resistance

12.1 Pesticides and Their Brief History

Pesticides are substances that used to prevent, mitigate, destroy or inhibit the growth of insect pests or unwanted animals such as rodents, snails, etc. The use of naturally occurring substances as pesticides were very common in olden times termed as 'biological poisons' including arsenic, pyrethrum, cyanide, sulfur, and lead. Elemental sulfur (S) and chemicals containing Arsenic (As) and lead (Pb) protected crops from insect pests starting from ancient Sumerian period to medieval times. In the nineteenth century, more natural compounds prepared from roots of tropical vegetables and chrysanthemums were researched upon for use as pesticide. The twentieth century witnessed a massive explosion in the World's population and food resources were in great demand for all. During World War II, scientific attempts directed toward the making and distribution of chemical weapons for mass destruction, consequently, steered toward the invention of "synthetic pesticides" for human welfare and increased agricultural resources.

The ill-famed DDT (dichlorodiphenyltrichloroethane) developed around 1939 showcases both the usefulness of such new pesticides and their innate dangers. DDT was invented by a Swiss chemist, as a persistent, insoluble and inexpensive weapon against insects. It soon gained popularity amongst home coming soldiers to ward off lice and eradicate mosquito/malaria problems in multiple nations. Only 20 years later, mankind became aware of its serious effects on after biomagnifications and led 86 countries to ban its use.

With dramatic escalations in fabrication and application of chemicals such as dithiocarbamic fungicides, BHC, DDT and 2,4-D, it was apparent that pesticide regulation and laws needed to be upgraded. Pesticides were found to be associated to a number of disorders and health problems, such as, neurologic and endocrine (hormone) system disorders, birth defects, cancer, and other diseases. The average American child (6–11 years old) were reported to possess obnoxious levels of the neurotoxic pesticides, e.g. organophosphates namely chlorpyrifos and methyl parathion [CDC (Centers for Disease Control and Prevention) data].

Addition of pesticides also has widespread effects on the microbial population, flora and fauna dwelling in the contaminating habitat (Table 12.1). Any natural or synthetic chemical when added to the soil select for the species that would be resistant and/or would be able to transform or degrade it. When PCBs (poly chlorinated biphenyls) were added to soil "mosaic pathways" evolved for the efficient metabolization of the xenobiotics by cooperative biochemical functioning, leading to a change in the overall microbial ecology (Abraham et al. 2002). The inherent potential of bacteria and bacterial enzymes for metabolism of added organophosphates, atrazine or PCBs (Abraham et al. 2002; Seffernick and Wachett 2001) has steered

Table 12.1 showing a list of pesticides classified on the type of pest with widely used examples

Pesticide class	Targeted pest(S)	Benefit during pest control	Examples
Acaricides/miticides	Mites	Stop sucking insect pests of plants or from animals,	Permethrin, dicofol, Nissorun 10WP
Algicides	Algae, marine plants, scum	Kill undesired algae	Yates bluestone copper Sulphate; Coptrol aquatic Algicide
Avicides	Birds	Protects crops from damage by birds	Strychnine, DRC-1339, Starlicide, Avitrol
Bactericides	Bacteria	Kill bacteria in desired locations	Validamycin, streptomycin
Disinfectants/ biocides/ antimicrobials	Microorganisms of various types, viruses	Eradicates microbes from target area, e.g., disinfection, sterilization, sanitization	Formaldehyde, GlutaraldehydeOrtho-phthalaldehyde
Fumigants	Nematodes, weed seeds, fungi, insects, etc	Exterminates undesired species from soil, commodities or space	Methyl bromide, phosphine
Fungicides	Fungi	Exterminates fungi causing plant diseases	Bravo Ultrex, Daconil Zn
Herbicides	Undesired plants (weeds)	Eradicates weeds that otherwise cause economic damage due to use of water, nutrients and light; visual nuisance eliminated	Nufarm simazine 900 DF herbicide; Farmoz simazine 900 WDG herbicide
Insecticides/ins. Growth regulators	Insects	Removes disease threat directed to humans & animals, contamination of merchandise/properties	SevinBonide captain Jack, Monterey garden insect spray
Moluscicides	Invertabrates, e.g., snails, slugs	Abolishes trouble or economic damage of invertebrates to valued plants or crops	Allicin, fentin
Piscicides	Fishes	Extremination of unsatisfactory fish from targeted waters	Rotenone, Fintrol
Rodenticides	Rodents	Abolishes disease to humans and damage to commodities/premises	Chlorophacinonediphacinone
Slimicides	Various lower plant/ animal forms, microbes	Checks slime formation of slime in aquatic environments	South Gippsland water copper Sulphate, Simagranz herbicide

our focus on using directed evolution (site directed mutagenesis and chromosome shuffling) for producing beneficial strains suitable for bioremediation and also industry. However, the change in ratios of the selective strains and altered dominance of few strains in the ecosystem might affect the beneficial microbes (involved in nutrient mobilization) and thus the functioning of the ecosystem on the whole.

Pesticides were question at the vanguard of the environmental movement that led to the launching of EPA. Rachel Carson's book "Silent Spring" was published in 1962 which sensationalized the dangers of DDT (and other pesticides) and resulted ingelling the populaces' anxieties on handling and application of chemicals issued that contaminated the air, water, wildlife and food supplies (and as found as residues in human tissues). The President's Science Advisory Committee (1963) framed a report on "The Use of Pesticides" which emphasized on reduction in the use of pesticides. In 1969,"HEW Secretary's Commission on Pesticides and Their Relationship to Environmental Health" also called Mrak Commission), released a report to emphasize upon the abolishing of DDT (except in essential cases of public health) due to its persistence and damaging effects on environment and ecology. The report also recommended the restricted use of other harmful pesticides to "essential uses" to limit the damages imposed by them to mankind and other life forms and vast environmental hazards (Mrak 1969, pp. 8–9).

In 1947, Federal Fungicide, Insecticide and Rodenticide Act (FIFRA) was imposed, when the synthetic organic pesticide industry was ready to thrive and blossom. FIFRA protected society from ineffective or dangerous products and USDA approved product labels prior to sale of products. Products were to be safe when used as directed by the label. However, this act was defective as it did not discontinue the sale orders for selling dangerous pesticides and restricted punishments for companies selling the harmful products. (Briggs 1992 p. 279) Such companies would acquire a "protest registration" and sell the product without proper registration and labelling by USDA (Briggs 1992 p. 279). Therefore, FIFRA was amended later on, and it was now mandatory to procure a federal registration number for registration of pesticides (1959), include forewarnings on labels (1961) and delete safety assertions from labels (1964).

FIFRA (Sec. 2) has defined a pesticide as: "(1) any substance or mixture of substances intended for preventing, destroying, repelling, or mitigating any pest, (2) any substance or mixture of substances intended for use as a plant regulator, defoliant or desiccant, and (3) any nitrogen stabilizer". In recent times certain microorganisms, such as bacteria have been found to be competent as pesticide active ingredients. Such natural microbial-pesticides inhibit and destroy pests, without showing any adverse effect on other organisms and/or do not give rise to harmful residues. Therefore such "safer" pesticides or 'biopesticides' are not governed by identical and rigorous registration obligations as chemical pesticides. Biochemicals, that are naturally occurring (or identical) chemicals may be considered as "safer" chemicals and therefore are granted expedited registration since they are naturally environment-friendly.

12.2 Pesticides: Broad Spectrum and Narrowspectrum

Pesticides are categorized as 'broad spectrum' and 'narrow spectrum' based on the types of pests that they are active against. Broad spectrum pesticides destroy or inhibit pests regardless of their species and include most organophosphate, neonicotinoid, carbamate insecticides and pyrethroid are indicated on the labels of commercial pesticides. Certain broad spectrum pesticides, such as chlorpyrifos, are made to act selectively when used in moderate concentrations. Broad spectra pesticides have the potential to impact large number of natural enemies and also beneficial ones. If the balance is destroyed and greater numbers of natural enemies survive as compared to beneficial ones, they will pose potential dangers for the environment. However, once applied they help to control harmful species in the following season and restrict the quantity of pesticides that is required to be re-applied.

Narrow spectrum pesticides or selective pesticides, can be potent against a certain kind of insect or family of insects. These chemicals are formulated to act against one kind of garden pests such e.g. ants, while the 'biocontrol' agents (predatory bugs in this case) that naturally feed on pests, remain untouched. This selective targeting encourages natural predators of pests to facilitate in combating the population of unwanted insects. However, narrow spectrum pesticides might not be safe for other applications exterior to the particular target. For example, Brodifacoum, a rodenticide, is poisonous to dogs that accidentally ingest it.

The farmer should be informed about the persistence of the pesticide and the frequency of reapplication required in each season. For example, Acetamiprid, is a broad-spectrum pesticide having long-lasting influence on natural enemies of the useful insects, but, required re-application to prevent harmful insects. On the other hand, Carbaryl bait is a narrow-spectrum pesticide that targets mites, does not show any influence on the natural enemies of the mites. The natural enemies will therefore control mites later on, in the season, without the need to reapply the same pesticide. Thus, the nonchalant use of pesticides (even specific ones) may challenge biodiversity and may result in the 'flare up' of unwanted weeds, damaging insects, and/or resistant pathogens.

Plants depend on beneficial microbes for availability of nutrients and to protect them from pathogens or degrade the compounds that might retard/inhibit their growth. Thus plant and soil microorganisms create a dynamic living system that is suited to either species for proper functionality. Multiple and repeated doses of pesticides cause reduction of soil diversity by abolishing the non-resistant strains and allowing dominance of only resistant species. Application of chemical fertilizers also impact soil microbial ecology as well as the plants. Unlike organic fertilizers, the inorganic ones do not provide nourishment to microbes; rather plants become adapted to higher levels of nutrients and water and microbes decline in numbers and indirectly cause increase in the numbers of pathogens (that are now ready to feed on plants).

12.3 Resistance to Increased and Repeated Application of Pesticides in Microbes and Insects

The ecosystem is balanced by the fine tuning between the crops and microbes. Grasses, trees or crops are assisted by soil microbes in obtaining water, solubilizing nutrients, protection from pests and pathogens, preventing nutrient losses and in breaking down inhibitory compounds that might retard growth. These microbes, in return, are benefitted by healthy development of plants inhabiting their niche and substances secreted from the plants root system. This synergistic development creates anenergetic living system that is undoubtedly disrupted by conventional systems that use pesticides, herbicides and fertilizers. The chemicals used to enhance plant growth may/may notcase detrimental effects on the soil microbes by killing or causing mutation pressure on them. While these chemicals probably target only specific species, their repeated use unavoidably annihilates the valuable microbes that otherwise contributed toward the balance of the ecosystem. Surviving microbes may become mutated or genetically altered such that they do not benefit the soil ecosystem in an attempt to become resistant to the applied chemical. The obliteration or modifications of first-level microorganisms affect the overall ecosystem of a given habitat from the beginning to the top-most level comprising of largest mammalian predators.

A review by Lo (2010) based on research carried out for 20 years showed that some pesticides stimulated the growth of microbes, but other pesticides had negative effects or no visible effects on the microbiota. Data showed that on one hand carbofuran stimulated the growth of anaerobic nitrogen fixers in flooded and non-flooded soil, however; on the other hand, another pesticide butachlor reduced *Azospirillum* and aerobic nitrogen fixers in non-flooded soil. Diuron and chlorotoluron had no effect on microbial populations although, linuron strongly affected microbes. Phosphorus(P)-containing herbicides e.g. glyphosate and insecticide e.g. methamidophos encouraged microbial growth in the soil, but other P-containing insecticide e.g. fenamiphos showed negative effects on bacteria responsible for nitrification. The adverse effect of pesticides to microbial populations has been exemplified similarly through other case studies. In a report by Clapperton and Regen (2012), it was shown that pesticides affected the soil biota, and soil ecosystem functioning specifically in a 'no-tillage' system. Increase in glyphosate (pesticide) levels in the rhizosphere attracted *Fusarium* (a fungus) and this phenomenon was consistent with the finding that over a short time period glyphosate actually stimulated fungal growth (Kremer et al. 2005, Araújo et al. 2003). During this time, consequently the ecological harmony between beneficial bacteria and fungi changed and root damage by the fungal pathogen further unbalanced populations and diversity in the rhizosphere community because pathogenic nematodes subsequently attracted to the areas of root damage. The plant reduces photosynthesis, and channelizes more energy to block the root damage. As a result, root exudates change in composition and subsequently influence the population, diversity, and activities/metabolism of the rhizosphere microorganisms. Consequently, the fungal-feeding nematodes (non-pathogenic to plants) increases in numbers opportunistically due to the flourishing fungi (non-mycorrhizal), and the ecosystem thereby attain a new balance. However, few plants might expire due to the combined effect of *Fusarium*, predatory nematodes, and other diseases.

Many bacteria are known to degrade pesticides as these bacteria are capable of using specific pesticides to meet their energy needs, i.e., using them as food. For example, bacterial strains isolated from triazine contaminated soil were capable of using triazine, simazine and atrazine for their energy source (Dinamarca et al. 2007). Triazine herbicides, e.g. atrazine, were introduced by man into the environment one-half century ago. Triazine compounds were initially found to be poorly biodegradable. Over several years after introduction, environmental half-lives of atrazine were measured again and they were found to be variable of 60–400 days. Evidence of nonspecific oxidation reactions, resulting in partial metabolism in soil, was obtained via soil metabolite and biochemical studies. More recently, environmental half-lives were re-calculated and were found to have decreased considerably; being only about 1–50 days. *Arthrobacter aurescens* TC1 is unique as it can utilize various s-triazines for both carbon and nitrogen sources. *A. aurescens* TC1 was isolated at a roadside spill site, where 1000 pounds of atrazine had contaminated 35 cubic yards of soil; the site was bioremediated to acceptable levels with approval by the U.S. Environmental Protection Agency (Shapir et al. 2007).

Paul et al. (2005a, b, 2006) reported the complete degradation of *p*-nitrophenol (PNP) that is formed due to the degradation of nitrophenolic organophosphate pesticides such as parathion and methyl parathion. They also reported the adaptation of some microorganisms for swimming toward PNP (chemotaxis) following a concentration gradient and degrading it completely. The persistence of insecticides and their effects on the rhizosphere microbes are more detrimental as compared to both herbicides and fungicides. Mostly insecticides are readily degraded by soil bacteria. Chlorpyrifos (Lorsban), a commonly used neurotoxin insecticide, may degrade within 20 days, (Lakshmi et al. 2008) and all negative effects of the chemical on soil bacteria and fungi impacted during that time recovers in few weeks (although, impact on soil fauna are comparatively more persistent). Chauhan et al. (2010) and Zhang et al. (2012) reported the oxidative degradation pathways of PNP by bacteria such *Pseudomonas* sp. and *Rhodococcus opacus* suggesting the complete removal of OP pesticides and its persistent residue (PNP) from soil.

Another interesting fact about the microbial utilization and consequently degradation of pesticides is exemplified by mineralization of the persistent chlorinated pesticide lindane. Indiscriminate use and extensive production of HCH (hexachlorocyclohexane) during the past 50 years created a serious environmental problem. Bacteria degraded HCH isomers were isolated from HCH contaminated soils procured from various different locations around the world (from the family Sphingomonadaceae). Interestingly, all these bacteria contain nearly identical *lin* genes (responsible for HCH degradation), which diverged to perform different catabolic reactions. The organization and diversity of *lin* genes studied amongst several sphingomonads, showed association with plasmids and IS6100. These element suggest that horizontal transfer of *lin* genes led to the distribution and diversity of

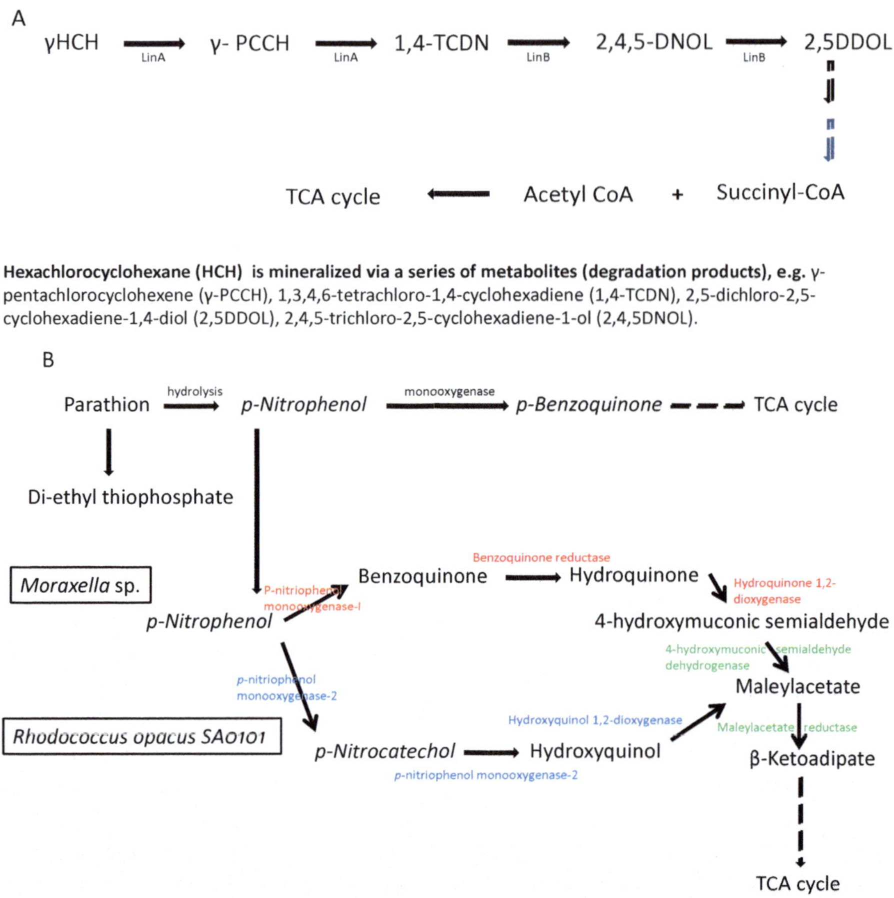

Fig. 12.1 (**a**) Degradation of HCH via *lin* gene(s) (**b**) Degradation of parathion via the formation of PNP which is degraded further via two pathways (**c**) Degradation of BPCs via the formation of various metabolites generated during the oxidation of diphenylmethane (upper) or benzophenone (lower) by isolated enzymes of biphenyl pathway or induced bacterial cells. (Adapted from J. Bacteriol. 2013; 195 (16): 3563–3574 and J. Soil Sci. Plant Nutr. 2010; 10 (3): 320–332) (**d**) Degradation of triazine by bacterial metabolism

microbes capable of HCH degradation. *Clostridium rectum*, *C. butyricum*, *C. pasteurianum*, *Citrobacter freundii*, *Desulfovibrio gigas*, etc. are anaerobic isolates that can catabolically break down one or more of the four HCH isomers (Lal et al. 2010). Striking similarities were disclosed in the mechanisms by which the aerobic HCH-degradative pathways evolved and those that have been found for other xenobiotics (Top and Springael 2005). Inevitable role of horizontal gene transfer amongst different species and insertion sequence-mediated transposition within the same species was predicted to be involved for a quick (50 years) assemblage of genes encoding a novel catabolic pathway (Lal et al. 2005). Figure 12.1 shows the

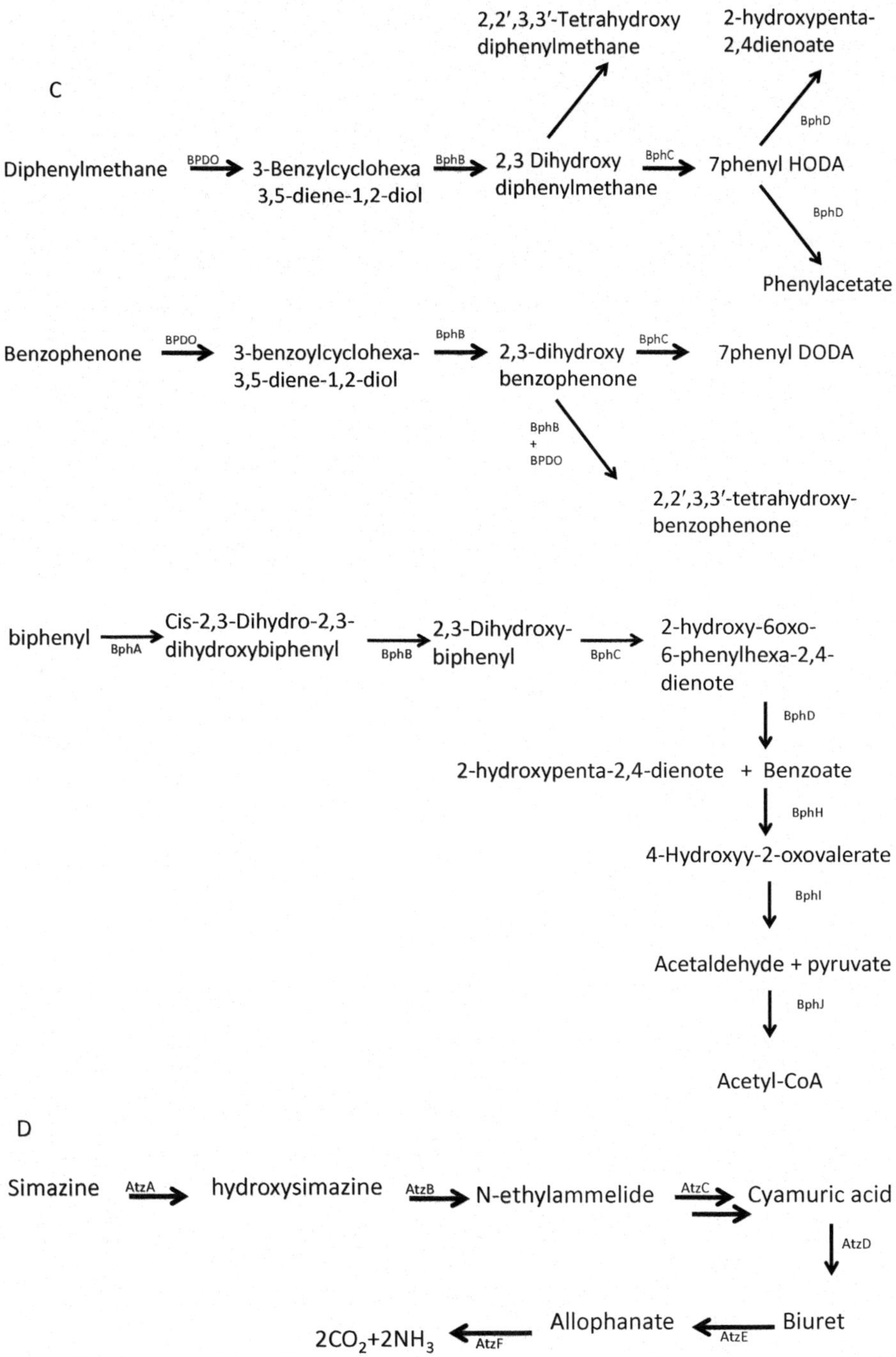

Fig. 12.1 (continued)

degradation of HCH via *lin* genes and parathion via the formation of PNP by several monooxygenases and oxido-reductases. These studies and more enforce the importance of evolution and adaptation of microbes for acquiring the capability of degrading pesticides for (a) survival in the presence of otherwise destructive xenobiotic compounds and (b) utilization of the pesticide as food (Table 12.2).

Similar or even more severe adverse effects are seen on the fauna due to repeated and indiscriminate use of pesticides of a long period. Repetitively using the same class of pesticides for controlling pests triggers unwanted variations in its gene pool resulting in another form of artificial selection, i.e. pesticide resistance. A small fraction of the pest population may survive exposure to the pesticide when it first applied to a particular habitat, due to the distinct genetic makeup of the pests. These genes are transmitted to the next generation for developing resistance. Subsequent applications of the same pesticide, selectively increases the number or ratio of less-susceptible individuals in the population. Repeated doses of pesticide application process of selection, the population gradually develops resistance to the pesticide. Across the word, about 500 species or more of insects, mites, and spiders have acquired certain level of pesticide endurance. The two spotted spider mite is a pest of most fruit crops and is disreputable for its remarkable capability of rapidly developing resistance to miticides (http://msue.anr.msu.edu/topic/grapes/integrated_pest_management/how_pesticide_resistance_develops).

By cultivating detoxifying bacteria (e.g. fenitrothion-degrading *Burkholderia* strains) in its gut, a pest called the bean bug (a major pest of soybean crops) can become instantly resistant to a common insecticide (Kikuchi et al. 2012). Their finding suggested that there is a possibility that the insecticide resistance may develop even in the absence of pest insects, and potentially move around horizontally between different pest insects and other organisms.

There are a number of means by which insects become resistant to insecticides that are used for crop protection and/or as public health products:

(i) Metabolic resistance

Resistant insects use their internal enzyme systems to break down insecticides faster as compared to susceptible strains and/or eliminate them from their system. Resistant strains may possess either elevated amounts or more efficient varieties of these enzymes and/or (ii) such enzyme systems may demonstrate a broader spectrum (degrade many different insecticides).

(ii) Target-site resistance

The target site is where the insecticide acts in any insect. This site may be genetically modified to (i) prevent the insecticide binding or (ii) interacting thus reducing or preventing the adverse pesticidal effects. Bacteria are capable of not only altering the enzyme targeted by antibiotics, but also by the use of enzymes to modify the antibiotic itself and thus neutralize it.

Table 12.2 shows a list of pesticides and microorganisms that degrade them along with some of the enzymes that mediate the degradation process

Xenobiotic pesticide	Pesticide class	Degrading microorganism	Enzyme involved	Ref
Coumaphos	Organophosphate insecticide	E coli (transformed)	Diethylthiophosphatase	Mansee et al. (2005)
Propachlor	Acetamide herbicide	*Pseudomonas* strain PEM1; Acinetobacter strain BEM2	Catechol oxygenase	Martín et al. (2000)
Chlorpyrifos	Organophosphate insecticide	Pseudomonas sp.	Chlorpyrifos hydrolase	Latifi et al. (2012)
HCH	Organochlorines insecticide	Sphingomonaspaucimobilis	Haloalkane dehalogenases	Lal et al. (2005)
DDT Endosulfan	Organochlorines insecticide	*P. aeruginosa; F. oryzihabitans*	Monooxygenase	Barragán et al. (2007)
Coumaphos, diethylphosphate and chlorferon	Organophosphate insecticide	*Escherichia coli (OPH), Pseudomonas monteilli* strain C11	Phosphotriesterase	Ha et al. (2009), Horne et al. (2002)
2,4-D DDT	Organochlorines herbicide and insecticide	*Pseudomonasfluorescens*	Oxidoreductases	Santacruz et al. (2005)
Methyl parathion/ tetrachlorvinphos	Organophosphate insecticide	Bacterial consortia	Hydrolase	Yañez-Ocampo et al. (2009)
Pyridaben	Acaricide (kills ticks and mites)	*Chryseobacterium, Acidovorax*	Not known	Yoon et al. (2010)
Cypermethrin	Synthetic pyrethroid	*Devosiayakushimensis, Pseudomonas aeruginosa* PAL106	Monooxygenase (CMO)	Naphade et al. (2012)

(iii) Penetration resistance

Resistant insects might absorb the toxin at a slower rate than susceptible insects. Resistance to penetration in insect prevents a wide range of insecticides because the outer cuticle acquires barriers to impede the absorption of the pesticides via the cuticle. Penetration resistance is frequently acquired in combination with other forms of pesticide resistance; reduced penetration consequently intensifies the overall effects of other resistance mechanisms.

(iv) Behavioural resistance

Resistant insects try to avoid the toxin (pesticides) as in the case of organochlorines, organophosphates, carbamates and pyrethroids. Insects consequently discontinue feeding in the incidence of certain insecticides, or abandon the area after spraying (for instance, they may move to the underside of a sprayed leaf, move deeper in the crop canopy or fly away from the target area).

12.4 Implication for Drug Resistance by Microbes

As in the case of pesticides, repeated or prolonged use of antibiotics causes bacteria to develop endurence to certain varieties of antibiotics. This means that the drugs are no moreactive at eradicating the bacteria and the infections become trickier to treat. If bacteria are resistant to more than one type of antibiotic, they are called multidrug-resistant organisms, or MDROs. There are several types of MDROs:c arbapenem-resistant Enterobacteriaceae (CRE), methicillin-resistant *Staphylococcus aureus* (MRSA), vancomycin-intermediate or vancomycin resistant *Staphylococcus aureus* (VISA/VRSA), and vancomycin-resistant enterococci (VRE). "Cross-resistant" bacteria are those that can survive in the presence of various antimicrobial molecules using similar defense mechanism(s). Bacteria transfer few genes or part of genetic material to other bacteria surviving in the same environment. Co-resistance is the property acquired when genetic information coding for various unrelated defense mechanisms is transferred in a single event and is also expressed in the 'new' bacterial host. This new strain conferred with the property of "co-resistance" to several antimicrobials (Todar 2008).

Examples of target-altering pathogens are *Staphylococcus aureus*, vancomycin-resistant Enterococci and macrolide-resistant *Streptococcus*, while examples of antibiotic-modifying microbes are *Pseudomonas aeruginosa* and aminoglycoside-resistant *Acinetobacter baumannii* (Fisher and Mobashey 2010). Kemboi et al. (2014) have shown the role played by horizontal gene transfer of antibiotic resistance genes from *Salmonella* spp. to *E. coli* there by conferring resistance to ampicillin, tetracyclin, nalidixic acid, etc. Some antibacterial drugs target only specific bacterial molecules (almost always proteins) and any mutation in these molecules

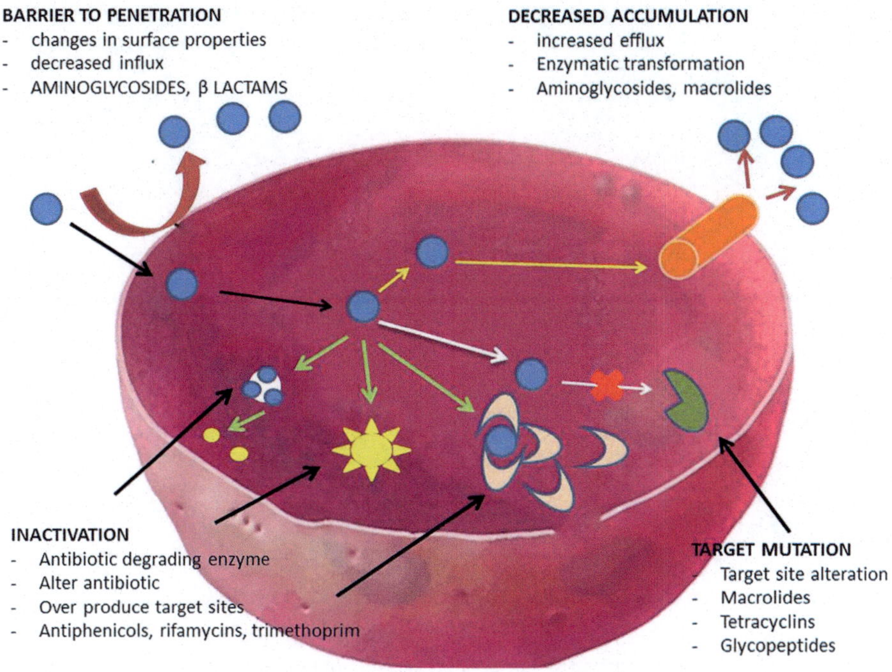

Fig. 12.2 shows the various mechanisms for developing resistance to drugs by pathogens

obstruct or reverse the critical effect of antimicrobial drugs, thereby resulting in antibiotic resistance (Fig. 12.2).

Recent works advocate that some antibiotics serve as signaling molecules at low concentrations that might be found in natural ecosystems, and therefore some of the genes encoding antibiotic resistance were originally selected for metabolic functioning or for signal transductions in their host cells. Higher concentration of antibiotics discharged in specific habitats (e.g., clinics/hospitals) due to irresponsible human activity has the potential to shift these metabolic roles (Martínez 2009) toward development of resistance to antibiotics/drugs. Horizontal gene transfer (Fig. 12.3) is the process in which an organism incorporates genetic material from another organism without being the offspring of that organism (Todar 2008). A resistant gene transfer in bacteria in actually the horizontal gene transfer where different bacteria of the different species (one having developed resistant gene) transfer resistant gene to the one having not developed resistance to antibiotics. Horizontal gene transfer (HGT) is a process where small amounts of DNA or genetic material can be transferred between individual bacteria of the same species or even between different species. Conjugation occurs when there is direct cell-cell contact between two bacteria (may/may not be evolutionarily close) and transfer of small pieces of DNA called plasmids takes place. This is thought to be the

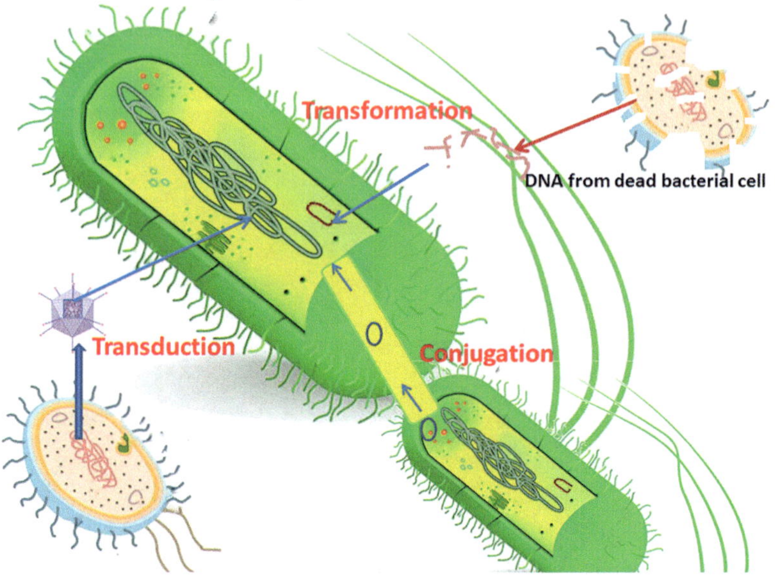

Fig. 12.3 showing the exchange of gene(s) via horizontal gene transfer for formation of novel pathways of pesticide degradation and/or building antimicrobial resistance

main mechanism of HGT. Acquisition of the resistance gene has been attributed to the increasing number of the resistant bacteria and the ability to acquire the resistance gene. Transduction via virus as the carrier of genetic information also leads to development of resistance genes. IS insertions has been reported to bring about mutations in bacterial genome leading to both drug resistance or improvising degradation of pesticides.

Several plasmids encode toxin–antitoxin systems (Paul et al. 2005b). If these plasmids contain antibiotic resistance genes (Moritz and Hergenrother 2007; Perichon et al. 2008; Sletvold et al. 2008), it may be highly likely that the microbes will become doubly protected and resistance against antimicrobials would persist.

A 'new generation' of pesticides or third generation of pesticides has now come into existence to prevent the growing risks of resistance and changes in the ecology of the contaminated risks. This class involves using insect hormones to provide insecticides that are more specific for their target pests. For example 'natalisin' impacts the reproductive function and mating behavior in insects and arthropods and has been used as an eco-friendly method for application instead of chemical pesticides. DuPont's Coragen' suspension concentrate and 'Ferterra' granular formulation are the successfully launched new generation insecticides in India.

12.5 Conclusion

The earth is contaminated with a large number of xenobiotic products that were initially aimed at human welfare such as pesticides (in soils and water) and pharmaceutical antimicrobials (in clinical set ups). Several bacteria capable of using xenobiotics (organic pesticides) as a carbon and/or nitrogen source, although it has been estimated that less than 1% of the bacterial species in soils are currently known (Paul et al. 2005a, b). This implies that a considerable proportion of the potential of bacterial metabolic capabilities to degrade dangerous chemicals is yet to be investigated and exploited for remediation purposes. The bacteria have also evolved over the years such that pesticides that were recalcitrant and escaped complete mineralization (HCH, DDT) are now being completely biodegraded due to the horizontal gene transfer across species. This unique phenomenon has also led to the increase in resistance of bacteria to antimicrobials too. Gene(s) exchanged between resistant and susceptible species have given rise to multi-drug resistant varieties that use several mechanisms to avoid antimicrobial compounds from acting upon them.

References

Abraham WR, Nogales B, Golyshin PN, Pieper DH, Timmis KN (2002) Polychlorinated biphenyl-degrading microbial communities in soils and sediments. Curr Opin Microbiol 5:246–253

Araújo ASF, Monteiro RTR, Abarkeli RB (2003) Effect of glyphosate on the microbial activity of two Brazilian soils. Chemosphere 52:799–804

Barragán HB, Costa C, Peralta J, Barrera J, Esparza F, Rodríguez R (2007) Biodegradation of organochlorine pesticides by bacteria grown in microniches of the porous structure of green bean coffee. Int Biodeterior Biodegrad 59:239–244

Briggs, SA (1992) Basic guide to pesticides: their characteristics and hazards. Taylor & Francis, Washington, DC

Chauhan A, Pandey G, Sharma NK, Paul D, Pandey J, Jain RK (2010) P-Nitrophenol degradation via 4-nitrocatechol in Burkholderia sp. SJ98 and cloning of some of the lower pathway genes. Environ Sci Technol 44:3435–3441

Clapperton J, Regen M (2012) Pesticide Effects on Soil Biology in No-till on the Plains, Leading Edge, Volume 8, Number 1 www.notill.org

Dinamarca MA, Cereceda BF, Fadic X, Seeger M (2007) Analysis of s-triazine-degrading microbial communities in soil using more probable number enumeration and tetrazolium-salt detection. Int Microbiol 10:209–215

Fisher JF, Shahriar M (2010) Enzymology of bacterial resistance. In: Comprehensive natural products II. Volume 8: enzymes and enzyme mechanisms. Elsevier, Boston, pp 443–201

Ha J, Engler CR, Wild J (2009) Biodegradation of coumaphos, chlorferon, and diethylthiophosphate using bacteria immobilized in Ca-alginate gel beads. Bioresour Technol 100:1138–1142

Horne I, Rebecca L, Harcourt TD, Sutherland RRJ, Oakeshott JG (2002) Isolation of a Pseudomonas monteilli strain with a novel phosphotriesterase. FEMS Microbiol Letters 206:51–55

Kemboi WW, Raphael W, Ramesh F (2014) Horizontal gene transfer of drug resistance genes between *Salmonella* and *Escherichia coli*. Int J Bioassays 3(06):3066–3072

Kikuchi Y, Hayatsu M, Hosokawa T, Nagayama A, Tago K, Fukatsu T (2012) Symbiont-mediated insecticide resistance. PNAS 109(22):8618–8622

Kremer RJ, Means NE, Kim S (2005) Glyphosate affects soybean root exudates and rhizosphere microorganisms. J Environ Analyt Chem 85(15):1165–1174

Lakshmi CV, Kumar M, Khanna S (2008) Biotransformation of chlorpyrifos and bioremediation of contaminated soil. Int Biodeterior Biodegrad 62:204–209

Lal R, Dogra C, Malhotra S, Sharma P, Pal R (2005) Diversity, distribution and divergence of *lin* genes in hexachlorocyclohexane-degrading sphingomonads. Trends Biotechnol 24:121–130

Lal R, Pandey G, Sharma P, Kumari K, Malhotra S, Pandey R, Raina V et al (2010) Biochemistry of microbial degradation of hexachlorocyclohexane and prospects for bioremediation. Microbiol Mol Biol Rev 74(1):58–80

Latifi AM, Khodi S, Mirzaei M, Miresmaeili M, Babavalian H (2012) Isolation and characterization of five chlorpyrifos degrading bacteria. Afr J Biotechnol 11:3140–3146. https://doi.org/10.5897/AJB11.2814

Lo CC (2010) Effect of pesticides on soil microbial community. J Environ Sci Health B 45:348–359

Mansee AH, Chen W, Mulchandani A (2005) Detoxification of the organophosphate nerve agent coumaphos using organophosphorus hydrolase immobilized on cellulose materials. J Ind Microbiol Biotechnol 32:554–560

Martín M, Mengs G, Plaza E, Garbi C, Sánchez M, Gibello A, Gutiérrez F, Ferrer E (2000) Propachlor removal by Pseudomonas strain GCH1 in a immobilized-cell system. Appl Environ Microbiol 66(3):1190–1194

Martínez JL (2009) The role of natural environments in the evolution of resistance traits in pathogenic bacteria. Proc R Soc B 276:2521–2530

Moritz EM, Hergenrother PJ (2007) Toxin-antitoxin systems are ubiquitous and plasmid encoded in vancomycin-resistant enterococci. PNAS 104:311–316

Mrak E (1969) Pesticides and their relationship to environmental health. U.S. Department of HEW, Washington, DC

Naphade SR, Durve AA, Bhot M, Varghese J, Chandra N (2012) Isolation, characterization and identification of pesticide tolerating bacteria from garden soil. Euro J Exp Bio 2(5):1943–1951

Paul D, Pandey G, Jain RK (2005a) Suicidal genetically engineered microorganisms for bioremediation: need and prospective. BioEssays 27:563–573

Paul D, Pandey G, Pandey J, Jain RK (2005b) Accessing microbial diversity for bioremediation and environmental restoration. Trends Biotechnol 23:135–142

Paul D, Singh R, Jain RK (2006) Chemotaxis of Ralstonia sp. SJ98 towards p-nitrophenol in soil. Environ Microbiol 8:1797–1804

Perichon B, Bogaerts LT, Frangeul L, Courvalin P, Galimand M (2008) Sequence of conjugative plasmid pIP1206 mediating resistance to aminoglycosides by 16S rRNA methylation and to hydrophilic fluoroquinolones by efflux. Antimicrob Agents Chemother 52:2581–2592

Santacruz G, Bandala E, Torres LG (2005) Chlorinated pesticidas (2,4 –D and DDT) biodegradation at high concentrations using immobilized *Pseudomonas fluorescens*. J Environ Sc Heal 40(4):571–583

Seffernick JL, Wackett LP (2001) Rapid evolution of bacterial catabolic enzymes: a case study with atrazine chlorohydrolase. Biochemistry 40:12747–12753

Shapir N, Mongodin EF, Sadowsky MJ, Daugherty SC, Nelson KE, Wackett LP (2007) Evolution of catabolic pathways: genomic insights into microbial s-Triazine metabolism. J Bacteriol 189:674–682

Sletvold H, Johnsen PJ, Hamre I, Simonsen GS, Sundsfjord A, Nielsen KM (2008) Complete sequence of *Enterococcus faecium* pVEF3 and the detection of an omega epsilon zeta toxin-antitoxin module and an ABC transporter. Plasmid 60:75–85

Todar K (2008) Textbook of bacteriology, 4th edn. University of Wisconsin press, Wisconsin

Top EM, Springael D (2005) The role of mobile genetic elements in bacterial adaptation to xenobiotic organic compounds. Curr Opin Biotechnol 14:262–269

Yañez OG, Penninckx M, Jiménez TGA, Sánchez SE, Ortiz HML (2009) Removal of two organophosphate pesticides employing a bacteria consortium immobilized in either alginate or tezontle. J Hazard Mater 168:1554–1561

Yoon C, Indiragandhi P, Anandham R, Cho S, Sa TM, Kim G (2010) Bacterial diversity and distribution from the whole mite extracts in Acaricide resistant and susceptible populations of two spotted spider mite-*Tetranychusurticae*. J Korean Soc Appl Biol Chem 53:446–457

Zhang S, Sun W, Xu L, Zheng X, Chu X, Tian J, Wu N, Fan Y (2012) Identification of the para-nitrophenol catabolic pathway, and characterization of three enzymes involved in the hydroquinone pathway, in Peudomonas sp. 1-7. BMC Microbiol 12:27. https://doi.org/10.1186/1471-2180-12-27

13 Antimicrobial Agents Used in Food Preservation or as Agricides and Effect on Microbes in Developing Antimicrobial Resistance

Soma Mukherjee, Nitin Dhowlaghar, and Wes Schilling

Abstract

Antimicrobial agents are used as preservatives in various foods to extend shelf-life and inhibit microbial growth. In addition, disinfectants are used to decontaminate and sanitize processing surfaces. Recently, there has been considerable interest in microbial adaptation and resistance to antimicrobials that are used in foods and disinfectants. However, the sources of antimicrobial resistance from the food and processing plant have not been well established. In this chapter, the effect of using antimicrobial agents in foods and food processing environments, including organics acids, alkaline agents, oxidative and quaternary ammonium compounds have been discussed along with the adaptation of microbial tolerance to these chemicals.

The effects of agricides, including pesticides, herbicides, and fungicides, on microflora is discussed with an emphasis on how these agricides affect soil microflora and their ability to develop resistance to the agricides that are used. Details are included on the mechanisms of how microbes develop resistance to antimicrobials. The majority of research that has been conducted on antimicrobial resistance pertains to pathogenic microflora. Human, plant, and pathogenic microflora are discussed and compared to help understand the phenomenon of chemical resistance. The acquired resistance can be identified using molecular tools but continued research is demanded to establish the best safe operating procedure for combating antimicrobial resistance.

S. Mukherjee (✉) · W. Schilling
Mississippi State University, Starkville, MS, USA
e-mail: soma.mukherjee75@gmail.com

N. Dhowlaghar
Washington State University, Pullman, WA, USA

S. M. Mandal, D. Paul (eds.), *Bacterial Adaptation to Co-resistance*,
https://doi.org/10.1007/978-981-13-8503-2_13

Keywords
Antimicrobials · Resistance · Gene transfer · Acquired resistance · Adaptation

13.1 Introduction

Discovery of antibiotics and incidence of antimicrobial resistance occurred at similar times in history. In 1928, Sir Alexander Fleming accidentally discovered antibiotics in his laboratory when he contaminated bacteria cells with Penicillin mold. In 1945, Sir Alexander Fleming warned that a sub-lethal dose of antibiotics can potentially be harmful and lead to resistance to the drug (Davies and Davies 2010). Antimicrobial resistance occurs due to the use of sublethal doses of antimicrobials in health care and agriculture environment. The demand for the production of food products with longer shelflife has necessitated the use of antimicrobials, antioxidants, and freezing, which have all contributed to the development of antimicrobial resistance (O'Neill 2015). The adequate therapeutic use of antibiotics in animals is essential for the treatment of infectious diseases. For animal welfare and food production, excessive and inappropriate use of antibiotics is concerning since adaptation of bacteria can lead to antibiotic inefficacy and increase the amount of infections that can occur (Van Boeckel et al. 2015). Antimicrobial resistant (AMR) bacteria are problematic in human health care and food production. If no action plan is taken to reduce microbial resistance, the estimated number of deaths due to antimicrobial resistance will rise to 10 million people per year by 2050 (based on United Nations report world population prospects and at the same time in BRICS (Brazil, Russia, India, China and South Africa) countries will show an increase up to 99% (O'Neill 2015).

Microbial exposure to low or sub-lethel or 'sub-therapeutic'levels of antibiotics trigger an increased incidence of antibiotic resistant bacteria in the clinical and agriculture sectors, due to an increase in the number of resistant genes as compared to similar microbes that have not been exposed to antibiotics (Allen 2014). This can lead to the proliferation of bacterial strains that cannot be controlled by antibiotics. AMR can occur when resistant genes/fragments of DNA are transmitted among bacteria through phages, plasmids and/or transposons, which results in horizontal gene transfer (HGT). Phages, plasmids, and transposons play a pivotal role in gene transfer as well as the mobility of the bacterium, which involves emergence, invasion and occupation. Three major areas through which HGT occur are environment, food (Jahan and Holley 2016) and animal and human digestive system. These areas also act as the bank of gene pool for the AMR genes which has been termed as "the resistome" (Forsberg et al. 2012). AMR in pathogenic bacteria demanded new antibiotic resistant drug formulations.

13.2 Food Production and Antimicrobials

Quaternary ammonium compounds, hypochlorides, peroxyacetic acid (PAA), chlorine dioxide and ionophores are commonly used in the food industry as sanitizers. These sanitizers are often rotated with respect to their use in order to inhibit AMR.

13.2.1 Quaternary Ammonium Compounds (QAC)

QAC's are used as disinfectants, sanitizers, fabric-softeners, and treatment of agricultural manure. Quaternary alkyl ammonium compounds (QUAAC) are water soluble compounds that are frequently found in waste water, and have contributed to AMR (Aase et al. 2000). QUAAC's are effective at controlling bacteria by exerting inter-communication with the molecules of microbial cell membrane. QUAAC's can effectively disrupt cell membrane integrity, which eventually results in cellular leakage (Bragg et al. 2014). Gram-negative bacteria are more easily controlled by QUAAC's in comparison to gram positive. In addition, QUAAC's are effective at destroying spores of bacteria and mycobacteria (White and McDermott 2001). Efflux pumps are transport proteins involved in the extrusion of toxic substrates (including all classes of clinically relevant antibiotics) from within cells into the external environment (Aase et al. 2000). QUAAC's may be contributing to AMR by inhibiting the efficacy of the energy driven efflux system thus causing multidrug resistance by bacterial strains (Suarez-Carmona et al. 2015). Various QAC genes have been identified in bacterial pathogens, which suggest that resistant genes are targeting many lipophilic cationic compounds from different chemical classes (Jaglic and Cervinkova 2012). Among other genes, *quc* was resistant to many beta-lactem antibiotics (Fuentes et al. 2005). QAC's can easily modify cell wall and reduce cell permeability, which may result in cross resistance between biocides and antibiotics (European Commission 2009; Fraise 2011).

Microorganisms acquire resistance to antibiotics through a combination of factors including adherence time or duration of contact with the QAC compounds, temperature and target organism. Compound concentration is a key factor with respect to AMR. Application of sublethal concentration of QUAAC's contribute to antimicrobial resistance to multiple compounds (Hegstad et al. 2010). Co-selection is conferred when the concentration of these compounds are much lower than the lethal concentration (Webber et al. 2015). This means that the AMR genes may be resistant to multiple antibiotics. This co-selection makes it difficult to reverse AMR. When BAC-tolerant *Salmonella enterica* serotype HvittingfossS41 was studied against active compounds and compared with wild type, this bacteria showed less susceptibility to many other antimicrobial compounds which confirms the acquisition of multiple drug resistance even though the bacteria was only exposed to one antibiotic (Condell et al. 2012). Similar changes were also observed in *Pseudomonas* species when exposed to contaminated sediment sample and sub-lethal concentrations of QUAAC. The effectiveness of antibiotics such as penicillin G, tetracycline and ciprofloxacin in controlling these resistant microorganisms was reduced, which was related to the ability of the microorganism to degrade these antibiotics by affecting the activity of the efflux pump (Tandukar et al. 2013).

13.2.2 Peracetic Acid (PAA)

PAA is a strong disinfectant that is used in food processing plants for sanitation and control of pathogens. PAA is relatively stable at recommended concentrations of 100–200 ppm and is useful for the removal of biofilm (Van der Veen and Abee 2011). Other desirable properties that make this component a replacer of chlorine is its inability to form foam, absence of phosphates, low corrosiveness, tolerance to hard water, and biodegradability. PAA is commonly used in immersion chillers and other locations on the processing plant in the poultry industry to help control *Salmonella* (Pérez-Varela et al. 2018). *Listeria monocytogenes* forms biofilms for protection against unfavorable conditions such as low and high-pH, high salt concentrations, heat, and cold. Persistent *L. monocytogenes* strains (adapted) have been identified as a source of contamination in food processing plants for 7–12 years (Sauders and Wiedmann 2007) and in food products up to 6 years (Nakamura et al. 2006). More recent data indicates that *L. monocytogenes* strains can persist for up to 10 years (Acciari et al. 2011). Therefore, if PAA is used in a processing plant to help control for *Listeria monocytogenes* or other pathogens, it is important to understand possible contamination sources, survival capacity of the pathogen(s), adequate sanitation processes, and the impact of sub-lethal conditions, including sub-lethal residual concentrations on food contact surfaces.

L. monocytogenes is a model test organism for stress adaptation studies due to its survival ability. To investigate the effect of stress adaptations of *L. monocytogenes,* strains on stainless steel (SS) coupons that are subjected to various stress treatments (NaCl and PAA) at different concentrations and following formation of biofilm, cells were exposed to hot water treatment for different durations (Belessi et al. 2011). PAA was proved to be the most effective treatment as a sanitizer. Similar results were also reported by mixed culture treatments when exposed to 160 ppm of PAA for 1 min with more than 3 log reduction (Somers and Lee Wong 2004). Improper cleaning and sanitation or limited accessibility to the biofilm (because of the insufficient contact time) have been hypothesized to contribute to stress adaptation (Pan et al. 2006). Acid-adapted cells at pH 5.0 were more resistant to PAA treatment than cells treated at pH 7.0. Acid adapted cells (pH 5.0) in a growth medium formed biofilm that was resistant to the PAA (2%) treatment. The possible explanation was substantiated by the maintenance of stress adaptation, the function of proton pump (Shabala et al. 2002) and induced cross-protection in treatments with higher concentrations of PAA (pH 3.6) (Tiganitas et al. 2009). However, acid adapted cells were not able to survive as well when inoculated on meat products. For example, acid-adapted cells inoculated on the surface of beef slices which were then treated with acid marinades (pH 2.5–4.3) and subsequently subjected to drying. The acid-adapted cells had reduced survivability when compared to the non-adapted cells (Calicioglu et al. 2002). The pH of the sanitizer is almost similar as the pH of the ambience that leads to the formation of biofilms. For example, acid resistant biofilm formation occurred in water (pH 6.8–7.0) or acid diluted (pH 3.1–4.7) fluids from a meat plant (Samelis et al. 2004).

13.2.2.1 Survival Kinetics of *L. monocytogenes*

Several molecular resistance systems have been proposed to explain the survival kinetics of *L. monocytogenes*, namely F_1F_0-ATPase complex, transcriptional regulators and decarboxylation reactions. These systems include glutamate and lysine decarboxylases and the production of ammonium ion and acetoin.

13.2.2.1.1 Transcriptional Regulators

The major role of Sigma factors (Transcriptional regulators) (which are also subunits of the prokaryotic RNA polymerase) are to identify a specific DNA sequence at promotor sites, which is influenced by factors that affect cell homeostasis. In *L. monocytogenes* and some other pathogens, the Sigma factor B (SigB) regulates gene expression that contribute to resistance towards adverse environmental conditions such as acid exposure, cold temperature, high salinity, higher temperature and oxidative stresses (Abram et al. 2008; Oliver et al. 2013). SigB also plays an important role in *L. monocytogenes* growth during stationary phase acid resistance. SigB also regulates gene expression that is related to virulence (Giotis et al. 2007). Over expression of SigB was in a human *L. monocytogenes* strain in soft cheese (Belessi et al. 2011). Other Sigma factors (Sig H, Mar R, Sig L) have been reported to protect *L. monocytogenes* in presence of organic acids (pH 6.0) (Mattila et al. 2012). Research needs to be continued to investigate different serovars as well as *L. monocytogenes* lineages to document the independent role of different sigma factors.

13.2.2.1.2 F_1F_0 ATPase and Glutamate Decarboxylase Acid Resistance System (GAD)

The GAD is responsible for controlling pH balance in different mechanisms in some bacteria. GAD produces one neutral compound γ-aminobutyrate (GABA). The enzyme is responsible for the production of GABA reduces acidic environment by decarboxylation reaction and produces GABA. The intercellular proton is utilized and acidic nature of cytoplasm is reduced. This GABA is transported through cell membrane by a GABA antiporter which is encoded as *gadC*. The synthesized ATP also acts to control the pH balance mediated by the F_1F_0 ATPase (Cotter et al. 2001).

Datta and Benjamin (1997) treated *L. monocytogenes* Scott A cells with an inhibitor (DCCD:N.N'-decyclohexylcarbodiimide) of the ATPase enzyme to evaluate the participation of the F_1F_0 ATPase system in *L. monocytogenes* resistance to acid. The cells treated with this compound were more susceptible to low pH (pH 3.5) when compared to control cells. The different survival rates were observed in different growth stages of the cells. Stationary growth phase cells exhibited 5 log survivals whereas exponential phase cells exhibited 2 log survivals. Later, Cotter et al. (2001) studied the participation of the F_1F_0-ATPase complex in low pH adapted cells. Low pH familiarized cells of *L. monocytogenes* (pH 5.5) exhibited 3 logs greater survival than the DCCD treated cells after 2 h of acid treatment at pH 3.5. Another acid resistant system is (2% inoculums were added and incubated for 16 h at pH 7.0 and then exposed to 1 h acid tolerance at pH 2.8. *L. monocytogenes* glutamate decarboxylase enzyme is encoded by five genes. Three genes are responsible for encoding the subunits of the enzyme and the other two encode the antiports. The existence of gene pairing was observed between *gad*D1T1 and *gad*D2T2 (Cotter et al. 2005). This operational process promote in the change of glutarate to γ-aminobutyrate (GABA) by cytoplasmic decarboxylation and by using existing proton within cellular space. GABA acts as an exporter from the cell through antiports that are located in the cell membrane. This change results in boost in the ionic balance of the cytoplasm (due to proton depletion) and a non-significant boost in extracellular pH (due to glutamate shift) (Booth et al. 2002). These authors reported that the GAD system is medium and strain-dependent. The observation of independent existence of GABA antiport for the accumulation of extracellular GABA led authors proposing the existence of two semi-independent systems: Extracellular GAD_e and the Intracellular GAD_i. The contribution of the extracellular system (GAD_e) in acid tolerance with GadD2 (this converts Glutamine in GABA) and GadD3 (Glt decarboxylase) have been reported. Complete understanding of individual role of these genes in other atmospheric conditions (acidic pH 3.5) have been confirmed by a whole set of mutated genes (Cotter et al. 2001).

13.2.2.2 Arginine Deiminase (ADI) System and Thiamine

This system is conjoined with all types of bacteria (Ryan et al. 2009) in relation to acid tolerance ability. Three enzyme systems have been identified in *L. monocytogenes*, namely arginine deiminase (ADI), catabolic ornithine carbamoyl-transferase (cOTC) and carbamate kinase (CK). This complete system induces acid tolerance which has been confirmed by these three operon-encoded proteins. These three proteins also take part in conversion of arginine. Thiamine also has an crucial act in acid resistance ability of *L. monocytogenes* (Madeo et al. 2012). ThiT plays an important role in the upward regulation or intake of thiamine. This is also an stimulative factor for more than one enzymes from important metabolic pathways more precisely associated with carbohydrate metabolism. The cells treated without thiamine were reported to show more sensitivity to acid treatment than the cells treated with sufficient amount of thiamine.

13.2.3 Chlorine

Chlorine is widely used as a cleaning compound (Krysinski et al. 1992) that is also used in drinking water (Sisti et al. 1998) at concentrations of 0.2–4 ppm to control microorganisms. Chlorine's antimicrobial mechanism is due to its high oxidation potential. Chlorine is used in food industry to reduce the microbial populations on the surface of raw fruits and vegetables and the sanitation of food-contact surfaces in processing plants (Kotula et al. 1997).

To study the sublethal concentration of chlorine on *L. monocytogenes,* stainless steel coupons were exposed to various concentrations of chlorine solution. Chlorine treatments of SS coupons reduced the attachment rate of the pathogen to the surface of the coupon (Lopes 1986). When the alkaline-adapted cells were exposed to the same sterile SS coupons, the cells showed increased attachment. This study further supports the previous results by Frank and Koffi 1990. After 8–12 days of consecutive chlorine treatment, cells exhibited resistance towards chlorine treatment. It can be concluded from this study that *L. monocytogenes* was able to adapt to chlorine stress during biofilm formation over a 12-day period (Soni et al. 2011). Ability of the cells to adapt to stress is dependent on environmental variables such as chlorine concentration, exposure time and the age of the culture. The pH of the chlorine solution is also important with respect to the ability of *L. monocytogenes* to adapt to stress (Bremer et al. 2002; Naïtali et al. 2009). Bremer et al. (2002) reported that the effective pH of chlorine to kill *L. monocytogenes* is less than 8.0. The capability of *L. monocytogenes* to resist adverse environmental condition shows similar trend of adaptation like many other microbes. Rapid transformation machinery can be developed by this strain to adapt to adverse conditions (Soni et al. 2011). In addition, *L. monocytogenes* can find refuge in conveyer belts, cracks in floors, and hard to reach places in machinery, which contributes to its adaptability and form biofilms (Nilsson et al. 2012).

13.2.3.1 Morphological Changes Due to Alkali Stress on *L. monocytogenes*

Low alkaline pHs (pH 7–8) affect the phagolysosome behavior (Segal et al. 1981) and cell morphology of *L. monocytogenes* in clinical and food samples (Vasseur et al. 2001). Giotis et al. (2007) evaluated morphological changes due to alkaline stress using scanning electron microscopy. Two types of cell cultures were obtained: a serotype 1/2 a strain and a derived mutant - sigma B factor deficient by deleting 1490 and 1788 of the sigma factor gene). After recovering cells from alkali treated and untreated suspensions, the cells were washed with phosphate buffer. Morphological data of the spheroid caps were recorded and these data were taken to measure cell volumes. No significant differences existed between equatorial radii of the two type's cell, but the calculated cell volume was greater for unstressed cells in comparison to stressed cells. The *L. monocytogenes* culture 10403S (the serotype 1/2 a strain) exposed to pH 9.5 showed filamentous growth compared to the control cells, and mutant strains exhibited similar filament and chain formation to the parent strain.

13.2.3.2 Molecular Change During Alkaline Adaptation of *L. monocytogenes*

Giotis et al. (2008) evaluated the possible proteomic and molecular changes in *L. monocytogenes* due to alkaline stress. The post translational modification of the stress survived cellular entities was evaluated. In preliminary experiments,

L. monocytogenes 10403S were grown in wide range of pH (5.0–9.0). The maximal tolerance pH of 9.5 was compared against pH 12.0 (lethal alkaline pH). Chloramphenicol was used to inhibit translation in model organism with the specific subunit of ribosome. A phenotypic change takes place during alkali adaptation in the presence of chloramphenicol due to the activation and induction of differentially expressed proteins. One dimensional electrophoresis data showed a marked change in newly synthesized protein at a very highly alkaline condition. The protein synthesis may play a crucial role in alkaline stress adaptation. 2D electrophoresis, mass spectrometry, and translational profiling indicated that the ATR (Alkali-Tolerance Response) mechanism helps the microorganism reduce excess alkalization and energy expenditure.

13.2.3.2.1 Molecular Change in Phosphate Binding Proteins

The major operon encoding genes of respiration were less activated during alkali adaptation. In contrast, glycerol metabolism pathway (dihydroxtacetone kinase, lmo1055) encoding genes were activated in alkali resistance cells. For a sudden shift in pH to 8.9 of the medium resulted in up regulation of the phosphate deficiency encoding genes (Atalla and Schumann 2003). Alkali phosphatase (lmo1870) was over expressed (~4.22 fold) when *L. monocytogenes* was shifted to an alkaline environment, that led to starvation (Matin et al. 1989). Phosphatase helps with solute transportation through the cell membrane. This allows *L. monocytogenes* to adapt to alkaline stress conditions (Whittam and Wheeler 1970). Addition of phosphate with certain molecules inhibit them to move through the cellular barrier and is promoted by the change in the transcriptional level genes.

13.2.3.2.2 Molecular Changes in Transport and Binding Proteins

After alkali stress, 87 genes were differentiated that are encoded for transport and binding protein. These genes contribute to the transportation of carbohydrate, organic acids, peptides, amino acids, and metals and help maintain cell fluid balance (osmoregulation), pH homeostasis, and signaling (Karpel et al. 1991).

13.2.3.2.3 Up Regulation of ATP Binding Cassette (ABC) Transporters

The ABC transporter system plays crucial part for the uptake and transportation of different components and solutes during multiplication in alkaline environment (Tomii and Kanehisa 1998). A significant number of ABC transporters have been identified with differential change in up and down regulated genes under alkaline stress conditions (Hyde et al. 1990; Ames et al. 1990; Davidson 2002). Transported peptides acidify cytoplasmic solutes through peptidase activity with free acidic amino acids as the proton source (Takami et al. 2002). Different type of heat shock proteins have been identified after alkali stress (dnaK, grpE and lmo0292), which have a similar structural pattern to HtrA, serine protease. Upregulation of chaperon proteins (clpE) exhibit ATPase activity and help the correct folding of denatured proteins. Other activity of this protein such as long term survival, cell division and virulence of *Listeria* at 42 °C has also been reported (Nair et al. 1999).

13.3 Antimicrobials and Livestock

Antimicrobials are used in agri-production (livestock) for various purposes including: (1) treatment of animal disease, (2) promoting animal growth (3) combat disease and prevent it from spreading to other livestock (metaphylaxis), and (4) to treat highly susceptible infections.

13.4 Ionophores

Ionophores are frequently used as food additives in the cattle and poultry industries to promote growth and also to prevent parasitic infection, known as coccidiosis (O'Neill 2015). Ionophores are not therapeutic antibiotics and therefore do not promote antibiotic resistance. They are used as a food additive to control coccidiosis and not as an antibiotic. Ionophores increase the amount of energy that is available from the diet. Many government authorities and pharma companies separate these ionophores recognize a distinct pathway action from antibiotic use regulations (Duax et al. 1996). This is because ionophores are not used as antibiotics in humans which have a different mode of action, ionophores can translocate across cell membranes and their resistance is highly specific. Europe has banned ionophores to promote growth, but the use to treat parasitic infection is still in practice. Ionophores are considered safer by most scientists since the development of resistance or tolerance is slow and thus cannot spread to other microbial populations (Simjee et al. 2012; Houlihan and Russell 2003). Houlihan and Russell (2003) also reported that ionophore resistance is likely an adaptation and not due to the acquisition of foreign genes.

13.5 Antibiotics

A portion of antibiotics that are ingested by humans is excreted (anywhere from 40–99%) in the urine and feces (Singer et al. 2016). A survey of wastewater treatment plants (WWTPs) (16 plants) in the UK revealed the presence of antibiotics in the water (Gardner et al. 2013). The reclaimed (with residual antibiotic) water is used to sprinkle irrigate crops, golf courses and landscapes. Purified excrement water, used for animal consumption, operational drain-water contain high levels of utilizable content of plant nutritional interest and can harbor microbes (Dodgen and Zheng 2016). Animal excreta also can contaminate soil microflora with antibiotic resistant genes. This has been established in a survey of animal excreta from animal farms in the Netherlands. Out of 80% of swine farms and 95% of cattle farms, 55% and 75% of the farms were positive for the presence of antibiotics (Berendsen et al. 2015). Oxytetracycline, doxycycline, and sulfadiazineare the most commonly found antibiotics that are isolated from animal feces. Other less commonly found antibiotics in animal feces include tetracycline, flumequine, lincomycin, and tylosin. The release of antibiotics into wastewater is accompanied by resistant genes. Researchers have reported that the origin of these resistant genes is the digestive tract of humans (Hu et al. 2013). The deposition of both antibiotic resistant genes and antibiotics in WWTPs can trigger the selection of new combinations of microbial resistance. This resistance can be transferred mobile genetic elements (Plasmids) and promote antibiotic resistance to multiple drugs by the microbial population (Singer et al. 2016). Bacterial SOS response can be initated by antibiotics, which causes bacteria to undergo hypermutation that creates genetic variability (Baharoglu and Mazel 2014). HGT is also stimulated by chemical, environmental, and abiotic sources (Warnes et al. 2012).

Intrinsic resistance is universally found within a bacterial species, and acquired resistance is obtained by mutations of intrinsic genes. Intrinsic resistance is carried by the species independently without any antibiotic stress or HGT (Cox and Wright 2013). The pattern and consequence of intrinsic gene transfer in pathogenic life entities are not clear (O'Neill 2015). Intrinsic resistance can be further increased through the up regulation of differential transcriptional factors which further up regulate cellular components that can withstand antibiotic effects (Sifaoui et al. 2001). These post translational changes are recognized as acquired resistance. Mobile genetic elements (MGE) contribute to acquired resistance. Three mechanisms govern this entire process of movement of hereditary units. First transformation occurs when bacteria can acquire unprotected DNA from the ambiance (Gaze et al. 2013). Conjugation occurs when a gene transfers through straightforward association (Lopatkin et al. 2016). In Transduction, genes move though bacteriophage to bacteria (Brown-Jaque et al. 2015). MGEs can freely transmit from cell to cell and can also integrate into intracellular MGEs. These transposable elements contain three important elements: transposons (Tn), insertion sequence (ISs) and integrons (In). Both gram positive and gram negative bacteria which contain integrative and conjugative genetic element -ICEs can transmit themselves or can utilize vectors for translocation. Mobile genetic elements contain carrying genes that can be characterized by phenotypes to the inheriting cell (Roberts and Mullany 2011).

13.6 Fungicides and Resistance

Almost three quarters of a million people die due to fungal infections every year (O'Neill 2015). No antifungal resistant fungicide has been reported as a potential threat to humans with the exception of azole based fungicides. Azole-based fungicidal (oral drug) components are used to treat *Aspergillus* infection. An increasing number of Azole-resistant fungal infections are expected in the future (O'Neill 2015). Azole based fungicides are used variously in different parts of the world. In Europe, 50% of total agricultural land for grain and grape cultivation are treated at least once in a year, whereas in the US is treated at 5% (Van Der Linden et al. 2013). It is clear that there is an association between the uses of fungicides and emergence of fungal resistance in that geographical area. For example, the use of fungicide is high in those areas. Every tulip bud is dipped in fungicides before plantation.

Approximately 7% of *Aspergillus fumigates* strains are azole-resistant, which is the highest rate that has been reported (Kano et al. 2014).

It is impractical to stop the use of fungicides since it would decrease staple food production. However, researchers can focus on the degradability of these compounds in the environment to reduce their persistence and resistance. In addition, scientists can work towards developing treatments that will be less susceptible to fungicidal resistance.

13.7 Conclusion

The transmission of AMR is a vast area linking medicine, plant cultivation, food and the interfaces within each other. In food for the transmission of AMR genes to occur, the bacterial presence in the food during production and to great extent food industries ensure that food is prepared in hygienic manner, should be devoid of pathogenic bacteria. Sub-lethal effects of sanitizers, fungicides and/or pesticides on microorganisms, their microbiome and ecosystem balance (rivers, sewage, irrigation system, hospital management system) has already been started impacting the effectiveness of antibiotics for treating pathogenic bacteria (Belessi et al. 2011). Out of 139 academic studies that were reviewed, 100 (70%) have linked antimicrobial resistance to the use of antimicrobials at insufficient concentration or environmental conditions that lead to microbial adaptations (Singer et al. 2016). The relevance of antimicrobial resistance and the contributing factors to the industrially used sanitizers or agricides can be roughly summarized in two high levels: (1) Sub-lethal effects of sanitizers, fungicides, pesticides on organisms, their microbiome and ecosystem balance (rivers, sewages, irrigation system, hospital management system) can impact the health and can affect economically important agriculture sector and food production system. (2) Relevant pathways for the dissemination of antibiotics, fungicides, disinfectants into animals, plants, food are under the Environmental protection agency.

References

Aase B, Sundheim G, Langsrud S, Rørvik LM (2000) Occurrence of and a possible mechanism for resistance to a quaternary ammonium compound in *Listeria monocytogenes*. Int J Food Microbiol 62(1–2):57–63

Abram F, Starr E, Karatzas KAG, Matlawska-Wasowska K, Boyd A, Wiedmann M et al (2008) Identification of components of the sigma B regulon in *Listeria monocytogenes* that contribute to acid and salt tolerance. Appl Environ Microbiol 74(22):6848–6858

Acciari VA, Torresi M, Migliorati G, Di Giannatale E, Semprini P, Prencipe V (2011) Characterisation of *Listeria monocytogenes* strains isolated from soft and semi-soft cheeses sampled in a region of Italy. Vet Ital 47(1):15–23

Allen HK (2014) Antibiotic resistance gene discovery in food-producing animals. Curr Opin Microbiol 19:25–29

Ames GFL, Mimura CS, Shyamala V (1990) Bacterial periplasmic permeases belong to a family of transport proteins operating from *Escherichia coli* to human: traffic ATPases. FEMS Microbiol Lett 75(4):429–446

Atalla A, Schumann W (2003) The pst operon of Bacillus subtilis is specifically induced by alkali stress. J Bacteriol 185(16):5019–5022

Baharoglu Z, Mazel D (2014) SOS, the formidable strategy of bacteria against aggressions. FEMS Microbiol Rev 38(6):1126–1145

Belessi CEA, Gounadaki AS, Psomas AN, Skandamis PN (2011) Efficiency of different sanitation methods on *Listeria monocytogenes* biofilms formed under various environmental conditions. Int J Food Microbiol 145:S46–S52

Berendsen BJ, Wegh RS, Memelink J, Zuidema T, Stolker LA (2015) The analysis of animal faeces as a tool to monitor antibiotic usage. Talanta 132:258–268

Booth IR, Cash P, O'byrne C (2002) Sensing and adapting to acid stress. Antonie Van Leeuwenhoek 81(1–4):33–42

Bragg R, Jansen A, Coetzee M, van der Westhuizen W, Boucher C (2014) Bacterial resistance to quaternary ammonium compounds (QAC) disinfectants. In: Infectious Diseases and Nanomedicine II. Springer, New Delhi, pp 1–13

Bremer PJ, Monk I, Butler R (2002) Inactivation of *Listeria monocytogenes/Flavobacterium* spp. biofilms using chlorine: impact of substrate, pH, time and concentration. Lett Appl Microbiol 35(4):321–325

Brown-Jaque M, Calero-Cáceres W, Muniesa M (2015) Transfer of antibiotic-resistance genes via phage-related mobile elements. Plasmid 79:1–7

Calicioglu M, Sofos JN, Samelis J, Kendall PA, Smith GC (2002) Destruction of acid-and non-adapted *Listeria monocytogenes* during drying and storage of beef jerky. Food Microbiol 19(6):545–559

CDC 2016 Multistate Outbreak of Listeriosis Linked to Frozen Vegetables http://www.cdc.gov/listeria/outbreaks/frozen-vegetables-05-16/index.html. Access 15 May 2018

Condell O, Iversen C, Cooney S, Power KA, Walsh C, Burgess C, Fanning S (2012) Efficacy of biocides used in the modern food industry to control *Salmonella*-links between biocide tolerance and resistance to clinically relevant antimicrobial compounds. Appl Environ Microbiol 78:3087–3097, AEM-07534

Cotter PD, Gahan CG, Hill C (2001) A glutamate decarboxylase system protects *Listeria monocytogenes* in gastric fluid. Mol Microbiol 40(2):465–475

Cotter PD, Ryan S, Gahan CG, Hill C (2005) Presence of GadD1 glutamate decarboxylase in selected *Listeria monocytogenes* strains is associated with an ability to grow at low pH. Appl Environ Microbiol 71(6):2832–2839

Cox G, Wright GD (2013) Intrinsic antibiotic resistance: mechanisms, origins, challenges and solutions. Int J Med Microbiol 303(6–7):287–292

Datta AR, Benjamin MM (1997) Factors controlling acid tolerance of *Listeria monocytogenes*: effects of nisin and other ionophores. Appl Environ Microbiol 63(10):4123–4126

Davies J, Davies D (2010) Origins and evolution of antibiotic resistance. Microbiol Mol Biol Rev 74(3):417–433

Davidson AL (2002) Not just another ABC transporter. Science 296(5570):1038–1040

Department of Health antimicrobial resistance strategy analytical working group

Dodgen LK, Zheng W (2016) Effects of reclaimed water matrix on fate of pharmaceuticals and personal care products in soil. Chemosphere 156:286–293

Duax WL, Griffin JF, Langs DA, Smith GD, Grochulski P, Pletnev V, Ivanov V (1996) Molecular structure and mechanisms of action of cyclic and linear ion transport antibiotics. Pept Sci 40(1):141–155

European Commission (2009) Scientific committee on emerging and newly identified health risks. Assessment of the antibiotic resistance effects of biocides. http://ec.europa

Forsberg KJ, Reyes A, Wang B, Selleck EM, Sommer MO, Dantas G (2012) The shared antibiotic resistome of soil bacteria and human pathogens. Science 337(6098):1107–1111

Fraise A (2011) Currently available sporicides for use in healthcare, and their limitations. J Hosp Infect 77(3):210–212
Frank JF, Koffi RA (1990) Surface-adherent growth of *Listeria monocytogenes* is associated with increased resistance to surfactant sanitizers and heat. J Food Prot 53(7):550–554
Fuentes DE, Navarro CA, Tantaleán JC, Araya MA, Saavedra CP, Pérez JM et al (2005) The product of the qacC gene of Staphylococcus epidermidis CH mediates resistance to β-lactam antibiotics in Gram-positive and Gram-negative bacteria. Res Microbiol 156(4):472–477
Gardner M, Jones V, Comber S, Scrimshaw MD, Coello-Garcia T, Cartmell E et al (2013) Performance of UK wastewater treatment works with respect to trace contaminants. Sci Total Environ 456:359–369
Gaze WH, Krone SM, Larsson DJ, Li XZ, Robinson JA, Simonet P et al (2013) Influence of humans on evolution and mobilization of environmental antibiotic resistome. Emerg Infect Dis 19(7):1–8
Giotis ES, McDowell DA, Blair IS, Wilkinson BJ (2007) Role of branched-chain fatty acids in pH stress tolerance in *Listeria monocytogenes*. Appl Environ Microbiol 73(3):997–1001
Giotis ES, Muthaiyan A, Blair IS, Wilkinson BJ, McDowell DA (2008) Genomic and proteomic analysis of the Alkali-Tolerance Response (AlTR) in *Listeria monocytogenes* 10403S. BMC Microbiol 8(1):102
Health Map 2018 Outbreak in Current location https://www.google.com/?gws_rd=ssl#q=HealthMap. Access 15 May 2018
Hegstad K, Langsrud S, Lunestad BT, Scheie AA, Sunde M, Yazdankhah SP (2010) Does the wide use of quaternary ammonium compounds enhance the selection and spread of antimicrobial resistance and thus threaten our health? Microb Drug Resist 16(2):91–104
Houlihan AJ, Russell JB (2003) The susceptibility of ionophore-resistant *Clostridium aminophilum* F to other antibiotics. J Antimicrob Chemother 52(4):623–628
Hu Y, Yang X, Qin J, Lu N, Cheng G, Wu N et al (2013) Metagenome-wide analysis of antibiotic resistance genes in a large cohort of human gut microbiota. Nat Commun 4:2151
Hyde SC, Emsley P, Hartshorn MJ, Mimmack MM, Gileadi U, Pearce SR et al (1990) Structural model of ATP-binding proteing associated with cystic fibrosis, multidrug resistance and bacterial transport. Nature 346(6282):362–365
Jaglic Z, Cervinkova D (2012) Genetic basis of resistance to quaternary ammonium compounds–the qac genes and their role: a review. Veterinarni Medicina 57(6):275–281
Jahan M, Holley RA (2016) Transfer of antibiotic resistance from *Enterococcus faecium* of fermented meat origin to *Listeria monocytogenes* and *Listeria innocua*. Lett Appl Microbiol 62(4):304–310
Kano R, Kohata E, Tateishi A, Murayama SY, Hirose D, Shibata Y et al (2014) Does farm fungicide use induce azole resistance in *Aspergillus fumigatus*? Sabouraudia 53(2):174–177
Karpel R, Alon T, Glaser G, Schuldiner S, Padan E (1991) Expression of a sodium proton antiporter (NhaA) in *Escherichia coli* is induced by Na+ and Li+ ions. J Biol Chem 266(32):21753–21759
Kotula KL, Kotula AW, Rose BE, Pierson CJ, Camp M (1997) Reduction of aqueous chlorine by organic material. J Food Prot 60(3):276–282
Krysinski EP, Brown LJ, Marchisello TJ (1992) Effect of cleaners and sanitizers on *Listeria monocytogenes* attached to product contact surfaces. J Food Prot 55(4):246–251
Lopatkin AJ, Huang S, Smith RP, Srimani JK, Sysoeva TA, Bewick S et al (2016) Antibiotics as a selective driver for conjugation dynamics. Nat Microbiol 1(6):16044
Lopes JA (1986) Evaluation of dairy and food plant sanitizers against *Salmonella typhimurium* and *Listeria monocytogenes* 1. J Dairy Sci 69(11):2791–2796
Madeo M, O'Riordan N, Fuchs TM, Utratna M, Karatzas KAG, O'Byrne CP (2012) Thiamine plays a critical role in the acid tolerance of *Listeria monocytogenes*. FEMS Microbiol Lett 326(2):137–143
Matin A, Auger EA, Blum PH, Schultz JE (1989) Genetic basis of starvation survival in non differentiating bacteria. Ann Rev Microbiol 43(1):293–314

Mattila M, Somervuo P, Rattei T, Korkeala H, Stephan R, Tasara T (2012) Phenotypic and transcriptomic analyses of Sigma L-dependent characteristics in *Listeria monocytogenes* EGD-e. Food Microbiol 32(1):152–164

Nair S, Frehel C, Nguyen L, Escuyer V, Berche P (1999) ClpE, a novel member of the HSP100 family, is involved in cell division and virulence of *Listeria monocytogenes*. Mol Microbiol 31(1):185–196

Naïtali M, Dubois-Brissonnet F, Cuvelier G, Bellon-Fontaine MN (2009) Effects of pH and oil-in-water emulsions on growth and physicochemical cell surface properties of *Listeria monocytogenes*: Impact on tolerance to the bactericidal activity of disinfectants. Int J Food Microbiol 130(2):101–107

Nakamura H, Tokuda Y, Sono A, Koyama T, Ogasawara J, Hase A et al (2006) Molecular typing to trace *Listeria monocytogenes* isolated from cold-smoked fish to a contamination source in a processing plant. J Food Prot 69(4):835–841

Nilsson RE, Latham R, Mellefont L, Ross T, Bowman JP (2012) MudPIT analysis of alkaline tolerance by *Listeria monocytogenes* strains recovered as persistent food factory contaminants. Food Microbiol 30(1):187–196

O'Neill J (2015) The review on antimicrobial resistance antimicrobials in agriculture and the environment: reducing unnecessary use and waste. Wellcometrust, London

Oliver HF, Orsi RH, Wiedmann M, Boor KJ (2013) σB plays a limited role in the ability of *Listeria monocytogenes* strain F2365 to survive oxidative and acid stress and in its virulence characteristics. J Food Prot 76(12):2079–2086

Pan Y, Breidt F, Kathariou S (2006) Resistance of *Listeria monocytogenes* biofilms to sanitizing agents in a simulated food processing environment. Appl Environ Microbiol 72(12):7711–7717

Pérez-Varela M, Corral J, Aranda J, Barbé J (2018) Functional characterization of AbaQ, a novel efflux pump mediating quinolone resistance in *Acinetobacter baumannii*. Antimicrob Agents Chemother 62(9):e00906–e00918

Roberts AP, Mullany P (2011) Tn 916-like genetic elements: a diverse group of modular mobile elements conferring antibiotic resistance. FEMS Microbiol Rev 35(5):856–871

Ryan S, Begley M, Gahan CG, Hill C (2009) Molecular characterization of the arginine deiminase system in *Listeria monocytogenes*: regulation and role in acid tolerance. Environ Microbiol 11(2):432–445

Samelis J, Kendall P, Smith GC, Sofos JN (2004) Acid tolerance of acid-adapted and nonadapted *Escherichia coli* O157: H7 following habituation (10°C) in fresh beef decontamination runoff fluids of different pH values. J Food Prot 67(4):638–645

Sauders BD, Wiedmann M (2007) Ecology of *Listeria* species and *L. monocytogenes* in the natural environment. Food Sci Technol. Marcel Dekker, New York, 161, 21

Segal AW, Garcia R, Goldstone H, Cross AR, Jones OT (1981) Cytochrome b-245 of neutrophils is also present in human monocytes, macrophages and eosinophils. Biochem J 196(1):363

Shabala L, Budde B, Ross T, Siegumfeldt H, Jakobsen M, McMeekin T (2002) Responses of *Listeria monocytogenes* to acid stress and glucose availability revealed by a novel combination of fluorescence microscopy and microelectrode ion-selective techniques. Appl Environ Microbiol 68(4):1794–1802

Sifaoui F, Arthur M, Rice L, Gutmann L (2001) Role of penicillin-binding protein 5 in expression of ampicillin resistance and peptidoglycan structure in *Enterococcus faecium*. Antimicrob Agents Chemother 45(9):2594–2597

Simjee S, Heffron AL, Pridmore A, Shryock TR (2012) Reversible monensin adaptation in *Enterococcus faecium*, *Enterococcus faecalis* and *Clostridium perfringens* of cattle origin: potential impact on human food safety. J Antimicrob Chemother 67(10):2388–2395

Singer AC, Shaw H, Rhodes V, Hart A (2016) Review of antimicrobial resistance in the environment and its relevance to environmental regulators. Front Microbiol 7:1728

Sisti M, Albano A, Brandi G (1998) Bactericidal effect of chlorine on motile Aeromonas spp. in drinking water supplies and influence of temperature on disinfection efficacy. Letters in applied microbiology 26(5):347–351.microbiology. *Current opinion in microbiology, 19*, v-vii

Somers EB, Lee Wong AC (2004) Efficacy of two cleaning and sanitizing combinations on *Listeria monocytogenes* biofilms formed at low temperature on a variety of materials in the presence of ready-to-eat meat residue. J Food Prot 67(10):2218–2229

Soni KA, Nannapaneni R, Tasara T (2011) The contribution of transcriptomic and proteomic analysis in elucidating stress adaptation responses of *Listeria monocytogenes*. Foodborne Pathog Dis 8(8):843–852

Suarez-Carmona M, Hubert P, Delvenne P, Herfs M (2015) Defensins: "Simple" antimicrobial peptides or broad-spectrum molecules? Cytokine Growth Factor Rev 26(3):361–370

Takami H, Takaki Y, Uchiyama I (2002) Genome sequence of Oceanobacillus iheyensis isolated from the Iheya Ridge and its unexpected adaptive capabilities to extreme environments. Nucleic Acids Res 30(18):3927–3935

Tandukar M, Oh S, Tezel U, Konstantinidis KT, Pavlostathis SG (2013) Long-term exposure to benzalkonium chloride disinfectants results in change of microbial community structure and increased antimicrobial resistance. Environ Sci Technol 47(17):9730–9738

Tiganitas A, Zeaki N, Gounadaki AS, Drosinos EH, Skandamis PN (2009) Study of the effect of lethal and sublethal pH and a_w stresses on the inactivation or growth of *Listeria monocytogenes* and *Salmonella* Typhimurium. Int J Food Microbiol 134(1–2):104–112

Tomii K, Kanehisa M (1998) A comparative analysis of ABC transporters in complete microbial genomes. Genome Res 8(10):1048–1059

Van Boeckel TP, Brower C, Gilbert M, Grenfell BT, Levin SA, Robinson TP et al (2015) Global trends in antimicrobial use in food animals. Proc Natl Acad Sci 112(18):5649–5654

Van Der Linden JW, Camps SM, Kampinga GA, Arends JP, Debets-Ossenkopp YJ, Haas PJ et al (2013) Aspergillosis due to voriconazole highly resistant *Aspergillus fumigatus* and recovery of genetically related resistant isolates from domiciles. Clin Infect Dis 57(4):513–520

Van der Veen S, Abee T (2011) Mixed species biofilms of *Listeria monocytogenes* and *Lactobacillus plantarum* show enhanced resistance to benzalkonium chloride and peracetic acid. Int J Food Microbiol 144(3):421–431

Vasseur C, Rigaud N, Hebraud M, Labadie J (2001) Combined effects of NaCl, NaOH, and biocides (monolaurin or lauric acid) on inactivation of *Listeria monocytogenes* and *Pseudomonas* spp. J Food Prot 64(9):1442–1445

Warnes SL, Highmore CJ, Keevil CW (2012) Horizontal transfer of antibiotic resistance genes on abiotic touch surfaces: implications for public health. MBio 3(6):e00489–e00412

Webber MA, Whitehead RN, Mount M, Loman NJ, Pallen MJ, Piddock LJ (2015) Parallel evolutionary pathways to antibiotic resistance selected by biocide exposure. J Antimicrob Chemother 70(8):2241–2248

White DG, McDermott PF (2001) Biocides, drug resistance and microbial evolution. Curr Opin Microbiol 4(3):313–317

Whittam R, Wheeler KP (1970) Transport across cell membranes. Annu Rev Physiol 32(1):21–60

Molecular Mechanisms of Action and Resistance of Antimalarial Drugs

14

Juveria Khan, Monika Kaushik, and Shailja Singh

Abstract

Drug-resistance in plasmodium parasites is manifested at large scale, putting all the malaria control efforts in vain with an urgent need of complementary strategies. A deeper understanding of mechanism of action of drugs, drug resistance and cross-resistance between drugs will pave a way to design an effective individualized drug for malaria-affected regions. This chapter summarizes the molecular mechanism of all currently available anti-malarial drugs and the factors playing significant role in the development and spread of resistance against the antimalarials.

Keywords

Plasmodium · Anti-malarial drug · Mechanism of action of drug · Drug resistance

14.1 Introduction

Malaria is caused by plasmodium parasite, a protozoan parasite. Infected female Anopheles mosquitoes carry the parasite for the transmission to the human beings. Five species of plasmodium responsible for malarial infection are *Plasmodium falciparum, Plasmodium ovale, Plasmodium vivax, Plasmodium malariae, and Plasmodium knowlesi* majorly in parts of Southeast Asia (Kantele and Jokiranta 2011). *Plasmodium falciparum* causes the most severe malaria. Malaria causes

Authors Juveria Khan and Monika Kaushik have equally contributed to this chapter.

J. Khan · M. Kaushik · S. Singh (✉)
School of Biotechnology, Jawaharlal Nehru University, New Delhi, India
e-mail: shailjasingh@mail.jnu.ac.in; shailja.singh@snu.edu.in; shailja.jnu@gmail.com

S. M. Mandal, D. Paul (eds.), *Bacterial Adaptation to Co-resistance*,
https://doi.org/10.1007/978-981-13-8503-2_14

infant mortality and death of pregnant women due to intrauterine growth retardation and abortion, respectively.

According to the WHO report of 2016, 212 million malaria cases with 429,000 deaths, especially among children were reported worldwide in 2015 despite having 21% decrease in the malarial incidence and 29% decrease in the malarial mortality rate globally from 2010 to 2015. 92% of deaths have taken place in the African region, followed by 6% deaths in South-East Asia region and 2% deaths in the Eastern Mediterranean Region. Some regions have shown admirable cut down in the malaria mortality rate. The malaria mortality rate has reduced by 58% in the Western Pacific Region, 46% in the South-East Asia Region, 37% in the Region of the Americas and 6% in the Eastern Mediterranean Region since 2010. In 2015, all the 53 countries of the European Region reported at least 1 year of zero cases of malaria (WHO 2016). Although this data of decrement is pleasing enough but the danger of malaria is still looming around. One of the major reasons for this dreadful disease besides ever-increasing mosquitoes, unavailability of vaccines and tropical climate is to overcome the disastrous challenge of drug resistance. At the same time, antimalarial drug development process is also lacking behind. In the period between 1975 and 1996, 1223 new drugs were registered but only 3 of them were antimalarial. Hence there is an urgent need for a rational antimalaria treatment policy.

The control methods of malaria include vector control, prophylactic drugs and vaccination. To date no completely effective vaccine is available for the control and prevention of malaria. Malarial chemotherapy plays a vital part in malaria control. Quinoline-containing antimalarials, such as quinine (QN), amodiaquine (AQ), chloroquine (CQ), primaquine (PQ) and mefloquine (MQ) are mainstays of chemotherapy against malaria which have long been used for clinical purpose. Among them chloroquine (CQ) rapidly became one of the most effective antimalarial drugs for first-line therapy after the Second World War owing to its highly effective, safe, well-tolerated and reasonably low cost features.

But within the period of ten to twelwe years, P. falciparum developed resistance against chloroquine in Colombia and the Cambodia-Thailand border region, spreading throughout the south Asia and South America. Resistance was not limited to these regions only but spread to Africa by the 1970 and soon the resistance rolled out at global level. Then as an alternative artemisinin-based combination therapy (ACTs) was introduced and still in use but it is heavy on pocket. CQ is the first line of treatment against P. vivax but resistance is emerging in P.vivax also (Price et al.). Different ACTs in use are Dihydroartemisinin/piperaquine, Artemether/lumefantrine, Artesunate–Mefloquine, Artesunate–Amodiaquine, Artesunate–Sulfadoxine–Pyrimethamine Artesunate–Pyronaridine, and recently approved Pyronaridine-tetraphosphate/artesunate (Pyramax) and piperaquinetetraphosphate/dihydroartemisinin (Eurartesim or PQ-DHA).

To conquer a war, one needs to be a thoughtful planner, well trained with one's tools of warfare: antimalarial drugs and not only this, knowing the strategies of enemy, understanding their tactics: anti-malarial drug resistance and exploiting their weaknesses are some important steps towards accomplishing success. In this chapter, we will discuss majority of drugs employed to cure malaria, their mode of action and the clever strategies of the plasmodium to overcome the action plan of different drugs i.e. antimalarial drug resistance. The broad understanding towards thesemechanisms will help us to enhance our action plan to cure our communities facing malarial trolls.

14.2 Classification of Antimalarials

There can be two ways to classify antimalarial drugs that are on the basis of their structure and the parasitic stages on which they act.

14.2.1 Parasite Stage Based

1. **Tissue schizonticides for causal prophylaxis:** Pyrimethamine and Primaquineare the drugs which act on tissue schizonticides. The clinical symptoms for malaria appear only when the infection reaches the erythrocyte stage. But thesedrugs act on the liver stage of the parasite. So theoretically, these drugs can prevent further development of infection but this is not practically possible since the infection can not be predicted before the clinical symptoms appear thus tissue schizonticides are more theoretical than practical.
2. **Tissue schizonticides for relapse prevention:** Malaria has the tendency of relapse on reactivation due to the hypnozoites present in *P. vivax* and *P. ovale* in the liver. Primaquine is the priliminary drug; pyrimethamine also shows such activity.
3. **Blood schizonticides:** These drugs are quite significant for anti-malarial chemotherapy since they act on the blood forms of the parasite thus terminate the clinical assault of malaria. Drugs like chloroquine, quinine, mefloquine, halofantrine, pyrimethamine, sulfadoxine, sulfones, and tetracyclines come under this category.
4. **Gametocytocides:** Drugs like Primaquine, Chloroquine and quinine act on gametocytes, killing the sexual form of parasites thus preventing the transmission of infection to mosquito. Quinine and Chloroquine have gametocytocidal activity against *P. vivax and P. malariae* but P. falciparum whereas Primaquine is active against all plasmodia, including P. falciparum.
5. **Sporontocides:** Primaquine and chloroguanide comes under this category. They inhibit the oocysts development in mosquito and thus ablate the transmission process.

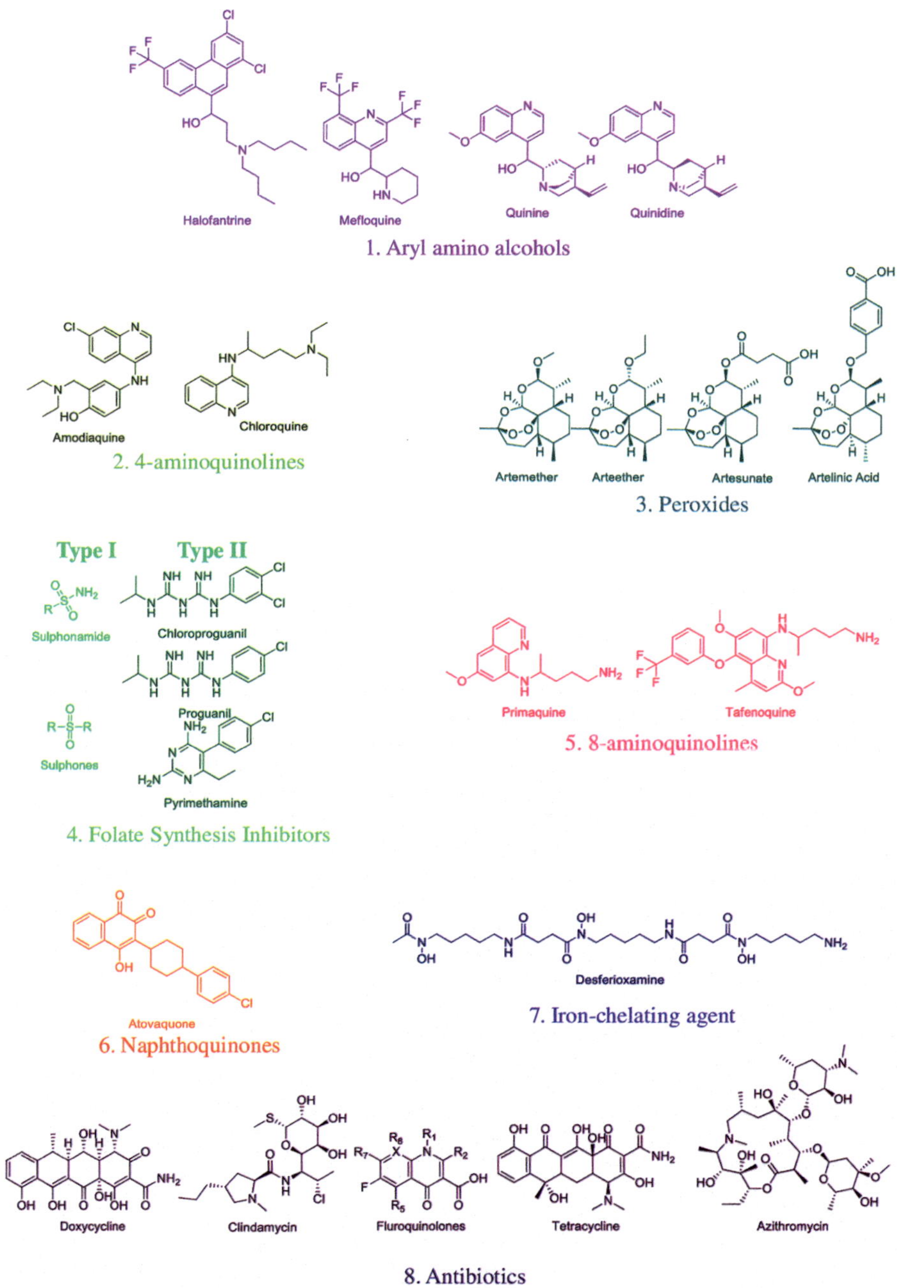

Fig. 14.1 Classification of Anti- Malarial drugs according to their structure

14.2.2 Structure Based

1. **Aryl amino alcohols:** Quinidine, mefloquine, quinine, halofantrine.
2. **4-aminoquinolines:** Amodiaquine, chloroquine.
3. **Endoperoxides:** Artemisinin derivatives – artemether, arteether, artesunate, artelinic acid
4. **Antifolates:** Type 1 – Dihydropteroate synthase inhibitor – sulphones, sulphonamides; Type 2 –dihydrofolatereductase inhibitor – biguanides like proguanil and chloroproguanil; diaminopyrimidine like pyrimethamine
5. **8-aminoquinolines:** WR238, 605, Primaquine
6. **Naphthoquinones:** Atovaquone
7. **Iron chelating agents:** Desferrioxamine
8. **Antibiotics:** Azithromycin, tetracycline, clindamycin, fluoroquinolones, doxycycline (Fig. 14.1)

14.3 4-Aminoquinolines: Chloroquine (CQ), Amodiaquine (AQ)

Both the CQ and AQ accumulate inside the parasite vacuole and intervene in the haem detoxification, which is a bi-product of haemoglobin degradation. Thus they share the similar mode of action.

14.3.1 Chloroquine

CQ is accumulated in the food vacuole of the parasite either by diffusion or protonation since CQ is a diprotic weak base and FV is acidic. It hinders with the detoxification of heme in the FV, leading to the death of the parasite death (Chou et al. 1980).

Cytoplasm of the erythrocyte is enriched with hemoglobin so upon digestion of the host's cytoplasm by the parasite, copious quantity of ferriprotoporphyrin IX (FPIX) is formed inside the FV. FPIX is toxic to the parasite as it permeabilizes the membrane causing changes in the pH leading to cell lysis (Chou and Fitch 1981; Chou and Fitch 1980, Fitch et al. 1982). The parasite defends itself from the toxic effects of FPIX by two ways: FPIX is polymerized to form hemozoin(HZ) inside the FV (Slater et al. 1991) or exits FV and undergoes degradation by glutathione present in the cytosol (Atamna and Ginsburg 1995). 70% of the FPIX is detoxified by cytosolic glutathione during the trophozoite stage, the stage most sensitive to CQ (Ginsburg et al. 1998). CQ is capable enough to counteract on both of the pathways in the following manner.

14.3.1.1 Mode of Action

A.Inhibition of polymerization of FPIX to HZ

In vitro studies have shown that the HZ formation is inhibited (Blauer and Akkawi 1997; Dorn et al. 1998) in the intact, CQ-treated, infected cells due to the binding of CQ to FPIX (Chou and Fitch 1993; Zhang et al. 1999). CQ is a weak base thus accumulates in the food vacuole which is acidic in nature. The force responsible for accumulation is the pH difference between the extracellular medium and the food vacuole. Polymerization of FPIX into HZ is inhibited leading to the accumulation of either free FPIX or its complex with CQ inside the food vacuole which is harmful to the plasmodium. Availability of FPIX is responsible for high affinity and saturable uptake of CQ, since the uptake of CQ is decreased upon the inhibition of FPIX (Bray et al. 1998) which deduces FPIX to be the receptor for CQ (Chou et al. 1980). Partition coefficient of FPIX in phospholipid membranes is 10^5 thus enabling its swift translocation (Light and Olson 1990a, b) which implies a pronounced competition between hemozoin formation and FPIX exit from the food vacuole illustrating its substantial egress from the vacuole. CQ enhances the solubility of FPIX in membrane (Ginsburg and Demel 1983), and thus increases its exit from the food vacuole, and now more of the CQ binds to FPIX in the cytosol. This is the point where the second mode of action of chloroquine comes into play.

B.Inhibition of glutathione mediated FPIX degradation

FPIX is released upon digestion of hemoglobin; around 50% of the FPIX in case of P.berghei (Burstein 1993) and 25–30% in case of P.falciparum (Ginsburg et al. 1998), is polymerized to hemozoin in the food vacuole (FV) whereas the rest of the FPIX escape out of the FV and is further degraded by glutathione (GSH) (Atamna and Ginsburg 1995). This implies that CQ not only inhibits FPIX polymerization by forming CQ-FPIX complex but also via some other way; In vitro studies have shown that CQ inhibits GSH dependent degradation of FPIX also. As soon as the *Plasmodium* is treated with CQ, a gradual dose and time dependent accumulation of FPIX is observed in the membrane fraction and a disturbance in the cation concentration is observed in infected cells (Chou and Fitch 1980). CQ enhances the FPIX-dependent permeabilization of membranes to sodium and potassium leading to FPIX accumulation which is responsible of parasite killing (Chou and Fitch 1981; Chou and Fitch 1980) The factors responsible for the cytotoxic effect of CQ (Krugliak and Ginsburg 1991) are: (a) generation of membrane-associated FPIX for increased permeability; (b) depletion of potassium in the cytosol to disturbthe cation homeostasis. In conclusion, both the cytosolic and vacuolar concentration of CQ is equally important to accomplish the action of the drug.

14.3.1.2 Resistance Development

In the late 1950s, the first CQ resistance case was reported in *Plasmodium falciparum*. Global expansion of resistance was soon observed forcing the development of new effective drugs, leading to the use of costly drugs i.e. artemisinin-based combination therapies (ACTs) used today. CQ is the first line treatment for *P. vivax* but even *P. vivax* is developing resistance to it (Price et al. 2014). The gene responsible for the development of CQ resistance (CQR) was identified by genetic studies. Later the gene and its role were confirmed by reverse genetic approaches (Fidock et al. 2000; Sidhu et al. 2002; Su et al. 1997). The intuitive function of *Plasmodium falciparum* Chloroquine Resistance Transporter (PfCRT) is not well established, although it is proposed to mediate the transfer of hemoglobin-derived peptides/amino acids from the FV (Martin and Kirk 2004). PfCRT is predicted to play different roles including chloride channel, a general activator, modulator of transport systems, a proton pump, a regulator of proton pumps and a proton-coupled transporter of cationic substrates (Juge et al. 2015).

The prominent mutation observed in the CQ resistant strain is found at position 76 in the transmembrane domain of PfCRT. Lysine at position 76 was found to mutate to different residues including threonine, asparagines and isoleucine. The mutation removes the positive charge from the substrate-binding site in the PfCRT due to which protonated CQ moves out of the FV (Blauer and Akkawi 1997).

Glutathione (GSH) is the main antioxidant present in the plasmodium and plays an important role in chloroquine resistant (Atamna and Ginsburg 1995; Ginsburg and Golenser 2003; Ginsburg et al. 1998). PfCRT carry out both the efflux of CQ and influx of GSH in the FV of parasite and thus converts the heme into hemozoin (Patzewitz et al. 2012).

14.4 Artemisnin

Artemisinin is a sesquiterpene lactone endoperoxide containing peroxide bridge, considered to be the key for its activity. Although artemisnin is widely used as antimalarial drug but the science behind its mode of action is still a matter of discussion (de Ridder et al. 2008; van Agtmael et al. 1999; Wang et al. 2010; Willcox 2009; Klonis et al. 2013).

It was discovered by a chineese group under Professor Youyou Tu in 1971. Artemisinin is the extract of herbal plant named, *Artemisia annua* (Tu 2011, Miller and Su 2011, Su and Miller 2015). It has low solubility in both oil and water. But synthetic derivatives of artemisinin including arteether, artemether, and artesunate developed later on, have higher solubility.

Later on artemsisnin was considered unsuitable as monotherapy so in 1984 it was first used in combination with other antimalarial drugs, known as Artimisinin Comination Therapy. The combination being more effective and less prone to resistance is used as first line of defence (Tu 2011, Brown 2010) since artemisnin and other conventional drugs have different modes of action (Li et al. 1984) so there is low chance for the parasite to develop mutations against two drugs simultaneously (Nosten and White 2007; White 1999).

14.4.1 Modes of Action

Artemisinin comes into action upon activation of endoperoxide bridge and releases free radicals. Following are two pathways for its activation:

1.The mitochondrial mediated degradation pathway (Sun et al. 2015).

The electron transport chain in mitochondria-activates artemisinin causing lipid peroxidation by generating reactive oxygen species leading to cytotoxicity and depolarization of the mitochondrial and plasma membrane (Sun et al. 2015, Wang et al. 2010, Mercer et al. 2010).

2.Heme- mediated degradation pathway (Sun et al. 2015).

In the heme-mediated pathway, two activation models are suggested

(a)Reductive scission model
(b)Open peroxide model

Both the models generate carbon-centered radical (O'Neill et al. 2010)

Heme is produced at both the ring and trophozoite stages of parasite de novo and by hemoglobin digestion respectively. However, the production of heme is higher via hemoglobin digestion rather than endogenous biosynthesis of heme, suggesting the major involvement of hemoglobin-derived heme for artemisinin activation (Wang et al. 2015, Klonis et al. 2011). Briefly, artemisinin forms complex with free heme and inhibits the formation of hemozoin (Chugh et al. 2013).

Several other proteins are also predicted to play part in the activation of artemisinin. Two cysteine proteases namely falcipain2a and falcipain3 (Xie et al. 2015), translationally controlled tumor protein (PfTCTP) (Bhisutthibhan et al. 1998; Eichhorn et al. 2012), PfATP6, an orthologous sarco/endoplasmic reticulum Ca2 + -ATPase (SERCA), (Eckstein-Ludwig et al. 2003). Apart from these targets, five other enzymes were also reported to act as potential targets of artemisinin, namely ornithine aminotransferase (OAT), pyruvate kinase (PyrK), L-lactate dehydrogenase (LDH), spermidine synthase (SpdSyn), and S-adenosylmethioninesynthetase (SAMS). Covalent modificationof these enzymes results in their irreversible malfunction (Delabays et al. 2001).

14.4.2 Resistance Development

Artemisinin has been used for over a decade as ACT to treat malaria. But recently, cases of artemisnin resistant in were observed in Southeast Asia, which has raised the concerns (Htut 2009; Witkowski et al. 2012; Ashley et al. 2014; Amaratunga et al. 2012). But interestingly no change in the earlier proposed targets (PfATP6 and PfTCTP) of artemisnin was observed (Afonso et al. 2006).

Resistance occurs primarily in the ring stage due to mutations in proteins including multidrug resistance proteins (van der Velden et al. 2015). and kelch propeller domain and (Afonso et al. 2006; Witkowski et al. 2013; Mok et al. 2014;

Takala-Harrison et al. 2012; Cheeseman et al. 2012; Ashley et al. 2014; Ariey et al. 2013; Isozumi et al. 2015; Straimer et al. 2014; Nyunt et al. 2014; Bayih et al. 2015).

14.5 Antifolates

Antifolates either act on dihydropteroate synthase (DHPS) (Brooks et al. 1994) or dihydrofolatereductase (DHFR) (Bzik et al. 1987). Both the enzymes are involved in the folate synthesis pathway. Drugs which are involved in the inhibition of DHPS are sulphones, sulphonamides, sulfadoxine and the drugs which inhibit DHFR activity are biguanides and diaminopyrimidine.

14.5.1 Targets in Folate-Biosynthesis Pathway

Folate needs to be synthesized in the diet since most of the enzymes involved in the folate biosynthesis pathway are not present in the mammalian system. So targeting enzymes of folate pathway can be a good strategy.

It has been found that inhibitors of DHPS results in the depletion of deoxythymidine triphosphate and thus reduces the production of DNA in *Plasmodium spp.* and interestingly there is no homologue of DHPS in *Homo sapiens*. (Schellenberg and Coatney 1961).

DHFR of *P. falciparum* is bifunctional unlike bacteria and eukaryotes (Bzik et al. 1987) Dihydrofolate synthase (DHFS) produces dihydrofolate which is converted to different folate derivaves by three enzymes including dihydrofolatereductase (DHFR), serine hydroxymethyltransferase (SHMT), and thymidylate synthase (TS).

Thus from the folate biosynthesis pathway, two enzymes can be targeted namely, DHFR and DHPS.

14.5.2 Available Drugs

Both the de novo (Ferone 1977) and salvage pathway (Krungkrai et al. 1989) are found to be involved in biosynthesis of folate in *Plasmodium spp.* But the predominant folate derivatives present in the mammalian host is a poor substrate for the two *Plasmodium* folate transporters (Salcedo-Sora et al. 2011) and secondly exogenous folate salvage pathway is not the primary source in *Plasmodium*. Folate synthesis pathway is involved in the synthesis of different metabolites in different pathways including purine, pyrimidine, amino acid biosynthetic, (Ferone 1977) and DNA replication (Schellenberg and Coatney 1961; Gutteridge and Trigg 1971; Newbold et al. 1982; Gritzmacher and Reese 1984; Triglia and Cowman 1999). Thus its inhibition can disturb the *Plasmodium* parasite abruptly. But the great agony is the rise of mutations observed in these genes, responsible for the resistance against antifolate drugs (Peterson et al. 1988).

Dihydropteroate synthase (DHPS) mediate combining of pteridine with para-amino benzoic acid (PABA) to form dihydropteroate. Sulfadoxine is a structural analog of PABA thus involved in the inhibition of DHPS (Nzila et al. 2000). However, since *Plasmodium* can take folate from its environment, many parasite isolates can survive even in the presence of DHPS inhibiting drug including sulfadoxinein. Nevertheless, DHPS is essential, as parasites with a non-functional *dhps* can not survive (Wang et al. 2004).

Pyrimethamine, a structural analogue of dihydrofolate inhibit DHFR, thus halts the production of folate derivatives made from both exogenous and *de novo* folate synthesis pathway.

Other derivatives of DHFR are WR99210, cycloguanil, chlorcycloguanil, pyrimethamine (Ferone et al. 1969; Milhous et al. 1985; Winstanley et al. 1995; Sirawaraporn et al. 1997; Nzila-Mounda et al. 1998) and trimethoprim in their decreasing order of potency (Ferone et al. 1969; Iyer et al. 2001).

14.5.3 Resistance Development

In the 1970s, antifolate resistance was observed within the fleeting time after they were introduced in South America and Southeast Asia (Le Bras and Durand 2003). Factors responsible for the development of resistance are Single-nucleotide polymorphisms (SNPs) and copy-number variation (CNV).

14.5.3.1 Single-Nucleotide Polymorphisms (SNPs)

Genetic alterations due to mutations in enzyme targets and transporter pumps are responsible for drug failure (Le Bras and Durand 2003). Resistance to antifolates arises after accumulation of point mutations, leading to increase in the levels of SP resistance *in vivo* (Le Bras and Durand 2003). The first mutation in *dhfr* is supposed to be S108 N in SP-exposed parasites which are followed by additional mutations in both target enzymes with constant drug pressure. Other identified mutations are A437G and K540E in *dhps* and N51I, C59R, S108 N in *dhfr* (Sridaran et al. 2010).

14.5.3.2 Copy-Number Variation (CNV) of GTP-Cyclohydrolase (GCH)

P. falciparum upregulate the expression of certain genes in order to compensate for the inhibitory activity of some drugs. Amplification of GTP-cyclohydrolase via copy number variation enhances the flux from the first step of the folate-synthesis pathway, thus increase the substrate for the downstream enzymes (Kidgell et al. 2006).

Increased GCH1 was not a very safe zone for parasite survival. In some parasite lines, it enhances pyrimethamine resistance while in others a decrease in resistance is observed when *gch1* was overexpressed at high levels, with a marked decrease in efficacy of downstream enzymes due to a combination of point mutations and drug pressure (Heinberg et al. 2013). The possible reason may be the accumulation of one of the folate intermediates which is involved in slowing down the downstream reactions leading to toxicity in parasites or inhibition of other essential enzymes (Kwon et al. 2008). Thus parasite needs to maintain a level of folate such that it's neither too high to compromise survival nor too low to cause deficiency (Fig. 14.2).

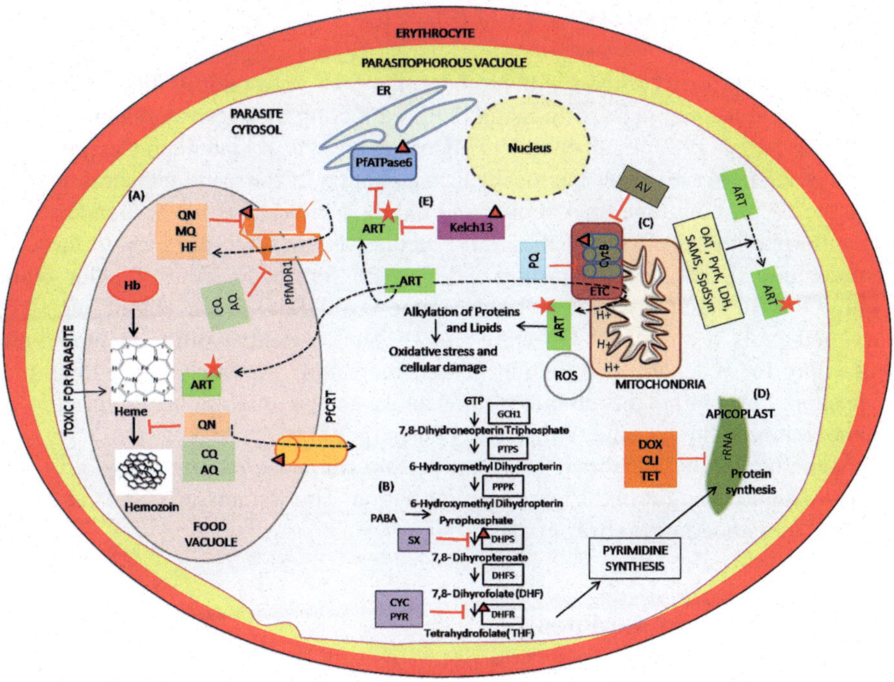

Fig. 14.2 Mechanism of Action and Resistance to Antimalarial Drugs: Targeted biological pathways **(A) Heme detoxification in Food Vacoule**: (i) 4- aminoquinoles: Chloroquine (CQ), Amodiaquine (AQ) and Artemisnin (ART): Artemisnin gets activated represented by (red star) binds to reactive heme and interferes with its detoxification by making complex with Heme and inhibits formation of hemezoin. (ii) Aryl amino alcohols Quinine (QN), Mefloquine (MQ), Halofantrine (HF) is sensitive to PfCRT and PfMDR1. Resistance occurs due to mutation (represented by red triangle) and copy number variation in gene pfcrt and pfmdr1 (transporter)

(B) Folate Biosynthetic pathway in cytosol: Folate Biosynthetic pathway enzymes Dihydropteroate synthase (DHPS) is sensitive to drug Sulphadoxine (SX) and dihydrofolate reductase (DHFR) is sensitive to Pyrimethamine (PYR) and Chloroproguanil (CYC). Parasite becomes resistant to SX due to mutation (represented by red triangle) in DHPS and to PYR, CYC due to mutation (represented by red triangle) and copy number variation in dhps and dhfr.

(C) Electron Transport chain in mitochondria: Activated Artemisnin (ART) (represented by star), Atovaquine (AV) and Primaquine (PQ) is sensitive to cytochrome b complex in electron transport chain via production of reactive oxygen species (ROS). Activated Artemisnin (ART) in mitochondria leads to alkylation of proteins and lipids which causes oxidative stress and cellular damage. Resistance to these drugs occurs due to mutation (represented by red triangle) in cytb gene.

(D) Protein synthesis pathway in Apicoplast: Antibiotics Doxycycline (DOX), Clindamycin (CLI), Tetracycline (TET) is sensitive to protein synthesis by targeting ribosomal RNA in Apicoplast.

(E) Other targeted molecules in Parasite: Activated Artemisnin (represented by red star) is sensitive to PfATP6 in Endoplasmic Reticulum. Resistance occurs due to mutation in pfatp6 gene or by Kelch 13 Propeller.

Other potential target of Artemisnin is pyruvate kinase (PyrK), S-adenosylmethionine synthetase (SAMS) spermidine synthase (SpdSyn), ornithine aminotransferase (OAT), and L-lactate dehydrogenase (LDH).

14.6 Arylmethanols

Arylmethanols form complex with protoporphyrin IX, thus inhibits the formation of the synthetic hemozoin i.e. beta-hematin thus affecting the growth of hematin crystals (Chou et al. 1980; Egan et al. 1997; Dorn et al. 1998) (Slater and Cerami 1992; Egan et al. 1994). But still the mechanism of action of the aryl methanols such as quinine (QN), quinidine, mefloquine and halofantrine (Hf) is not wellunderstood. The morphological effects of these aryl methanols are quite different to those of chloroquine. Hemoglobin transport vesicles are formed in case of chloroquine within the parasite (Yayon et al. 1984; Hoppe et al. 2004), while absent in case of arylmethanols (Famin and Ginsburg 2002). Thus, scientist differ in their view regarding the mechanism of action of aryl methanols that in place of inhibiting hemazoin formation, the quinoline methanols act by inhibition of ingestion of hemoglobin by the parasite (Famin and Ginsburg 2002).

Mutations in the membrane transport protein cause the resistance against these drugs with no changes in the hemozoin formation. Thus hemozoin still can be considered as a good target (Bray et al. 2005).

14.6.1 8-Aminoquinolines

8-Aminoquinoline is a derivative of quinoline with amine group attached at the eighth position of quinoline There are three members in the family of 8-aminoquinoline i.e. primaquine, tafenoquine and pamaquine. They act on hypnozoites and has been used as prophylactic agent for malaria. The use of pamaquine has been stopped but primaquine is widely used against p. vivax and P. ovale and tefenoquine is in phase III of clinical trials.

14.6.2 Mode of Action

Primaquine is an Food and Drug Administration (FDA) approved drug, (Hill et al. 2006). Its mode of action is not completely unraveled. It generates the oxygen free radical and disrupt the electron transport chain of respiration in parasite (Hill et al. 2006). Primaquine forms reactive molecules via via cytochrome P450 and causes both the lysis of blood and inhibition of parasite growth (Pybus et al. 2013). It has multiple effects from hypnozoite elimination in *P. vivax* and *P. ovale to gametocytocidaal against* falciparum malaria. Although it shows weak activity on asexual stage (very weak for *P. falciparum*). It has already been 60 years of using Primaquine but it still play an important role as antimalarial. (Vale et al. 2009).

Despite such a successful history, primaquine pose serious hazards and safety issues as well, such as haemolysis in glucose-6-phosphate dehydrogenase (G6PD) deficient individuals (Ramos Junior et al. 2010). G6PD is an X-linked genetic disorder andquite common in tropical areas with frequencies varying from 3 to 30%. Thus restricting its use for some groups like pregnant women (Fernando et al. 2011). Medical practitioners are often at variance regarding recommendations and prescription practice of this drug. The 8-aminoquinolines play significant role in reducing infectivity to mosquitos. Studies have shown that primaquine together with artemisinin combination treatment (ACT) is very effective against gametocytes (WHO 2013).

Primaquine is currently used to treat vivax malaria at three different stages: patients with confirmed parasitaemia for blood as well as initial liver stage and for terminal prophylaxis (Hill et al. 2006) i.e. treatment given at the end of the exposure period. This step is taken as a precaution since vivax malaria often relapses due to the presence of hypnozoites.

14.6.3 Resistance Development

There is no evidence of resistance in the liver stage activity but reduced enzyme activity is observed in people with CYP2D6 genetic polymorphisms indicationg reduced efficiency of primaquine (Pybus et al. 2012, Bennett et al. 2013).

Primaquine resistance is hard to determine due to the following reasons.

1. Primaquine is always given combination with a blood schizontocide, and to determine the actual drug lackingefficacy is difficult (Gascon et al. 1994).
2. Dose proportional to body weight is an important factor before declaring treatment failure (Takeuchi et al. 2010).
3. Primaquine is given to remove hypnozoites which are responsible for relapse. But it is difficult to to distinct relapse from re-infection. Since, a patient in malaria endemic region might get re-infected with P. vivax which can be misunderstood as relapse.

14.7 Metal Chelators

There are no free non-mutable metals available as drug target till date except heme. Copper, iron and zinc are the three important metals involved in different cellular processes like electron transfer reactions, iron metabolism, transport, DNA-protein interactions etc. These metals can be toxic if homeostasis is disturbed. Metal chelators are effective at high concentrations with a slow rate of clearance.

14.7.1 Iron Chelators

One of the host defense mechanism is to sequester the iron in order to fight infectious diseases (Cabantchik et al. 1999). This short-term exhaustion is harmful for *Plasmodium* with very less effect on host system. When *Plasmodium* culture is treated with iron chelators, growth of the parasite is inhibited, most effectively on trophozoites. (Raventos-Suarez et al. 1982; Atkinson et al. 1991). In vivo studies have shown that when mice infected with *P. chabaudi* is given diet, deficient in iron, the mortality and parasitemia was found to get reduced (Harvey et al. 1985). Iron chelators act on various enzymes containing iron interfere with the artemisinins but otherwise represent a strategy of targeting multiple enzymes containing iron. Sequestration of iron is a common host defense strategy to fight infectious agents. It is shown to play an essential role in different experiments like dietary iron deficiency protects mice against *P. chabaudi*, decreasing mortality and parasitemia Iron chelators inhibit liver stage as well as erythrocyte stage malaria and number of merozoites released from the liver is enhanced on iron supplementation (Loyevsky et al. 1999). Therefore *Plasmodium* growth can be inhibited*in vivo* by iron chelation and iron deficiency.

14.7.2 Mode of Action

In vitro studies reflected that iron chelators act against both the liver and erythrocytes stages. They possibly act by inhibiting ribonucleotide reductase from DNA synthesis function (Mabeza et al. 1999). Siderophores, hemophores or receptors for iron transport have not been found in *Plasmodium*. It even lacks a functional heme-oxygenase which is involved in hemoglobin digestion in order to release heme iron instead causes sequesteration of heme into heme crystal in food vacuole preventing re-use of iron (Scholl et al. 2005). A chelatable non-ferritin pool with low molecular weight was reported to be present in erythrocytes involved in providing bioavailable iron (Loyevsky et al. 1999). But the mechanism to obtain iron for intrahepatocyte*Plasmodium* is yet to decipher. The tricarboxylic acid cycle actively (Krungkrai et al. 1999; Krungkrai et al. 2000) produces succinyl Co-A for the synthesis of heme during gametocytogenesis, (Vaidya and Mather 2009). During sexual stages in *P. falciparum,* expression of several genes including cytochrome b, involved in heme biosynthesis is enhanced several-fold (Learngaramkul et al. 1999; Young et al. 2005). Although the relationship between gametocytes and iron requirements is yet to unravelin detail but can be a good target.

14.7.3 Drawbacks

Iron chelators like deferoxamine (DFO) have not shown very impressive results and are also difficult to administer (Thuma et al. 1998) buthave shown antimalarial activity. DFO has half-life of 20 minutes. It can not be taken orally rather administererd via injection or infusion (Iyer et al. 2001). Deferiprone has half-life of 1–2 hours only so multiple doses are needed to be taken on a daily basis. It is taken orally but may cause fatal agranulocyotisis (Neufeld 2006). Thus these drugs are not pharmacologically suitable as antimalarial and lack safety.

14.8 Naphthoquinones: Atovaquone

Atovaquone disrupt the mitochondrial membrane potential by inhibiting electron transport via acting on cytochrome *bc*1 complex. Cytochrome bc1 complex is present in the mammalian system also but the drug shows antimalarial activity at nanomolar concentration, thus harmless for mammalian system. (Srivastava et al. 1997)

14.8.1 Mechanism of Action

Sequence *ef* loop of cytochrome b present in both the *P. yoelii* and *R. capsulatus* are highly similar, thus both the species are sensitive to atovaquone in nanomolar concentration whereas mammalian bc1 complex is resistant to atovaquone. Iron-sulfur protein is involved in the transfer of electron from ubihydroquinone to cytochrome *c*1. But in presence of atovaquone, the iron-sulfur protein domain of the Rieske [2Fe-2S] protein is stuck in a cytochrome *b*-binding conformation due to the interaction of atovaquone with the Qo site of cytochrome *bc*1 complexes, thus, halting the immobilization of electron from the [2Fe-2S] cluster to cytochrome *c*1. In summary, there can be two possible mechanism of inhibition. First, inhibition of substrate (ubihydroquinone) binding due to atovaquone binding within the Qo site and second, by inhibiting or slowing down the electron transfer since binding affect the mobility of the cluster domain (Mather et al. 2005).

14.8.2 Mechanism of Resistance Development

Resistance has developed against atovaquone due to the mutation in the *ef* loop of cytochrome b. but in paralla with resistance, a ten-fold decreament in the efficacy of cytochrome *bc*1 complex is also found, but with no prominent effect on the parasite growth. (Srivastava et al. 1999). Thus affinity for both the substrate, ubihydroquinone and inhibitor, atovaquone is reduced in the mutated strains. The enzyme activity to immobilize the [2Fe-2S] cluster domain is lost and the IC50 also gets doubled. Similar mutations were observed with atovaquone resistance in clinical isolates (Fivelman et al. 2002; Korsinczky et al. 2000) and in experimental models of rodent malaria (Syafruddin et al. 1999).

14.9 Other Drugs with Antimalarial Activity

There are few drugs which are not yet part of standard antimalarial regimen but shows antimalarial activity including tetracycline, azithromycin, clindamycin norfloxacin and ciprofloxacin etc.

14.9.1 Tetracycline

In the early 1940s, a broad spectrum family was discovered namely tetracyclines. These were found to be active against protozoa including plasmodium. In 1950 it was observed that tetracycline was effective against uncomplicated malaria caused by *P. falciparum* and *P. vivax*.

Tetracycline is produced from cycline, a synthetic antibiotic. Bacteria from the genus *Streptomyces* naturally produce cycline (Tan et al. 2011). Tetracycline, doxycycline and minocycline come in the same class of antibiotics. Tetracycline has bacteriostatic action and it acts by inhibiting protein synthesis in bacteria; they have a wide spectrum of activity (Gialdroni Grassi 1993).

14.9.1.1 Mechanism of Action

Cyclines inhibit bacterial protein synthesis (Roberts 1996). They bind to 30S ribosomal protein and to various ribonucleic acids in the 16S ribosomal RNA although its mechanism of action is not properly understood. *Plasmodium* has three types of ribosomes: mitochondrial, plastid and nuclear. Several studies have shown that tetracycline inhibits the de novo synthesis of thymine and cytosine by blocking the synthesis of mitochondrial protein and reducing the catalytic activity of mitochondrial enzyme including dihydroorotate dehydrogenase (Blum et al. 1984; Kiatfuengfoo et al. 1989; Budimulja et al. 1997; Prapunwattana et al. 1988). Not only this, the synthesis of nucleotides and deoxynucleotides is also inhibited in *P. falciparum* (Yeo et al. 1997). During *in vitro* studies of minocycline on *P. falciparum* strain, it was shown that it reduced the transcription of genes present in mitochondria and apicoplast (Lin et al. 2002). Antibiotics are found to have slow mode of action so they should be prescribed in combination with a faster acting drug. But some adverse effects were observed in case of pregnant women and children less than 8 year of age on treatment with doxycycline.

14.9.2 Doxycycline

It has dual action i.e. a slow acting blood schizonticidal and as prophylactic drug in regions with chloroquine and multi-drug resistant against *P. falciparum* malaria.

14.9.3 Clindamycin

It binds to 50s subunit of ribosomes and thus inhibits protein synthesis. It can be given in combination with drug resistant malaria along with quinine. But it has some adverse effects like pseudomembrane colitis and skin rashes.

14.9.4 Fluoroquinolones

Their results are inconsistent but both ciprofloxacin and norfloxacin have shown anti-malarial activity both in vitro and in vivo.

14.9.5 Azithromycin

Azithromycin act as a prophylactic agent against chloroquine resistant stain *P. falciparum* infection (Table 14.1).

Table 14.1 Antimalarial Drugs Mechanism of Action and Resistance

Class	Drug Name	Life cycle stage of parasite	Mechanism of action	Mechanism of resistance	
				Cellular mechanism	Molecular mechanism
Aryl amino alcohols	Quinine/ quinidine	- blood Schizonticides - gametophyte Schizonticides	Inhibition by production of toxic heme adducts by accumulating in digestive vacuole of parasite.	Production of efflux transporter Point mutation and copy number variation in Pfmdr1.	Mutation in Pfcrt(K76 T,K76I,K76 N)
	Mefloquine	- blood Schizonticides -Sporontocides	Inhibition by production of toxic heme adducts by accumulating in digestive vacuole of parasite. It also has cytosolic mode of action.	Point mutation and copy number variation in Pfmdr1 that lead to accumulation of drug in digestive vacuole away from cytosolic site of action.	Mutation in Pfcrt(K76 T,K76I,K76 N)
	Halofantrine	- blood Schizonticides -Sporontocides	Inhibition by production of toxic heme adducts by accumulating in digestive vacuole of parasite.	Point mutation and copy number variation in Pfmdr1.	Mutation in Pfcrt(K76 T,K76I,K76 N)
4 - aminoquinoles	Chloroquine	- blood Schizonticides - gametophyte Schizonticides	Inhibition by production of toxic heme adducts by accumulating in digestive vacuole of parasite.	Point mutation in Pfmdr1, Pfcrt.	Mutation in Pfcrt(K76 T,K76I,K76 N)
	Amodiaquine	- blood Schizonticides	Inhibition by production of toxic heme adducts by accumulating in digestive vacuole of parasite.	Point mutation and copy number variation in Pfmdr1, Pfcrt.	

(continued)

Table 14.1 (continued)

Class	Drug Name	Life cycle stage of parasite	Mechanism of action	Mechanism of resistance	
				Cellular mechanism	Molecular mechanism
Anti-folates	TYPE –I: Sulfadoxine	- blood Schizonticides - liver Schizonticides	Inhibits the Plasmodium dihydropteroatesynthetase (PfDHPS) enzyme.	Point mutation in Pfdhps. Copy number variation	Pf*dhps*(A437G, K540E)
	TYPE- II: Pyrimethamine/ Chloroproguanil	- blood Schizonticides - liver Schizonticides -Sporontocides	Inhibits the Plasmodium dihydrofolatereductase (PfDHFR) enzyme. It inhibits folate biosynthesis	Point mutation in Pfdhfr binding site Copy number variation.	*Pfdhfr*(N51I, C59R, S108 N, I164L).
8 - aminoquinolines	Primaquine	- gametophyte Schizonticides - liver Schizonticides -Sporontocides	Production of reactive oxygen species (ROS) and interfering with electron transport in mitochondria.	Not known	CYP2D6 genetic polymorphisms
Antibiotics	Tetracycline	- blood Schizonticides	Inhibitors of protein synthesis pathway by binding to 30s rRNA		
	Doxycyclin/ Clindamycin/	- blood Schizonticides - liver Schizonticides	Inhibitors of protein synthesis pathway by binding to 23 s rRNA		
ENDOPEROXIDES	Artemisinin / Artemether/ Arteether/ Artesunate	- blood Schizonticides - gametophyte Schizonticides	Inhibition by production of toxic heme adducts by accumulating in digestive vacuole of parasite. Production of reactive oxygen species (ROS).	Polymorphism in K13, Point mutation in pfatpase6. Point mutation and change in copy number in pfmdr1.	Polymorphism of 86 N, 184F, and 1246D alleles of Pfmdr1.
Naphthoquinones	Atovaquone	- blood Schizonticides - liver Schizonticides - Sporontocides	Inhibits mitochondrial electron transport in cytochrome bc1 complex of parasite.	Nucleotide and point mutation in cytb gene.	Y302C

14.10 Combination Therapy

Increasing resistance poses a threat to treat malaria, relinquishing conventional monotherapy less effective. WHO suggests the use of artemisnin-based combination therapy (ACT) in areas where *P.falciparum* infection is prominent to tackle occurrence of resistance and improving treatment outcome. Well known combinations used are: Artemether/lumefantrine, Dihydroartemisinin/piperaquine, Artesunate–Amodiaquine, Artesunate–Mefloquine, Artesunate–Pyronaridine, Artesunate–Sulfadoxine–Pyrimethamine and recently approved Pyronaridine-tetraphosphate/artesunate (Pyramax) and piperaquinetetraphosphate/dihydroartemisinin (Eurartesim or PQ-DHA). ACTs are used for the treatment of acute, uncomplicated malaria, giving a rapid decreament in parasitaemia with a three-day regimen, thus improving compliance and reducing the chances of relapse through the slower elimination of the partner component (WHO 2011).

Based on pharmacokinetic knowledge, the concern of cardiac saftety in malaria patients can be eliminated by providing PQ/DHA. Artemisnin derivatives are very potent and decrease the number of parasite upto a factor of 10^4 per intraerythrocytic developmental cycle thus leaving few parsites for the combiniatorial drug to kill. Artemisnin have the ability to kill gametophytes as well as interrupting malarial transmission making it a preferable drug in epidemic cases. Artemisnin derivatives are very effective in treating non-immune patients who may readily develop complicated malaria because of its immediate onset of therapeutic effect but their termination elimination time is very less approximately 2 hours. (Keating 2012) Thus, in order to spare from the reoccurrence monotherapy should be avoided and combination therapy (ACT) shoud be taken for uncomplicated malaria (WHO report 2011).

Artemisnin are consumed orally in patients with uncomplicated malaria. Artemisnin cannot be metabolized to DHA but it is still used as a primary antimalarial. Artemisnin does not interfere with parasitic hepatic stages and thus lack prophylactic activity. They cannot be utilized as chemopreventives as they have a short half life. Artemisnin affects early gametophyte stages of parasite development and has ability to interfere with mosquito transmission (Kumar and Zheng 1990).

14.10.1 Different ACTs in Use Are Following

14.10.1.1 Artemether/Lumefantrine

It is most useful and effective in all endemic regions except for the regions where MDR is widespread including Cambodia and the border regions of Thailand (Cibulskis et al. 2011). Artemether is a methyl-ether derivative of artemisinin wheras lumefantrine is a fluorine derivative with its activity against blood stage (Woodrow et al. 2005). It is an effective combination with a very few unfavourable incidents documented. Lumefantrine is not available as a monotherapy. Artemether/lumefantrine combination is consumed with fatty food so that optimal concentration of plasma drug can be achieved (Haynes 2001). It is a matter of concern when combination is used as Stand-by emergency treatment (SBET), because of occurrence of nausea and declination of food by malaria patients.

14.10.1.2 Dihydroartemisinin/Piperaquine

In 2011 European Medicines Agency approved Dihydroartemisinin/piperaquine (DHP) for treating symptomatic and uncomplicated malaria in all including adults, children and infants older than 6 months or above 5 kg. For more than 10 years, DHP has been used for treating malaria in endemic regions under the name Artekin. A malaria patient mostly feel nauseated and anorectic but the advantage dihydroartemisinin/piperaquine has is that it is taken on fasting.

14.10.1.3 Artesunate-Amodiaquine

It has been found to be an effective therapy in regions where 28-day cure rates exceeded 80% with themonotherapy of amodiaquine (Adjuik et al. 2002; Martensson et al. 2005; Durrani et al. 2005). Amodiaquine is a 4-aminoquinoline derivative of chloroquine but shows efficacy against chloroquine resistant strains of *P.falciparum* with exception in some parts of East and Southern Africa. Amodiaquine is converted to desethylamodiaquine upon oral administration and shows its antimalarial activity.

14.10.1.4 Artesunate-Mefloquine

Mefloquine is a quinoline methanol belonging to the class of arylamino alcholos (Palmer et al. 1993). Use of mefloquine as a monotherapy for treating malaria has led to rapid spread of resistance majorily due to copy number variation and mutation of PfMDR1 gene (Price et al. 1999). Evidence have suggested that initial implementation of the lowerdose of mefloquine as a monotherapy causes resistance but with implentation of higher dose in combination with ACT is less likely to cause resistance (Simpson et al. 2000). This improves bioavailability and reduces vomiting (Simpson et al. 1999). There has always been uncertainity about using mefloquine for pregnant women but some observations from Thailand showed increase in stillbirth risk after the use of mefloquine at any stage of pregnancy.

14.10.1.5 Artesunate-Pyronaridine

Pyronaridine, a synthetic antimalarial was earlier taken as monotherapy, developed in China. It structurally resembles amodiaquine though it is relatively more potent towards resistant parasites. Its pharmacokinetic features are not completely known but similar to other drugs of this category, it has substantial distribution with slow elimination. The mechanism of action and mechanism of expected resistance have not been identified but is expected to be the same as that in their general class of classification. Artesunate–Pyronaridine combination is well sustainable, useful and efficacious.

14.10.1.6 Artesunate-Sulfadoxine-Pyrimethamine

Sulfadoxine is absorbed slowly and remains in the system for a longer duration wheras pyrimethamine acts against folic acid pathway. This combination of

Artesunate and Pyrimethamine act synergistically with rare side effects in boththe adults and children with uncomplicated malaria. The trio-combination of Artesunate–Sulfadoxine–Pyrimethamine has been monitired widely in adults and children with uncomplicated malaria and is quite effective in regions where 28-day cure rates with sulfadoxine–pyrimethamine alone exceed 80% (Adjuik et al. 2004; Barnes et al. 2006; von Seidlein et al. 2000).

14.11 Cross Resistance

Drug resistance is itself a big troll which is getting worst with the development of cross resistance and multi-drug resistance. Cross resistance occurs among the drugs that belongs to the same chemical family or have the same mode of action whereas multidrug resistance is the phenomenon when resistance is developed against two or more drugs belonging to different classes and thus having a different mechanism of action. In case of *P. falciparum*, both the cross resistance and multidrug resistance are evident at wide scale. As a result of drug resistance, parasites are not completely removed from the bloodstream or sometimes their removal is delayed leading to the production and transmission of gametocytes with resistant genotype. Some drugs show incomplete cross-resistance like pyrimethamine and cycloguanil in all the three strains including *P. falciparum*, (Milhous et al. 1985; Winstanley et al. 1995) *P. gallinaceum*, (Rollo 1952) and *P. berghei* (Thompson and Bayles 1968) observed *in vitro* and *in vivo* as well (Vestergaard Olsen 1983). Other drugs like pyrimethamine-proguanil (Clyde and Shute 1957), and sulfonamides-sulfones (Peters 1975) also show incomplete resistance. Several studies have shown that the degree of cross-resistance in resistant clones varies (Jones 1953) like resistance induced by cycloguanil is "broader" than the resistance induced by pyrimethamine (Thompson and Bayles 1968) As far as multidrug resistance is concerned drugs like quinine, mepacrine and proguanil possess MDR in the Central American strain of *Plasmodium* (Earle et al. 1948).

14.12 Conclusion

Malaria is an important public health challenge. It poses substantial risk to billions worldwide. Thus controlling malaria is a big challenge and great requirement. But the deprivation of efficasious drugs and the continuing evolution of resistance make it a daunting task. The combination of improved knowledge about the mechanism of action of different antimalarial drugs and the mechanism of development of resistance against these drugs will improve our understanding and broaden our ways to tackle this burden of development of new and effective drugs. The advent of drug-resistant parasites severely spoil efforts to control malaria and must be curtailed by using complementary strategies. In this chapter we have discussed the mode of action of the available drugs, molecular mechanisms of drug resistance and understanding of cross-resistance between drugs so that our undersatnding to rationally design an individualized potent antimalarial drug can be developed. Not onlyis this, but the different combination of ACT which is the most prevalent treatment nowadaysalso discussed. But the parasitic system advances with the advancement of the medication. Now, the parasite has developed cross resistance, which is forcing to fall back to the drawing board. To counteract all these factors, a comprehensive understading behind the mechanism of available drugs and the parasitic mode of developing resistance need to be understood in proper detail.

References

Adjuik M, Agnamey P, Babiker A, Borrmann S, Brasseur P, Cisse M, Cobelens F, Diallo S, Faucher JF, Garner P, Gikunda S, Kremsner PG, Krishna S, Lell B, Loolpapit M, Matsiegui PB, Missinou MA, Mwanza J, Ntoumi F, Olliaro P, Osimbo P, Rezbach P, Some E, Taylor WR (2002) Amodiaquine-artesunate versus amodiaquine for uncomplicated Plasmodium falciparum malaria in African children: a randomised, multicentre trial. Lancet 359:1365–1372

Adjuik M, Babiker A, Garner P, Olliaro P, Taylor W, White N (2004) Artesunate combinations for treatment of malaria: meta-analysis. Lancet 363:9–17

Afonso A, Hunt P, Cheesman S, Alves AC, Cunha CV, Do Rosario V, Cravo P (2006) Malaria parasites can develop stable resistance to artemisinin but lack mutations in candidate genes atp6 (encoding the sarcoplasmic and endoplasmic reticulum Ca2+ ATPase), tctp, mdr1, and cg10. Antimicrob Agents Chemother 50:480–489

Amaratunga C, Sreng S, Suon S, Phelps ES, Stepniewska K, Lim P, Zhou C, Mao S, Anderson JM, Lindegardh N, Jiang H, Song J, Su XZ, White NJ, Dondorp AM, Anderson TJ, Fay MP, Mu J, Duong S, Fairhurst RM (2012) Artemisinin-resistant Plasmodium falciparum in Pursat province, western Cambodia: a parasitc clearance rate study. Lancet Infect Dis 12:851–858

Ariey F, Witkowski B, Amaratunga C, Beghain J, Langlois AC, Khim N, Kim S, Duru V, Bouchier C, Ma L, Lim P, Leang R, Duong S, Sreng S, Suon S, Chuor CM, Bout DM, Menard S, Rogers WO, Genton B, Fandeur T, Miotto O, Ringwald P, Le Bras J, Berry A, Barale JC, Fairhurst RM, Benoit-Vical F, Mercereau-Puijalon O, Menard D (2013) A molecular marker of artemisinin-resistant Plasmodium falciparum malaria. Nature 505:50–55

Ashley EA, Dhorda M, Fairhurst RM, Amaratunga C, Lim P, Suon S, Sreng S, Anderson JM, Mao S, Sam B, Sopha C, Chuor CM, Nguon C, Sovannaroth S, Pukrittayakamee S, Jittamala P, Chotivanich K, Chutasmit K, Suchatsoonthorn C, Runcharoen R, Hien TT, Thuy-Nhien NT, Thanh NV, Phu NH, Htut Y, Han KT, Aye KH, Mokuolu OA, Olaosebikan RR, Folaranmi OO, Mayxay M, Khanthavong M, Hongvanthong B, Newton PN, Onyamboko MA, Fanello CI, Tshefu AK, Mishra N, Valecha N, Phyo AP, Nosten F, Yi P, Tripura R, Borrmann S, Bashraheil M, Peshu J, Faiz MA, Ghose A, Hossain MA, Samad R, Rahman MR, Hasan MM, Islam A, Miotto O, Amato R, Macinnis B, Stalker J, Kwiatkowski DP, Bozdech Z, Jeeyapant A, Cheah PY, Sakulthaew T, Chalk J, Intharabut B, Silamut K, Lee SJ, Vihokhern B, Kunasol C, Imwong M, Tarning J, Taylor WJ, Yeung S, Woodrow CJ, Flegg JA, Das D, Smith J, Venkatesan M, Plowe CV, Stepniewska K, Guerin PJ, Dondorp AM, Day NP, White NJ (2014) Spread of artemisinin resistance in Plasmodium falciparum malaria. N Engl J Med 371:411–423

Atamna H, Ginsburg H (1995) Heme degradation in the presence of glutathione. A proposed mechanism to account for the high levels of non-heme iron found in the membranes of hemoglobinopathic red blood cells. J Biol Chem 270:24876–24883

Atkinson CT, Bayne MT, Gordeuk VR, Brittenham GM, Aikawa M (1991) Stage-specific ultrastructural effects of desferrioxamine on Plasmodium falciparum in vitro. Am J Trop Med Hyg 45:593–601

Barnes KI, Little F, Smith PJ, Evans A, Watkins WM, White NJ (2006) Sulfadoxine-pyrimethamine pharmacokinetics in malaria: pediatric dosing implications. Clin Pharmacol Ther 80:582–596

Bayih AG, Getnet G, Alemu A, Getie S, Mohon AN, Pillai DR (2015) A unique plasmodium falciparum K13 gene mutation in Northwest Ethiopia. Am J Trop Med Hyg 94:132–135

Bennett JW, Pybus BS, Yadava A, Tosh D, Sousa JC, Mccarthy WF, Deye G, Melendez V, Ockenhouse CF (2013) Primaquine failure and cytochrome P-450 2D6 in Plasmodium vivax malaria. N Engl J Med 369:1381–1382

Bhisutthibhan J, Pan XQ, Hossler PA, Walker DJ, Yowell CA, Carlton J, Dame JB, Meshnick SR (1998) The Plasmodium falciparum translationally controlled tumor protein homolog and its reaction with the antimalarial drug artemisinin. J Biol Chem 273:16192–16198

Blauer G, Akkawi M (1997) Investigations of B- and beta-hematin. J Inorg Biochem 66:145–152

Blum JJ, Yayon A, Friedman S, Ginsburg H (1984) Effects of mitochondrial protein synthesis inhibitors on the incorporation of isoleucine into Plasmodium falciparum in vitro. J Protozool 31:475–479

Bray PG, Mungthin M, Ridley RG, Ward SA (1998) Access to hematin: the basis of chloroquine resistance. Mol Pharmacol 54:170–179

Bray PG, Ward SA, O'neill PM (2005) Quinolines and artemisinin: chemistry, biology and history. *Curr Top Microbiol Immunol* 295:3–38

Brooks DR, Wang P, Read M, Watkins WM, Sims PF, Hyde JE (1994) Sequence variation of the hydroxymethyldihydropterin pyrophosphokinase: dihydropteroate synthase gene in lines of the human malaria parasite, Plasmodium falciparum, with differing resistance to sulfadoxine. Eur J Biochem 224:397–405

Brown GD (2010) The biosynthesis of artemisinin (Qinghaosu) and the phytochemistry of Artemisia annua L. (Qinghao). Molecules 15:7603–7698

Budimulja AS, Syafruddin, Tapchaisri P, Wilairat P, Marzuki S (1997) The sensitivity of Plasmodium protein synthesis to prokaryotic ribosomal inhibitors. Mol Biochem Parasitol 84:137–141

Burstein A (1993) Panic as a posttraumatic stressor. Am J Psychiatry 150:842

Bzik DJ, Li WB, Horii T, Inselburg J (1987) Molecular cloning and sequence analysis of the Plasmodium falciparum dihydrofolate reductase-thymidylate synthase gene. Proc Natl Acad Sci U S A 84:8360–8364

Cabantchik ZI, Moody-Haupt S, Gordeuk VR (1999) Iron chelators as anti-infectives; malaria as a paradigm. FEMS Immunol Med Microbiol 26:289–298

Cheeseman IH, Miller BA, Nair S, Nkhoma S, Tan A, Tan JC, Al Saai S, Phyo AP, Moo CL, Lwin KM, Mcgready R, Ashley E, Imwong M, Stepniewska K, Yi P, Dondorp AM, Mayxay M, Newton PN, White NJ, Nosten F, Ferdig MT, Anderson TJ (2012) A major genome region underlying artemisinin resistance in malaria. Science 336:79–82

Chou AC, Fitch CD (1980) Hemolysis of mouse erythrocytes by ferriprotoporphyrin IX and chloroquine. Chemotherapeutic implications. J Clin Invest 66:856–858

Chou AC, Fitch CD (1981) Mechanism of hemolysis induced by ferriprotoporphyrin IX. J Clin Invest 68:672–677

Chou AC, Fitch CD (1993) Control of heme polymerase by chloroquine and other quinoline derivatives. Biochem Biophys Res Commun 195:422–427

Chou AC, Chevli R, Fitch CD (1980) Ferriprotoporphyrin IX fulfills the criteria for identification as the chloroquine receptor of malaria parasites. Biochemistry 19:1543–1549

Chugh M, Sundararaman V, Kumar S, Reddy VS, Siddiqui WA, Stuart KD, Malhotra P (2013) Protein complex directs hemoglobin-to-hemozoin formation in Plasmodium falciparum. Proc Natl Acad Sci U S A 110:5392–5397

Cibulskis RE, Aregawi M, Williams R, Otten M, Dye C (2011) Worldwide incidence of malaria in 2009: estimates, time trends, and a critique of methods. PLoS Med 8:e1001142

Clyde DF, Shute GT (1957) Resistance of Plasmodium falciparum in Tanganyika to pyrimethamine administered at weekly intervals. Trans R Soc Trop Med Hyg 51:505–513

de Ridder S, van der Kooy F, Verpoorte R (2008) Artemisia annua as a self-reliant treatment for malaria in developing countries. J Ethnopharmacol 120:302–314

Delabays N, Simonnet X, Gaudin M (2001) The genetics of artemisinin content in Artemisia annua L. and the breeding of high yielding cultivars. Curr Med Chem 8:1795–1801

Dorn A, Vippagunta SR, Matile H, Jaquet C, Vennerstrom JL, Ridley RG (1998) An assessment of drug-haematin binding as a mechanism for inhibition of haematin polymerisation by quinoline antimalarials. Biochem Pharmacol 55:727–736

Durrani N, Leslie T, Rahim S, Graham K, Ahmad F, Rowland M (2005) Efficacy of combination therapy with artesunate plus amodiaquine compared to monotherapy with chloroquine, amodiaquine or sulfadoxine-pyrimethamine for treatment of uncomplicated Plasmodium falciparum in Afghanistan. Tropical Med Int Health 10:521–529

Earle DP, Berliner RW, Taggart JV, Zubrod CG, Welch WJ, Bigelow FS, Kennedy TJ, Shannon JA (1948) Studies on the chemotherapy of the human malarias. X. the suppressive antimalarial effect of Paludrine. J Clin Invest 27:130–133

Eckstein-Ludwig U, Webb RJ, Van Goethem ID, East JM, Lee AG, Kimura M, O'neill PM, Bray PG, Ward SA, Krishna S (2003) Artemisinins target the SERCA of Plasmodium falciparum. Nature 424:957–961

Egan TJ, Ross DC, Adams PA (1994) Quinoline anti-malarial drugs inhibit spontaneous formation of beta-haematin (malaria pigment). FEBS Lett 352:54–57

Egan TJ, Mavuso WW, Ross DC, Marques HM (1997) Thermodynamic factors controlling the interaction of quinoline antimalarial drugs with ferriprotoporphyrin IX. J Inorg Biochem 68:137–145

Eichhorn T, Winter D, Buchele B, Dirdjaja N, Frank M, Lehmann WD, Mertens R, Krauth-Siegel RL, Simmet T, Granzin J, Efferth T (2012) Molecular interaction of artemisinin with translationally controlled tumor protein (TCTP) of Plasmodium falciparum. Biochem Pharmacol 85:38–45

Famin O, Ginsburg H (2002) Differential effects of 4-aminoquinoline-containing antimalarial drugs on hemoglobin digestion in Plasmodium falciparum-infected erythrocytes. Biochem Pharmacol 63:393–398

Fernando SD, Rodrigo C, Rajapakse S (2011) Chemoprophylaxis in malaria: drugs, evidence of efficacy and costs. Asian Pac J Trop Med 4:330–336

Ferone R (1977) Folate metabolism in malaria. Bull World Health Organ 55:291–298

Ferone R, Burchall JJ, Hitchings GH (1969) Plasmodium berghei dihydrofolate reductase. Isolation, properties, and inhibition by antifolates. Mol Pharmacol 5:49–59

Fidock DA, Nomura T, Talley AK, Cooper RA, Dzekunov SM, Ferdig MT, Ursos LM, Sidhu AB, Naude B, Deitsch KW, Su XZ, Wootton JC, Roepe PD, Wellems TE (2000) Mutations in the P. falciparum digestive vacuole transmembrane protein PfCRT and evidence for their role in chloroquine resistance. Mol Cell 6:861–871

Fitch CD, Chevli R, Banyal HS, Phillips G, Pfaller MA, Krogstad DJ (1982) Lysis of Plasmodium falciparum by ferriprotoporphyrin IX and a chloroquine-ferriprotoporphyrin IX complex. Antimicrob Agents Chemother 21:819–822

Fivelman QL, Butcher GA, Adagu IS, Warhurst DC, Pasvol G (2002) Malarone treatment failure and in vitro confirmation of resistance of Plasmodium falciparum isolate from Lagos, Nigeria. *Malar J* 1:1

Gascon J, Gomez ARCE, E J, Menendez C, Valls ME, Corachan M (1994) Poor response to primaquine in two cases of Plasmodium vivax malaria from Guatemala. Trop Geogr Med 46:32–33

Gialdroni Grassi G (1993) Tetracyclines-extending the atypical spectrum. Int J Antimicrob Agents 3(Suppl 1):S31–S46

Ginsburg H, Demel RA (1983) The effect of ferriprotoporphyrin IX and chloroquine on phospholipid monolayers and the possible implications to antimalarial activity. Biochim Biophys Acta 732:316–319

Ginsburg H, Golenser J (2003) Glutathione is involved in the antimalarial action of chloroquine and its modulation affects drug sensitivity of human and murine species of Plasmodium. Redox Rep 8:276–279

Ginsburg H, Famin O, Zhang J, Krugliak M (1998) Inhibition of glutathione-dependent degradation of heme by chloroquine and amodiaquine as a possible basis for their antimalarial mode of action. Biochem Pharmacol 56:1305–1313

Gritzmacher CA, Reese RT (1984) Protein and nucleic acid synthesis during synchronized growth of Plasmodium falciparum. J Bacteriol 160:1165–1167
Gutteridge WE, Trigg PI (1971) Action of pyrimethamine and related drugs against Plasmodium knowlesi in vitr. Parasitology 62:431–444
Harvey PW, Bell RG, Nesheim MC (1985) Iron deficiency protects inbred mice against infection with Plasmodium chabaudi. Infect Immun 50:932–934
Haynes RK (2001) Artemisinin and derivatives: the future for malaria treatment? Curr Opin Infect Dis 14:719–726
Heinberg A, Siu E, Stern C, Lawrence EA, Ferdig MT, Deitsch KW, Kirkman LA (2013) Direct evidence for the adaptive role of copy number variation on antifolate susceptibility in Plasmodium falciparum. Mol Microbiol 88:702–712
Hill DR, Baird JK, Parise ME, Lewis LS, Ryan ET, Magill AJ (2006) Primaquine: report from CDC expert meeting on malaria chemoprophylaxis I. Am J Trop Med Hyg 75:402–415
Hoppe HC, Van Schalkwyk DA, Wiehart UI, Meredith SA, Egan J, Weber BW (2004) Antimalarial quinolines and artemisinin inhibit endocytosis in Plasmodium falciparum. Antimicrob Agents Chemother 48:2370–2378
Htut ZW (2009) Artemisinin resistance in Plasmodium falciparum malaria. N Engl J Med 361:1807–1808. author reply 1808
Isozumi R, Uemura H, Kimata I, Ichinose Y, Logedi J, Omar AH, Kaneko A (2015) Novel mutations in K13 propeller gene of artemisinin-resistant Plasmodium falciparum. Emerg Infect Dis 21:490–492
Iyer JK, Milhous WK, Cortese JF, Kublin JG, Plowe CV (2001) Plasmodium falciparum cross-resistance between trimethoprim and pyrimethamine. Lancet 358:1066–1067
Jones SA (1953) Experiment to determine if a proguanil-resistant strain of P. falciparum would respond to large doses of pyrimethamine. Br Med J 1:977
Juge N, Moriyama S, Miyaji T, Kawakami M, Iwai H, Fukui T, Nelson N, Omote H, Moriyama Y (2015) Plasmodium falciparum chloroquine resistance transporter is a H+-coupled polyspecific nutrient and drug exporter. Proc Natl Acad Sci U S A 112:3356–3361
Kantele A, Jokiranta TS (2011) Review of cases with the emerging fifth human malaria parasite, Plasmodium knowlesi. Clin Infect Dis 52:1356–1362
Keating GM (2012) Dihydroartemisinin/Piperaquine: a review of its use in the treatment of uncomplicated Plasmodium falciparum malaria. Drugs 72:937–961
Kiatfuengfoo R, Suthiphongchai T, Prapunwattana P, Yuthavong Y (1989) Mitochondria as the site of action of tetracycline on Plasmodium falciparum. Mol Biochem Parasitol 34:109–115
Kidgell C, Volkman SK, Daily J, Borevitz JO, Plouffe D, Zhou Y, Johnson JR, Le Roch K, Sarr O, Ndir O, Mboup S, Batalov S, Wirth DF, Winzeler EA (2006) A systematic map of genetic variation in Plasmodium falciparum. PLoS Pathog 2:e57
Klonis N, Creek DJ, Tilley L (2013) Iron and heme metabolism in Plasmodium falciparum and the mechanism of action of artemisinins. Curr Opin Microbiol 16:722–727
Klonis N, Crespo-Ortiz MP, Bottova I, Abu-Bakar N, Kenny S, Rosenthal PJ, Tilley L (2011) Artemisinin activity against Plasmodium falciparum requires hemoglobin uptake and digestion. Proc Natl Acad Sci U S A 108:11405–11410
Kong, K-F (2005) Characterization of the amp genes involved in the regulation of β-lactamase expression in *Pseudomonas aeruginosa* (2005). ProQuest ETD Collection for FIU. AAI3206025. https://digitalcommons.fiu.edu/dissertations/AAI3206025
Korsinczky M, Chen N, Kotecka B, Saul A, Rieckmann K, Cheng Q (2000) Mutations in Plasmodium falciparum cytochrome b that are associated with atovaquone resistance are located at a putative drug-binding site. Antimicrob Agents Chemother 44:2100–2108
Krugliak M, Ginsburg H (1991) Studies on the antimalarial mode of action of quinoline-containing drugs: time-dependence and irreversibility of drug action, and interactions with compounds that alter the function of the parasite's food vacuole. Life Sci 49:1213–1219
Krungkrai J, Webster HK, Yuthavong Y (1989) De novo and salvage biosynthesis of pteroylpentaglutamates in the human malaria parasite, Plasmodium falciparum. Mol Biochem Parasitol 32:25–37

Krungkrai J, Burat D, Kudan S, Krungkrai S, Prapunwattana P (1999) Mitochondrial oxygen consumption in asexual and sexual blood stages of the human malarial parasite, Plasmodium falciparum. *Southeast Asian J Trop Med Public Health* 30:636–642

Krungkrai J, Prapunwattana P, Krungkrai SR (2000) Ultrastructure and function of mitochondria in gametocytic stage of Plasmodium falciparum. Parasite 7:19–26

Kumar N, Zheng H (1990) Stage-specific gametocytocidal effect in vitro of the antimalaria drug qinghaosu on Plasmodium falciparum. Parasitol Res 76:214–218

Kwon YK, Lu W, Melamud E, Khanam N, Bognar A, Rabinowitz JD (2008) A domino effect in antifolate drug action in Escherichia coli. Nat Chem Biol 4:602–608

Le Bras J, Durand R (2003) The mechanisms of resistance to antimalarial drugs in Plasmodium falciparum. Fundam Clin Pharmacol 17:147–153

Learngaramkul P, Petmitr S, Krungkrai SR, Prapunwattana P, Krungkrai J (1999) Molecular characterization of mitochondria in asexual and sexual blood stages of Plasmodium falciparum. Mol Cell Biol Res Commun 2:15–20

Li GQ, Arnold K, Guo XB, Jian HX, Fu LC (1984) Randomised comparative study of mefloquine, qinghaosu, and pyrimethamine-sulfadoxine in patients with falciparum malaria. Lancet 2:1360–1361

Light WR 3rd, Olson JS (1990a) The effects of lipid composition on the rate and extent of heme binding to membranes. J Biol Chem 265:15632–15637

Light WR 3rd, Olson JS (1990b) Transmembrane movement of heme. J Biol Chem 265:15623–15631

Lin Q, Katakura K, Suzuki M (2002) Inhibition of mitochondrial and plastid activity of Plasmodium falciparum by minocycline. FEBS Lett 515:71–74

Loyevsky M, Sacci JB, Boehme P, Weglicki W, John C, Gordeuk VR (1999) Plasmodium falciparum and Plasmodium yoelii: effect of the iron chelation prodrug dexrazoxane on in vitro cultures. Exp Parasitol 91:105–114

Mabeza GF, Loyevsky M, Gordeuk VR, Weiss G (1999) Iron chelation therapy for malaria: a review. Pharmacol Ther 81:53–75

Martensson A, Stromberg J, Sisowath C, Msellem MI, Gil JP, Montgomery SM, Olliaro P, Ali AS, Bjorkman A (2005) Efficacy of artesunate plus amodiaquine versus that of artemether-lumefantrine for the treatment of uncomplicated childhood Plasmodium falciparum malaria in Zanzibar, Tanzania. Clin Infect Dis 41:1079–1086

Martin RE, Kirk K (2004) The malaria parasite's chloroquine resistance transporter is a member of the drug/metabolite transporter superfamily. Mol Biol Evol 21:1938–1949

Mather MW, Darrouzet E, Valkova-Valchanova M, Cooley JW, Mcintosh MT, Daldal F, Vaidya AB (2005) Uncovering the molecular mode of action of the antimalarial drug atovaquone using a bacterial system. J Biol Chem 280:27458–27465

Mercer AE, Copple IM, Maggs JL, O'neill PM, Park BK (2010) The role of heme and the mitochondrion in the chemical and molecular mechanisms of mammalian cell death induced by the artemisinin antimalarials. J Biol Chem 286:987–996

Milhous WK, Weatherly NF, Bowdre JH, Desjardins RE (1985) In vitro activities of and mechanisms of resistance to antifol antimalarial drugs. Antimicrob Agents Chemother 27:525–530

Miller LH, Su X (2011) Artemisinin: discovery from the Chinese herbal garden. Cell 146:855–858

Mok S, Ashley EA, Ferreira PE, Zhu L, Lin Z, Yeo T, Chotivanich K, Imwong M, Pukrittayakamee S, Dhorda M, Nguon C, Lim P, Amaratunga C, Suon S, Hien TT, Htut Y, Faiz MA, Onyamboko MA, Mayxay M, Newton PN, Tripura R, Woodrow CJ, Miotto O, Kwiatkowski DP, Nosten F, Day NP, Preiser PR, White NJ, Dondorp AM, Fairhurst RM, Bozdech Z (2014) Drug resistance. Population transcriptomics of human malaria parasites reveals the mechanism of artemisinin resistance. Science 347:431–435

Neufeld EJ (2006) Oral chelators deferasirox and deferiprone for transfusional iron overload in thalassemia major: new data, new questions. Blood 107:3436–3441

Newbold CI, Boyle DB, Smith CC, Brown KN (1982) Stage specific protein and nucleic acid synthesis during the asexual cycle of the rodent malaria Plasmodium chabaudi. Mol Biochem Parasitol 5:33–44

Nosten F, White NJ (2007) Artemisinin-based combination treatment of falciparum malaria. Am J Trop Med Hyg 77:181–192
Nyunt MH, Hlaing T, Oo HW, Tin-Oo LL, Phway HP, Wang B, Zaw NN, Han SS, Tun T, San KK, Kyaw MP, Han ET (2014) Molecular assessment of artemisinin resistance markers, polymorphisms in the k13 propeller, and a multidrug-resistance gene in the eastern and western border areas of Myanmar. Clin Infect Dis 60:1208–1215
Nzila AM, Mberu EK, Sulo J, Dayo H, Winstanley PA, Sibley CH, Watkins WM (2000) Towards an understanding of the mechanism of pyrimethamine-sulfadoxine resistance in Plasmodium falciparum: genotyping of dihydrofolate reductase and dihydropteroate synthase of Kenyan parasites. Antimicrob Agents Chemother 44:991–996
Nzila-Mounda A, Mberu EK, Sibley CH, Plowe CV, Winstanley PA, Watkins WM (1998) Kenyan Plasmodium falciparum field isolates: correlation between pyrimethamine and chlorcycloguanil activity in vitro and point mutations in the dihydrofolate reductase domain. Antimicrob Agents Chemother 42:164–169
O'neill PM, Barton VE, Ward SA (2010) The molecular mechanism of action of artemisinin--the debate continues. Molecules 15:1705–1721
Palmer KJ, Holliday SM, Brogden RN (1993) Mefloquine. A review of its antimalarial activity, pharmacokinetic properties and therapeutic efficacy. Drugs 45:430–475
Patzewitz EM, Salcedo-Sora JE, Wong EH, Sethia S, Stocks PA, Maughan SC, Murray JA, Krishna S, Bray PG, Ward SA, Muller S (2012) Glutathione transport: a new role for PfCRT in chloroquine resistance. Antioxid Redox Signal 19:683–695
Peters W (1975) The chemotherapy of rodent malaria, XXII. The value of drug-resistant strains of P. berghei in screening for blood schizontocidal activity. Ann Trop Med Parasitol 69:155–171
Peterson DS, Walliker D, Wellems TE (1988) Evidence that a point mutation in dihydrofolate reductase-thymidylate synthase confers resistance to pyrimethamine in falciparum malaria. Proc Natl Acad Sci U S A 85:9114–9118
Prapunwattana P, O'sullivan WJ, Yuthavong Y (1988) Depression of Plasmodium falciparum dihydroorotate dehydrogenase activity in in vitro culture by tetracycline. Mol Biochem Parasitol 27:119–124
Price RN, Cassar C, Brockman A, Duraisingh M, van Vugt M, White NJ, Nosten F, Krishna S (1999) The pfmdr1 gene is associated with a multidrug-resistant phenotype in Plasmodium falciparum from the western border of Thailand. Antimicrob Agents Chemother 43:2943–2949
Price RN, Von Seidlein L, Valecha N, Nosten F, Baird JK, White NJ (2014) Global extent of chloroquine-resistant Plasmodium vivax: a systematic review and meta-analysis. Lancet Infect Dis 14:982–991
Pybus BS, Marcsisin SR, Jin X, Deye G, Sousa JC, Li Q, Caridha D, Zeng Q, Reichard GA, Ockenhouse C, Bennett J, Walker LA, Ohrt C, Melendez V (2012) The metabolism of primaquine to its active metabolite is dependent on CYP 2D6. Malar J 12:212
Pybus BS, Sousa JC, Jin X, Ferguson JA, Christian RE, Barnhart R, Vuong C, Sciotti RJ, Reichard GA, Kozar MP, Walker LA, Ohrt C, Melendez V (2013) CYP450 phenotyping and accurate mass identification of metabolites of the 8-aminoquinoline, anti-malarial drug primaquine. Malar J 11:259
Ramos Junior WM, Sardinha JF, Costa MR, Santana MS, Alecrim MG, Lacerda MV (2010) Clinical aspects of hemolysis in patients with P. vivax malaria treated with primaquine, in the Brazilian Amazon. Braz J Infect Dis 14:410–412
Raventos-Suarez C, Pollack S, Nagel RL (1982) Plasmodium falciparum: inhibition of in vitro growth by desferrioxamine. Am J Trop Med Hyg 31:919–922
Roberts MC (1996) Tetracycline resistance determinants: mechanisms of action, regulation of expression, genetic mobility, and distribution. FEMS Microbiol Rev 19:1–24
Rollo IM (1952) Daraprim-experimental chemotherapy. Trans R Soc Trop Med Hyg 46:474–484. discussion, 498–508
Salcedo-Sora JE, Ochong E, Beveridge S, Johnson D, Nzila A, Biagini GA, Stocks PA, O'neill PM, Krishna S, Bray PG, Ward SA (2011) The molecular basis of folate salvage in Plasmodium falciparum: characterization of two folate transporters. J Biol Chem 286:44659–44668

Schellenberg KA, Coatney GR (1961) The influence of antimalarial drugs on nucleic acid synthesis in Plasmodium gallinaceum and Plasmodium berghei. Biochem Pharmacol 6:143–152

Scholl PF, Tripathi AK, Sullivan DJ (2005) Bioavailable iron and heme metabolism in Plasmodium falciparum. Curr Top Microbiol Immunol 295:293–324

Sidhu AB, Verdier-Pinard D, Fidock DA (2002) Chloroquine resistance in Plasmodium falciparum malaria parasites conferred by pfcrt mutations. Science 298:210–213

Simpson JA, Price R, Ter Kuile F, Teja-Isavatharm P, Nosten F, Chongsuphajaisiddhi T, Looareesuwan S, Aarons L, White NJ (1999) Population pharmacokinetics of mefloquine in patients with acute falciparum malaria. Clin Pharmacol Ther 66:472–484

Simpson JA, Watkins ER, Price RN, Aarons L, Kyle DE, White NJ (2000) Mefloquine pharmacokinetic-pharmacodynamic models: implications for dosing and resistance. Antimicrob Agents Chemother 44:3414–3424

Sirawaraporn W, Sathitkul T, Sirawaraporn R, Yuthavong Y, Santi DV (1997) Antifolate-resistant mutants of Plasmodium falciparum dihydrofolate reductase. Proc Natl Acad Sci U S A 94:1124–1129

Slater AF, Cerami A (1992) Inhibition by chloroquine of a novel haem polymerase enzyme activity in malaria trophozoites. Nature 355:167–169

Slater AF, Swiggard WJ, Orton BR, Flitter WD, Goldberg DE, Cerami A, Henderson GB (1991) An iron-carboxylate bond links the heme units of malaria pigment. Proc Natl Acad Sci U S A 88:325–329

Sridaran S, Mcclintock SK, Syphard LM, Herman KM, Barnwell JW, Udhayakumar V (2010) Anti-folate drug resistance in Africa: meta-analysis of reported dihydrofolate reductase (dhfr) and dihydropteroate synthase (dhps) mutant genotype frequencies in African Plasmodium falciparum parasite populations. Malar J 9:247

Srivastava IK, Rottenberg H, Vaidya AB (1997) Atovaquone, a broad spectrum antiparasitic drug, collapses mitochondrial membrane potential in a malarial parasite. J Biol Chem 272:3961–3966

Srivastava IK, Morrisey JM, Darrouzet E, Daldal F, Vaidya AB (1999) Resistance mutations reveal the atovaquone binding domain of cytochrome b in malaria parasites. Mol Microbiol 33:704–711

Straimer J, Gnadig NF, Witkowski B, Amaratunga C, Duru V, Ramadani AP, Dacheux M, Khim N, Zhang L, Lam S, Gregory PD, Urnov FD, Mercereau-Puijalon O, Benoit-Vical F, Fairhurst RM, Menard D, Fidock DA (2014) Drug resistance. K13-propeller mutations confer artemisinin resistance in Plasmodium falciparum clinical isolates. Science 347:428–431

Su XZ, Miller LH (2015) The discovery of artemisinin and the Nobel prize in physiology or medicine. Sci China Life Sci 58:1175–1179

Su X, Kirkman LA, Fujioka H, Wellems TE (1997) Complex polymorphisms in an approximately 330 kDa protein are linked to chloroquine-resistant P. falciparum in Southeast Asia and Africa. Cell 91:593–603

Sun C, Li J, Cao Y, Long G, Zhou B (2015) Two distinct and competitive pathways confer the cellcidal actions of artemisinins. Microb Cell 2:14–25

Syafruddin D, Siregar JE, Marzuki S (1999) Mutations in the cytochrome b gene of Plasmodium berghei conferring resistance to atovaquone. Mol Biochem Parasitol 104:185–194

Takala-Harrison S, Clark TG, Jacob CG, Cummings MP, Miotto O, Dondorp AM, Fukuda MM, Nosten F, Noedl H, Imwong M, Bethell D, Se Y, Lon C, Tyner SD, Saunders DL, Socheat D, Ariey F, Phyo AP, Starzengruber P, Fuehrer HP, Swoboda P, Stepniewska K, Flegg J, Arze C, Cerqueira GC, Silva JC, Ricklefs SM, Porcella SF, Stephens RM, Adams M, Kenefic LJ, Campino S, Auburn S, Macinnis B, Kwiatkowski DP, Su XZ, White NJ, Ringwald P, Plowe CV (2012) Genetic loci associated with delayed clearance of Plasmodium falciparum following artemisinin treatment in Southeast Asia. Proc Natl Acad Sci U S A 110:240–245

Takeuchi R, Lawpoolsri S, Imwong M, Kobayashi J, Kaewkungwal J, Pukrittayakamee S, Puangsa-Art S, Thanyavanich N, Maneeboonyang W, Day NP, Singhasivanon P (2010) Directly-observed therapy (DOT) for the radical 14-day primaquine treatment of Plasmodium vivax malaria on the Thai-Myanmar border. Malar J 9:308

Tan KR, Magill AJ, Parise ME, Arguin PM (2011) Doxycycline for malaria chemoprophylaxis and treatment: report from the CDC expert meeting on malaria chemoprophylaxis. Am J Trop Med Hyg 84:517–531

Thompson PE, Bayles A (1968) Reciprocal cross resistance between cycloguanil hydrochloride and pyrimethamine in Plasmodium berghei infections in mice. J Parasitol 54:588–593

Thuma PE, Mabeza GF, Biemba G, Bhat GJ, Mclaren CE, Moyo VM, Zulu S, Khumalo H, Mabeza P, M'hango A, Parry D, Poltera AA, Brittenham GM, Gordeuk VR (1998) Effect of iron chelation therapy on mortality in Zambian children with cerebral malaria. Trans R Soc Trop Med Hyg 92:214–218

Triglia T, Cowman AF (1999) The mechanism of resistance to sulfa drugs in Plasmodium falciparum. Drug Resist Updat 2:15–19

Tu Y (2011) The discovery of artemisinin (qinghaosu) and gifts from Chinese medicine. Nat Med 17:1217–1220

Vaidya AB, Mather MW (2009) Mitochondrial evolution and functions in malaria parasites. Annu Rev Microbiol 63:249–267

Vale N, Moreira R, Gomes P (2009) Primaquine revisited six decades after its discovery. Eur J Med Chem 44:937–953

Van Agtmael MA, Eggelte TA, Van Boxtel CJ (1999) Artemisinin drugs in the treatment of malaria: from medicinal herb to registered medication. Trends Pharmacol Sci 20:199–205

van der Velden M, Rijpma SR, Russel FG, Sauerwein RW, Koenderink JB (2015) PfMDR2 and PfMDR5 are dispensable for Plasmodium falciparum asexual parasite multiplication but change in vitro susceptibility to anti-malarial drugs. Malar J 14:76

Vestergaard OLSEN V (1983) Unsuccessful effect of sulfadoxine-pyrimethamine in the treatment of falciparum malaria in Tanzania. Ugeskr Laeger 145:751

Von Seidlein L, Milligan P, Pinder M, Bojang K, Anyalebechi C, Gosling R, Coleman R, Ude JI, Sadiq A, Duraisingh M, Warhurst D, Allloueche A, Targett G, Mcadam K, Greenwood B, Walraven G, Olliaro P, Doherty T (2000) Efficacy of artesunate plus pyrimethamine-sulphadoxine for uncomplicated malaria in Gambian children: a double-blind, randomised, controlled trial. Lancet 355:352–357

Wang P, Wang Q, Aspinall TV, Sims PF, Hyde JE (2004) Transfection studies to explore essential folate metabolism and antifolate drug synergy in the human malaria parasite Plasmodium falciparum. Mol Microbiol 51:1425–1438

Wang J, Huang L, Li J, Fan Q, Long Y, Li Y, Zhou B (2010) Artemisinin directly targets malarial mitochondria through its specific mitochondrial activation. PLoS One 5:e9582

Wang J, Zhang CJ, Chia WN, Loh CC, Li Z, Lee YM, He Y, Yuan LX, Lim TK, Liu M, Liew CX, Lee YQ, Zhang J, Lu N, Lim CT, Hua ZC, Liu B, Shen HM, Tan KS, Lin Q (2015) Haem-activated promiscuous targeting of artemisinin in Plasmodium falciparum. Nat Commun 6:10111

White N (1999) Antimalarial drug resistance and combination chemotherapy. Philos Trans R Soc Lond Ser B Biol Sci 354:739–749

WHO (2011). World Malaria Report 2011

WHO (2013). World Malaria Report 2013

WHO (2016). World Malaria Report 2016

Willcox M (2009) Artemisia species: from traditional medicines to modern antimalarials–and back again. J Altern Complement Med 15:101–109

Winstanley PA, Mberu EK, Szwandt IS, Breckenridge AM, Watkins WM (1995) In vitro activities of novel antifolate drug combinations against Plasmodium falciparum and human granulocyte CFUs. Antimicrob Agents Chemother 39:948–952

Witkowski B, Amaratunga C, Khim N, Sreng S, Chim P, Kim S, Lim P, Mao S, Sopha C, Sam B, Anderson JM, Duong S, Chuor CM, Taylor WR, Suon S, Mercereau-Puijalon O, Fairhurst RM, Menard D (2012) Novel phenotypic assays for the detection of artemisinin-resistant

Plasmodium falciparum malaria in Cambodia: in-vitro and ex-vivo drug-response studies. Lancet Infect Dis 13:1043–1049

Witkowski B, Khim N, Chim P, Kim S, Ke S, Kloeung N, Chy S, Duong S, Leang R, Ringwald P, Dondorp AM, Tripura R, Benoit-Vical F, Berry A, Gorgette O, Ariey F, Barale JC, Mercereau-Puijalon O, Menard D (2013) Reduced artemisinin susceptibility of Plasmodium falciparum ring stages in western Cambodia. Antimicrob Agents Chemother 57:914–923

Woodrow CJ, Haynes RK, Krishna S (2005) Artemisinins. Postgrad Med J 81:71–78

Xie SC, Dogovski C, Hanssen E, Chiu F, Yang T, Crespo MP, Stafford C, Batinovic S, Teguh S, Charman S, Klonis N, Tilley L (2015) Haemoglobin degradation underpins the sensitivity of early ring stage Plasmodium falciparum to artemisinins. J Cell Sci 129:406–416

Yayon A, Timberg R, Friedman S, Ginsburg H (1984) Effects of chloroquine on the feeding mechanism of the intraerythrocytic human malarial parasite Plasmodium falciparum. J Protozool 31:367–372

Yeo AE, Edstein MD, Shanks GD, Rieckmann KH (1997) Potentiation of the antimalarial activity of atovaquone by doxycycline against Plasmodium falciparum in vitro. Parasitol Res 83:489–491

Young JA, Fivelman QL, Blair PL, De La Vega P, Le Roch KG, Zhou Y, Carucci DJ, Baker DA, Winzeler EA (2005) The Plasmodium falciparum sexual development transcriptome: a microarray analysis using ontology-based pattern identification. Mol Biochem Parasitol 143:67–79

Zhang J, Krugliak M, Ginsburg H (1999) The fate of ferriprotorphyrin IX in malaria infected erythrocytes in conjunction with the mode of action of antimalarial drugs. Mol Biochem Parasitol 99:129–141

Management and Control of Antimalarial Drug Resistance

15

Amrita Chakrabarti, Vigyasa Singh, and Shailja Singh

Abstract

Resistance to antimalarial drugs is a great threat to malaria elimination efforts and adds resurgency of malaria incidences and deaths. Rapid and indiscriminate use of drugs in the treatment of malarial infections has led to the emergence and spread of resistant strains. With condensed efforts for monitoring drugs resistance in clinical setup integrated by *in vitro* drug susceptibility assays and analysis of resistance markers can help in malaria eradication policy. In this chapter we discussed the demographical distribution of malaria throughout the world along with its clinical manifestations. We have also emphasized the mechanism of action underlying activity of FDA approved antimalarial drugs and also described effect of wide spectrum usage of antimalarials and combinational therapy against the deadly disease. Various factors leading to emergence of resistance including the mutation rate of the parasite for their bio-adaptation, the overall parasite load, strength of drug selection and treatment compliance. We summarize the chapter stating the number of probabilities which defines the future of antimalarial drug resistance and cumulative approaches to combat it.

Authors Amrita Chakrabarti and Vigyasa Singh have been equally contributed to this chapter.

A. Chakrabarti
Department of Life Sciences, School of Natural Sciences, Shiv Nadar University, Uttar Pradesh, India
e-mail: ac221@snu.edu.in

V. Singh · S. Singh (✉)
Special Centre for Molecular Medicine, Jawaharlal Nehru University, New Delhi, India
e-mail: shailja.jnu@gmail.com

S. M. Mandal, D. Paul (eds.), *Bacterial Adaptation to Co-resistance*,
https://doi.org/10.1007/978-981-13-8503-2_15

Keywords
Malaria · Anti-malarial drugs · Drug resistance · Combination therapy · Drug development

15.1 Introduction

Resistance to any drug, chemical or toxin is a ramification of a gradual, progressive change in a gene pool of a population from various generations due to factors such as mutations or natural selection and proves to be a prominent response towards selective pressures focused on any living organism (Davies and Davies 2010). Drug responsiveness of individual organism varies with fitness that enhances the survival potential. Although endogenous in nature, the immune system of an organism plays a significant role in transmittal of drugs within the system (Belkaid and Hand 2014) and confronts similar problems as in external drug delivery system (Yingchoncharoen et al. 2016). The term drug resistance is associated with the depreciation in the efficacy of a drug or a small molecule used in medication against certain microbes, parasites or cancers such as an antimicrobial or antiparasitic or an antineoplastic in curing an ailment or a disease condition (Demain and Sanchez 2009).

Drug-resistant traits are often rooted by successive progenitors resulting in each consecutive population more drug-resistant than the previous one. To avoid the emergence of such condition in a particular population one of the ways is to use a drug that block sexual reproduction or cell division or horizontal gene transfer in the entire target population (Wilson et al. 2016).

In the current world amongst the several tropical parasitic diseases, Malaria continues to be one of the most pernicious infectious diseases associated with very high mortality and morbidity rates (Ponts and Roch 2013). It is a hematoprotozoan parasitic infection which is transmitted by specific species of female anopheline mosquitoes (White 2004). With an approximately 212 million infections and 445,000 deaths as reported by WHO in 2016, malaria has been severely affecting countries of African and South-East Asian subcontinent which includes Republic Of Congo, Ivory Coast, Angola, Uganda, Kenya and India (Hotez and Kamath 2009). There has been a resurgent threat of malaria in contemporary world because of the seemingly boundless burden it has foisted on poverty stricken countries especially in the tropics (Frenk 2006). The attempts at the control of this deadly disease have stumbled after the collapse of the global eradication campaign in the 1960s. Primarily toddlers and preschoolers in sub-Saharan Africa and South Asia are most affected among other children. Control of this deadliest disease has traditionally relied on two armaments: restrain of the population of female anopheline mosquito vector through eviction of breeding sites, insecticides usage and deterrence of physical contact of mosquitoes with humans with effective case management. The physical association with human beings can be impeded by the practice of screens usage and mosquito bed net, particularly impregnated with insecticide or indoor residual spraying. The long term anticipation for the third armament, an effective malaria vaccine, has kept all of us in despair for not being materialized (White 2004).

Antimalarials drugs has played pivotal role in case management (mainly chloroquine, and currently used artemisinin-based combinatorial therapy along with sulfadoxine-pyrimethamine [SP] (Eastman and Fidock 2009). These drugs are mostly prescribed along with antipyretics which are widespread in use. But the major concern and obstacle towards successful treatment of the infection is the appearance of resistance towards antimalarial drugs which significantly contributes in global malaria-related mortality. Drug resistance introduced by antimalarials resulted in a deferment or failure to eliminate asexual stages of parasites from the blood stream. The gametocytes then produced from these parasites are mainly responsible for transmission of resistant genotype (Lin et al. 2010). In the latest scenario, pathogens have trained themselves to change their patterns of virulence and introduction of such trained pathogens in the environment is needed to be monitored.

According to the definition provided by WHO in 1967 for drug resistance against parasitic diseases; parasite strain have ability to survive or multiply despite the absorption of a drugs given in particular quantity of doses than those usually recommended but within the tolerance of the subject (Hyde 2007; World Health Organization 2015). As the pharmacokinetic profile of the antimalarials differ widely among individual entities, the definition of resistance is also altered based on the concentration profile of the effective drug (Arrow et al. 2004). Thus this issue of drug resistance and failure in its treatment is strongly associated to dose, time and concentration. Poor patient consent in respect to either drug dosage or extent of treatment duration along with individual variations of pharmacokinetics represent an important role in failure of therapy. Other factors involve in the failure are poor drug quality, drug synergy, inadequate absorption, rapid elimination and poor biotransformation of prodrug (Kay et al. 2014).

The utmost reason behind the emergence of resistance relies on extensive misuse by taking improper regime for the medication and deployment of these antimalarial drugs. This drug resistance further led to an immense selection pressure on malaria parasites to evolve distinct mechanisms of resistance. This also describes the underlying reason for two fold increase of malaria-owing child mortality in eastern and southern Africa (Cui et al. 2015).

To overcome the drug resistance many strategies are being pursued much familiar approaches like use of novel drug, natural entities with antimalarial attributes, targeted drug delivery systems and combination therapy have been undertaken by clinicians. Recognizing the requirement to accelerate the advancement in diminishing the global burden of malaria, WHO developed the Global Technical Strategy for Malaria 2016–2030 (GTS) (Hemingway et al. 2016). It sets out an aspect for boosting the momentum towards malaria elimination. The WHO blueprint to fight against malaria is further integrated with the Roll Back Malaria advocacy plan, Action and investment to defeat malaria 2016–2030 (AIM). Collectively, these documents provide a strong affirmation towards the requirement for universal access to interference for malaria prevention, diagnosis and treatment. It is the necessity that all countries should hasten efforts towards malaria elimination; and its surveillance should be the crux of the intervention (Huijben and Paaijmans 2018).

15.2 Disease Incidence and Demographical Distribution of Malaria

15.2.1 Occurrence and Geographical Dispersion of the Disease

According to WHO reports Malaria transmission occurs in five regions namely South–East Asian, Eastern Mediterranean, The Americas, Western Pacific Region and Africa. Apart from these 5 documented regions, the sixth one that is the European Union has been omitted from the list of malaria transmission regions; as it has proclaimed as malaria free region by WHO due to the recent registered reports suggesting zero endemic cases of malaria in recent years. According to 2015 WHO reports, there were 212 million recent incidents of malaria worldwide and an approximated 438,000 deaths (range 148–304 million). In 2016, the Global estimation accounted 3.3 billion people in 91 countries and territories are at risk of being infected with malaria and 1.2 billion are highly susceptible of developing the disease (>1 in 1000 possibility of acquiring malaria in a year). In the international forum the disease was marked by 306,000 under-five deaths out of which 292,000 were African children. The global disease burden is distributed unevenly with an approximation of 92% malaria driven death in African subcontinent, accompanied by 5% death in the South-East Asian region and 2% detected in the Eastern Mediterranean Region (World Health Organization 2016).

To our astonishment this deadly disease is not just confined to poor tropical regions of the world, even the developed countries such as United States are in clinch of the serious and fatal bouts of malaria. Almost all the recent travelers in United States are infected by malaria parasite and roughly 1700 cases of malaria are reported every year in the United States. First- and second-generation migrants from malaria-endemic countries returning to their native places are more likely to become infected with malaria due to inappropriate malaria prevention measures. U.S had recorded 63 outbreaks of locally transmitted mosquito-borne malaria between the years of 1957 and 2015; in such condition, regional mosquitoes become infected by feeding on blood of persons carrying malaria parasites (acquired in endemic areas) and then transmit malaria to local residents. Thus, there is a constant threat of reintroduction of malaria in the United States. During the period of 1963–2015, 97 cases of transfusion-transmitted malaria were reported in the United States. Recently the infection was classified as severe in 279 patients (16%) out of which 10 died. This was the largest number of deaths of malaria infected patients ever recorded in the United States since 2001. Of these 36 cases were among pregnant women, eight (22%) of which were severe, all of whom although survived (Cullen et al. 2016).

15.2.2 Diversity in Malarial Parasites and Its Anopheline Vectors

Among, the four species of plasmodium infecting human beings, *Plasmodium falciparum*, accounts for most severe illnesses and deaths from malaria. The present distribution of human-pathogenic *Plasmodium* species displays prevalence of *P. falciparum* and is estimated to cause 85% of infections acquired in Sub-SaharanAfrica,73% in Central America and the Caribbean, 24% in South America, 14% in Oceania, and 9% in Asia, while the distribution of *P. vivax* is concentrated in the few parts of Africa, covering Djibouti, Ethiopia, Eritrea, Sudan and Somalia which together accounted for 5% of total infection. Infections attributed to *P. vivax* accounted for 86% acquired in Oceania, 80% in Asia, 67% in South America and 22% in Central America and the Caribbean. Both *P. falciparum* and *P. vivax* are common in western Pacific and south-eastern Asia. Although *P. malariae* may appear in all malarious areas, its prevalence is normally low. In tropical Africa, people have encountered incidence of co-infection of both *P. falciparum* and *P. malariae*. *P. ovale* is widespread mainly in tropical Africa whereas *P. knowlesi* infection occurs only in definite forested areas of South-East Asia. Pregnant women have developed susceptibility towards *P. falciparium* in malaria endemic countries and the parasite has a profound effect on the birth weight of newborn which in turn decreases the baby's chances of survival. Other pathological symptoms include increased parasite sequestration in placenta, intrauterine growth retardation, and premature interruption of pregnancy, child mortality, increased maternal death and also development of congenital malaria. Thus at high transmission zones maternal and new born mortality are at peak (Molina-Cruz and Barillas-Mury 2014).

The retrospective and prospective view towards understanding malaria burden in India is quite important as it contributes 77% of the total malaria in Southeast Asian region of the world. India contributes towards the largest population which is at risk of malariain the world, with 85% living in malarious endemic zones (Kumar et al. 2007).

The epidemiological scenario of malaria in India is challenging because of geo-ecological diversity, several ecotypes including urban malaria and a vast distribution of nine anopheline vectors transmitting three Plasmodial species: *P. falciparum, P. vivax,* and *P. malariae.* Diverse transmission intensities extending from unstable to hyperendemic form, creates a challenging task to eliminate *Anopheles culicifacies* which has a wide distribution in rural areas, being the principal vector of rural malaria. *An. stephensi* is the dominant urban vector whereas *An. fluviatilis* is rampant in the hills and foothills; *An. minimus*, *An. nivipes*, *An. philippinen-sis* along with *An.dirus* are vectors in the northeast part of India. *An. sun-daicus* is confined to Andaman and Car Nicobar Islands with *An. annularis* and *An. varuna* as secondary vectors with varied distribution pattern.

In India, the state of Orissa contributes the most towards malaria. It has a population of 36.7 million (3.5%) but it shares 25% of a total of 1.5–2 million reported annual malaria cases, 39.5% of *P. falciparum* malaria, and 30% of deaths caused by malaria in India (Source NVBDCP, India). Similarly, other states such as West Bengal, Jharkhand, Chhattisgarh and Karnataka too contributed in malaria mortality. These states are partially inhabited by ethnic tribes living in the forest ecosystems. There is existence of meso to hyper endemic conditions of malaria with the predominance of *P. falciparum* to the magnitude of 90% or even more (Kumar et al. 2007). According to WHO report of 2018, India which is accounted for 4% of global malaria burden in 2017 has made significant progress towards the disease elimination by the end of 2018. It has accounted for 24% reduction in registered cases of malaria in the recent years which is quite good news itself.

15.3 Clinical Manifestations of Malaria and Anti-Malarial Drug Treatment

There are various pathophysiological mechanisms behind plasmodium infection and is understood to varying degrees. Clinical manifestations of malaria include fever, headache, chills and sweating triggered by dispensation of plasmodia from the mature schizonts (one of the asexual stages of the parasite in human host) into the blood stream. Without appropriate treatment or failure of an active immune response primed by repeated previous malaria infections leads to increase in the number of parasites with every 48 h of its reproductive cycle. A mature infection may comprise of 10^{12}circulating plasmodia in the bloodstream. Children suffering with severe form of malaria occasionally develop the following symptoms such as respiratory hardships in relation to metabolic acidosis, severe anemia or cerebral malaria. Multi-organ infection is prominent in adults. Asymptomatic infections may occur to people residing in malaria endemic areas who have developed partial immunity towards it. In conjugationto first- and second-line of antimalarial drug treatments which are commonly used in medication; collateral and auxiliary care measures (e.g., blood transfusions, intravenous fluids, supplemental oxygen and anti-seizure medication) may be necessary in case of severe manifestations. The aims of treatment are to reduce the rate of morbidity of an acute episode and mortality caused by the long-term deficits from malaria and to eradicate the infection entirely in order to avoid recurrence (Shaukat et al. 2012).

Pharmacokinetics and pharmacodynamics are two important aspects for any antimalarial drug development. The word pharmacokinetics refers to the process through which the drugs or compounds are absorbed, metabolized, distributed and excreted out of the body of an individual. Pharmacokinetic features of antimalarial drugs differ considerably among diverse individuals. Some responses are genetically determined, some depends on health status while other on dietary factors. Pharmacodynamics on the other hand refers to the way the drug acts on the target. The principal function of antimalarial drugs with respect to uncomplicated form of malaria is to prevent parasite multiplication and to reduce total parasite load in the body. Proliferation of parasite in non-immune individual is very high (at multiplication rates of 6–20 per 48 h cycle). Antimalarial compounds which exert maximum effect reduce the parasite number 10–10,000-fold. Thus the efficacy of drug depends on Pharmacodynamics (Brocks and Mehvar 2003).

15.3.1 Classification of Anti Malarials According to Their Mode of Action and Chemical Structure

- According to anti-malarial activity the classification is as follows (Fig. 15.1):
  - **Blood schizonticides:** Drugs under this category are known for showing their activity on the blood stage of the parasite and by that they help toabort clinical incursion of the pathogen. They account for their substantial role in anti-malarial chemotherapy. These include artemisinin, quinine, chloroquine, mefloquine, halofantrine, pyrimethamine, tetracyclines, sulfadoxine, sulfones etc. (Obaldia et al. 1997).
  - **Tissue schizonticides for seminal prophylaxis:** Drugs of this kind act primarily on the tissue stage of the plasmodia which grow inside the liver and commence the erythrocytic forms. Development of the affliction can be obviated by arresting this stage. This activity is shown by Pyrimethamine and Primaquine. Nevertheless since the prediction of the infection is impossible before the onset of clinical symptoms, this mode of therapy is much appropriate theoretically than practically (F. et al. 2010).
  - **Tissue schizonticides for averting relapse:** They target the hypnozoites of P. vivax and P. ovale in the liver which cause relapse of symptoms when reacti-

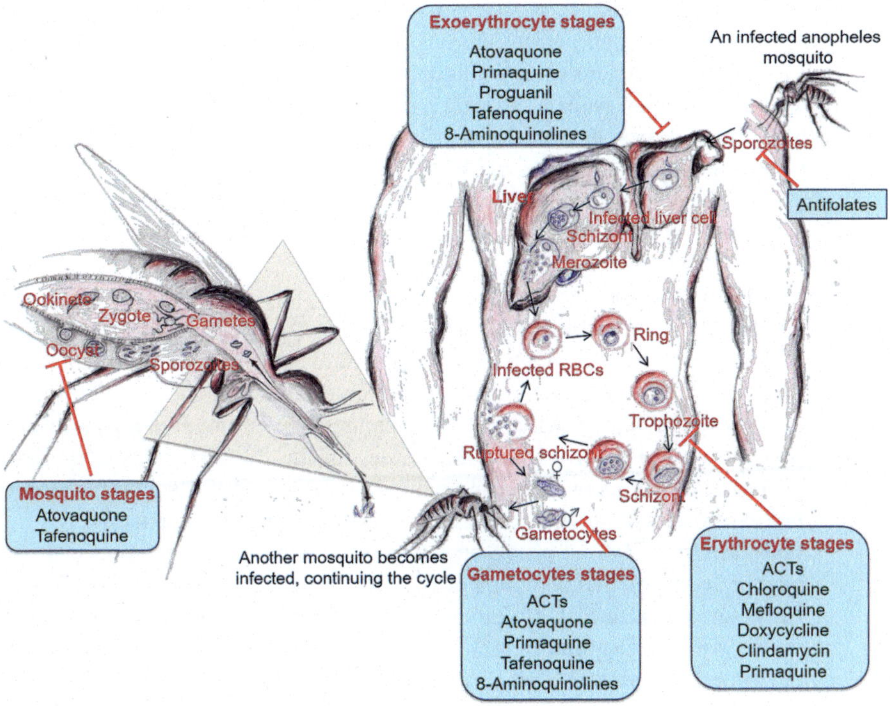

Fig. 15.1 Life cycle of malaria parasite and site of action of antimalarial drugs

vated. Primaquine act as the prototype drug; pyrimethamine also encompasses such activity (Davidson et al. 1981).

- **Gametocytocides:** The drugs under this category damage the sexual forms of the parasite in the blood stage and thereby prevent transmission of the infection to the vector. and Quinine and chloroquineshow gametocytocidal activity against P. vivax and P. malariae, but not against P. falciparum. Whereas, Artemisinin and Primaquine have demonstrated gametocytocidal activity against all plasmodia parasites, including P. falciparum (Sun et al. 2014).
- **Sporontocides:** The development of oocysts in the mosquito are obviated by such drugs and thus ablate the transmission. Primaquine and chloroguanide fall under this category (Rieckmann et al. 1968).
- Structural classification for the anti-malarial drug is as follows:
  (a) **Aryl amino alcohols:** Quinine, quinidine (cinchona alkaloids), piperaquine, halofantrine, amodiaquine, mefloquine, lumefantrine, tafenoquine
  (b) **4-aminoquinolines:** Chloroquine, amodiaquine.
  (c) **8-aminoquinolines:** WR238, 605,Primaquine
  (d) **Naphthoquinones:** Atovaquone
  (e) **Folate synthesis inhibitors:** Type 1 are competitive inhibitors of dihydropteroate synthase such assulphonamidesand sulphones.
  Type 2 blocks dihydrofolate reductase. This category contains biguanides like proguanil and chlorproguanil; diaminopyrimidine like pyrimethamine.
  (f) **Antimicrobials:** Tetracycline, doxycycline, clindamycin, azithromycin, fluoroquinolones
  (g) **Peroxides:** Artemisinin (Qinghaosu) derivatives and analogues such as artemether, artesunate, arteether, dihydroartemisinin
  (h) **Iron chelating agents:** Desferrioxamine
  (i) **Polyphenolic compounds:** Ellagic acid

Mechanism of action of few important drugs.

(a) **Aryl Amino Alcohols**

- **Quinine**: It is the chief alkaloid from bark of cinchona and depicts good efficacy in the management of severe form of falciparum related malaria in infected areas with known resistance to chloroquine (Achan et al. 2011). It also shows gametocytocidal activity against *P. vivax* and *P. malariae*. Being a weak base in nature, it is concentrated in the food vacuoles of *P. falciparum*. It functions by inhibiting heme polymerase, thereby allows accumulation of its substrate, heme which is very toxic for malaria parasite.
- **Halofantrine**: Halofantrineis structurally related to quinine however it is a phenanthrene methanol which was first developed in1960s by the Walter Reed Army Institute of Research. Its mechanism of action is almost similar in comparison to quinine and chloroquine. It forms toxic complexes with ferriprotoporphyrin IX that damages the membrane of the plasmodia. This synthetic antimalarial is quite effective against multi drug resistant *P.falciparum*malaria (Bryson and Goa 1992).

- **Mefloquine:** It was born as a consequence of research into novel anti malarials during the Vietnam war and hence considered as a 'new' drug. However, there was no further progress obtained for this drug. Mefloquine produces swelling of food vacuoles of *P. falciparum.* It forms toxic complexes with free heme group that damages membrane and interact with other parasite components. It has good activity against the blood forms of falciparum malaria (Palmer et al. 1993).
- **Lumefantrine**: Being an aryl alcohol, it is related to quinine, mefloquine and halofantrine. It is however devoid of cardiac toxicity as compared to halofantrine. It is being used in combination therapy with artemether and has shown excellent result against uncomplicated falciparum malaria in children and adults (Stover et al. 2012).

(b) **4 - Aminoquinolines**

- **Chloroquine**: It is the most widely used anti-malarial drug to administer all types of malarial infections. It is also the cheapest antimalarial agent available in market and has low toxicity. The characteristic feature of chloroquine is its capability to concentrate itself from a range of nanomolar levels inside red blood cells but outside the parasite to levels of one million times (millimolar level) higher in the acid vacuole of parasite inside an erythrocyte (Krogstad and Schlesinger 1987). It works by interfering with heme dimerization as it forms drug –heme complex which is much more stable than heme –heme complex. Chloroquine blocks the parasitic enzyme heme polymerase that acts in the detoxifying biochemical process within the malaria parasite which transform the toxic heme into non-toxic hemozoin (malaria pigmentation) therebyculminating the process by aggregation of toxic heme within the parasite. The other mechanism may include intercalation of the drug with parasite DNA.

(c) **8 - Aminoquinolines**

- **Primaquine:** It plays an important role in abrogating plasmodia multiplication by effectively killing its gametocidal stage, including *P.falciparum*. It is an essential co-drug with chloroquine. However, its mode of actionis not well established. It may act by producing reactive oxygen species or may interfere with the electron transport system in the parasite. It also has a direct effect against the dormant tissue forms of *P. Vivax* and *P. Ovale*malaria, and thus offers free radical cure and avert relapses. Chloroquine plus Primaquine remains the first-liner administrative drugs for radical cure of vivax and ovale malaria in most regions (Galappaththy et al. 2013).

(d) **Naphthoquinones**

- **Atovaquone:** Being synthetic in nature, Atovaquone is a hydroxy derivative of naphthoquinonedeveloped in the early 1980s. It has a lipophilic moiety which interferes with mitochondrial electron transport system and thereby prevents ATP formation and hampers the pyrimidine biosynthesis in plasmodia. It typically targets cytochrome bc1 complex and breach the membrane potential and blocks cellular respiration (Srivastava et al. 1997). It is very useful against plasmodias well as Toxoplasma and Pneumocystis carinii (Nixon et al. 2013).

(e) **Antifolates**

- **Combination of Sulfadoxine and Pyrimethamine:** Pyrimethamine and cycloguanil (biguanides) interfere with folic acid biosynthesis pathway by blocking the parasite enzyme known as dihydrofolate reductase-thymidylate synthase (DHFR). Thereby it inhibits synthesis of purines and pyrimidines, which are crucial for DNA synthesis, cellular division and multiplication. This causes failure of nuclear division at the time of schizogony in erythrocytes and liver.

  Sulfonamides inhibit the parasite enzyme dihydropteroate synthase (DHPS) which helps in usage of para- amino-benzoic acid in the synthesis of dihydropteric acid. It abrogates the earlier step in the folic acid pathway. Thus, the combination of these two classes of drugs offers two step synergic blockade of parasite division (Nyunt et al. 2010).

(a) **Antimicrobials**

Many Antimicrobials have been tested for their potential anti -malarial effects. Few of them such as Tetracyclines, Clindamycin and Azithromycin are in use worldwide to combat malaria.

- **Tetracyclines:** One of the first antibiotics used by the mankind and still continued to mark its usefulness in broad range of microbial and parasitic infections including malaria. These are bacteriostatic agents which acts by binding to the 30s ribosome subunit, further inhibiting protein synthesis. They are effective against a wide range of organisms including aerobic and anaerobic gram positive and gram-negative bacteria, *Rickettsia*, *Coxiella burnetii*, *Mycoplasma, Ureaplasma, Chlamydia, Legionella*, *Spirochaetes*, *Brucella* and *Helicobacter pylori.* This is a useful antibiotic for treatment of drug resistant *P. falciparum* malaria. It acts relatively slow and hence should always be combined with a faster acting drug like quinine. Due to its adverse effects on bones and teeth, its usage by children below 8 years of age and by pregnant women are avoided (Gaillard et al. 2015).
- **Clindamycin**: It inhibits the protein synthesis by binding to the 50s subunit of ribosomes. It is used for drug resistant malaria along with quinine. Adverse effects include pseudo membrane colitis and skin rashes (Lell and Kremsner 2002).
- **Azithromycin:** Azithromycin has anti-malarial activity and found to be useful as a causal prophylactic agent, against chloroquine resistant *P. falciparum* infection (Van Eijk and Terlouw 2007).

(b) **Peroxides**

- **Artemisinins:** It is a known sesquiterpene lactone peroxidase which is characterized by endoperoxide bridge in its chemical structure unlike any other known antimalarial drug (Brossi et al. 1988). This bridge is necessary for its antimalarial activity. Three derivatives are available in market, known as dihydroartemisinin (DHA); a closely related, oil-soluble methyl ether compound, artemether and the water-soluble hemi-succinate derivative artesunate. It has much clinical importance due to the fact thatartemisinins are metabolized to DHA inside body and this form has comparable antimalarial activity. Artemisinin treatment causes lipid peroxidation especially in the presence of heme. The activity of the drug is characterized by its inhibitory role of*P falciparum*-encoded sarcoplasmic-endoplasmic reticulum calcium ATPase (PfATP6) (Eckstein-Ludwig et al. 2003). These compounds are the fastest acting antimalarial available. These drugs start acting within the range of 12 hours. These prominent properties of the drug are very useful in managing complicated *P. falciparum* malaria. These drugs are also effective against the chloroquine resistant strains of *P. falciparum*. It is proved to be effective at killing the broadest range of asexual stages of the parasite, ranging from medium sized rings-stage to early schizonts and thus inhibits the development of the trophozoites, preventing the progression of the disease. Young circulating parasites are killed before they sequester in the deep microvasculature. Artemisinin compounds have been reported to reduce gametocytogenesis by their action on early (stage I-III) gametocytes.

(c) **Iron Chelators**

- **Desferrioxamine (Df):** The drug shows stage dependent antimalarial activity. It works as cytostatic drug for a range of asexual stages from early ring stage to early trophozoite stage whereas concomitantly it works as cytocidal for mature trophozoite and early schizonts of the parasite. The cytocidal impact of DF appears to correspond with the period of maximum DNA synthesis by the plasmodia thus suggesting ribonucleotide synthetase the most likely target for blockage. This is very interesting observation as the iron which is bound to an active site of tyrosyl residues, as in ribonucleotide reductase, is directly available to chelators (Hershko 1989). Preferably, DF can also act by chelating the low molecular weight iron pool in the parasite' causing interference with the ironsupply to various parasite metabolites (Hershko and Peto 1988). Drug combination with this type of iron chelators might lead to improved antimalarial performance as the combination will enhance the permeability into parasitized RBCs (Bunnag et al. 1992).

(d) **Polyphenolic Compounds**

- **Ellagic Acid:** The compound is recently found in various plant products and has antioxidant properties. It has been characterized as a potent *Pf*GluPho (glucose-6-phosphate dehydrogenase 6-phosphogluconolactonase) inhibitor. In *P. falciparum*, *Pf*GluPho enzyme plays key role in pentose phosphate pathway as the first two reactions of the PPP are catalyzed by this bifunctional enzyme. Ellagic acid inhibits the production of ROS and thus acts on trophozoite and schizonts stages of erythrocytic cycle of *P. falciparum,* which are known to coincide with the maximal generation of ROS during oxidative metabolism of the parasites. It also plays important role in changing cellular redox potential by depleting glutathione which further induces *Plasmodium* death. Thus it acts as potent antiplasmodial compound (Soh et al. 2009).

15.3.2 The Efficacy of Artemisinin Combination Therapy (ACTs) in Malaria Therapy

Combination drug treatment approaches are prevalent in dealing with many infectious diseases such as TB and HIV infection and in general, it is also relevant in case of malaria. The logical explanation behind the usage of ACT is that the probability of plasmodia simultaneously acquiring resistance due to emergence of genetic mutations by two different drugs with varied modes of action is much reduced than the chance of pathogen developing resistance to single drugs. In the present scenario, there are a number of ACTs being approved and tested in distinct *P. falciparum*-endemic regions (Cui and Su 2009). For the treatment of uncomplicated form of *falciparum* malaria in adults and children, Artemether–lumefantrine (Coartem) regimen is frequently used in a fixed-dose oral combination therapy (Adjei et al. 2009). Its excellent effectiveness against *P. falciparum* malaria has been confirmed in multiple clinical trials. Artesunate–mefloquine combination has been broadly used in Southeast Asia. DHA–piperaquine (Artekin) is yet another fixed-dose drug conjugates, majorly formulated in tablets and is commercially feasible among many Asian countries. Later on in this century, due to the appearance of resistance towards Chloroquine;Piperaquine was developed as a replacement of it and is used widely in China. Numerous clinical trials have demonstrated that this ACT with a 3-day regimen was highly effective and well tolerated (Zwang et al. 2009). Recently a fixed-dose artesunate–pyronaridine combination also showed excellent efficacy against uncomplicated form of *falciparum* malaria in children (Wongsrichanalai and Sibley 2013). In addition, several other ACTs – such as artesunate–sulfadoxine–pyrimethamine (SF), artesunate–amodiaquine and artesunate–chlorproguanil–dapsone have been developed and are under clinical trials. In many areas, ACTs depicted profound effectivity against *falciparum* malaria, with cure rates exceeding 90% (Anvikar et al. 2012). Clinical reports of ACTs in children also proved to have an immense effect in controlling the disease phenotype. Artemisinin and ACTs, together perform well against *Plasmodium vivax* malaria (Gogtay et al. 2013).

The underlying strategy of ACT development suggests in ACT, the partner drugs must be structurally independent of each other, with slow *in vivo* elimination profile and should be able to target parasites which have not yet developed resistance. Although the inclusion of artemisinin derivatives can improve the efficacy of few conventional antimalarial drugs in regions where parasites have developed high-level resistance towards these drugs, reintroduction of these antimalarial agents in ACTs iscontroversial or questionable. Although the use of artesunate–mefloquine is widely deployed in parts of Thailand and Cambodia yet high-level resistance to mefloquine is quite widespread. In China, extensive use of piperaquine became the major cause of parasites becoming more 'resistant' to the drug (Nosten and White 2007), but DHA–piperaquine is still very crucial in treating malaria parasites. The resistance towards amodiaquine and chloroquine is highly comparable and the dosage of Fansidar (SF) in treating *falciparum* malaria is waning in many African countries. ThusACTs with these partner drugs are still paving their way towards modern combinatorial therapy. Although deployment of such ACTs with a failing partner drug cause reversalof resistance towards partner drug as validated in the case of artesunate–mefloquine in Thailand, it must be cautioned since the effectiveness of ACTs might get compromised with the use of an inappropriate partner drug (Cui and Su 2009).

15.4 Current Status of Drug-Resistant Malaria and Detection of Resistance

15.4.1 Origin of Drug Resistance

In early 1950s when steps were taken to csontrol and eradicate malaria, the programs showed exciting results in several countries but thereafter the initiative to eliminate malaria failed due to emergence of drug resistance in parasites (Fig. 15.2) (Farooq and Mahajan 2004; Gero and Upston 1992).

There are various aspects which influence the development of resistance to the antimalarial drugs. CQ increases pH in the food vacuole and inhibits hemozoin synthesis. Mutation in the gene of *P. falciparum* chloroquine resistance

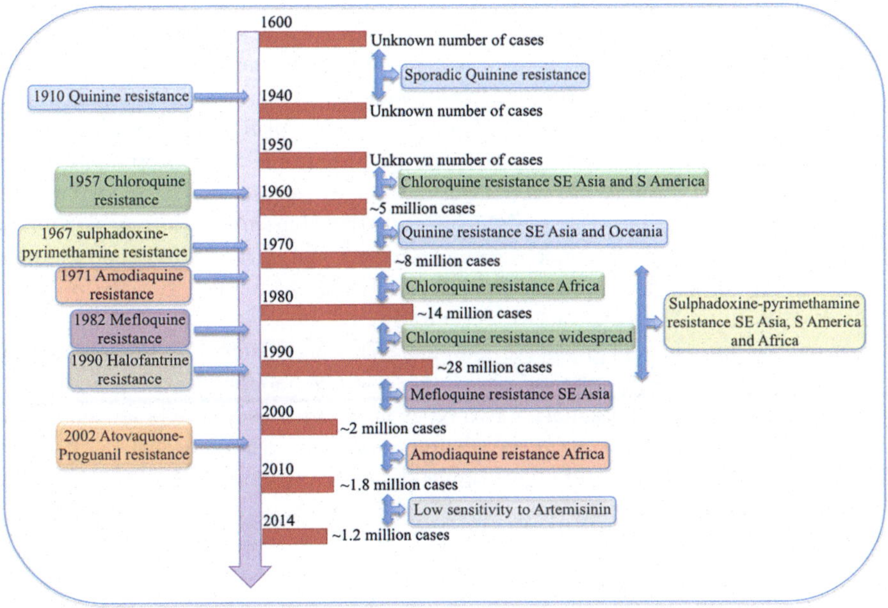

Fig. 15.2 Timeline of first confirmed report (left) and developing periodic (right) of antimalarial drug resistance with number of cases

transporter- a transmembrane protein (pfcrt) on chromosome 7, localized to the surface of the food vacuole mediates leakage of CQ out of the food vacuole (Riordan et al. 1985). Resistance to SP occurs because of point mutations in DHFR (Dihydrofolate reductase) and DHPS (Dihydropteroate synthase) enzymes in which the binding affinity of the drugs to these enzymes is greatly reduced. Mutation in the gene PfATP6 influence susceptibility/sensitivity to artemisinins. It is sarcoplasmic endoplasmic reticulum calcium dependent ATPase (SERCA) like transporter which is inhibited by Fe2+ accumulation after artemisinin treatment. Artemisinin allow cleavage of the peroxide bridge that result in the enzyme inhibition and parasite death (Eckstein-Ludwig et al. 2003).

15.4.2 Detection and Testing of Drug-Resistance

Proper diagnosis of malaria infection is very critical because symptoms of complicated malaria may develop suddenly in patients leaving to increase in fatality (Vaidya Kuladeepa 2012). The gold standard technique for malaria diagnosis is the identification of parasites in a Giemsa stained blood smear. Other widely used and more sensitive methods for detection include Rapid diagnostic tests (RDTs), Quantitative Buffy Coat (QBC) test and polymerase chain reaction (PCR) (Armstrong 1999; Moody 2002).

Determination of drug resistance in malaria parasites whether against gametocytocides, schizonticides can be analyzed by either *in-vitro* or *in-vivo* drug susceptibility tests (Wernsdorfer and Payne 1991). In 1990, a modified protocol has been implemented by WHO. This screening involved relatively shorter (7–14 days) follow-up period that included clinical, parasitological, hematocrit and fever assessment on Day 0, 3, 7 and 14. The underlying idea is that reappearance of parasites in peripheral blood within 14 days of treatment is more likely due to recrudescence than reinfection (Schapira et al. 1988). As infected people are susceptible to anaemia, in-vivo proceedings allowed investigation of hematological recovery after malaria therapy (Bloland et al. 1993). Unfortunately, these methodologies could not be fully standardized because of various external factors such as differential immunity of hosts, difference in absorption/metabolism of particular drug and potential misclassification of reinfections as recrudescence.

Inspite of immune and physiopathological status of the host, *in-vitro* assays helps in determination of the phenotype of the parasite independently (Farooq and Mahajan 2004). Previously mainly two types of in vitro assays are practiced; one is WHO schizont maturation assay and the other one isotopic micro test (Rieckmann et al. 1978). These tests require technological advancement and relatively expensive that makes them potentially less suitable for regular antimalarial drug efficacy assessment.

15.5 Factors Associated with Drug Resistance of Malaria

Accumulating evidences have revealed various factors relevant to the emergence, spreading and ramification of the deadly malaria. Among such factors those associated to drug resistance include myriad human behavior due to civil disturbances, socio-economic problems, increased mobility of populations, environmental and climatic changes. Besides these, certain biological factors such as pharmacokinetics of the antimalarial drug and interrelation of host, parasite and vector biology might play crucial role in understanding disease resistance. Finally, unsuccessful treatment modalities can also be one of the factors for development of disease (O'Meara et al. 2010).

15.5.1 Biological Influences on Resistance

Based on the published documents related to the parasite responsiveness towards in-vitro pharmacokinetic profiles of antimalarial drugs, it is always thought that a residual number of parasites are able to survive the treatment (Wernsdorfer 1991) which are further removed by the host immune system under normal conditions. Sometimes, post treatment factors leads to ineffective immune response towards clearing the residual parasite populations which also play a key role in the survival of parasite.

Also, in the case of immune-compromised individuals such as young children, parasites repeat many cycles through populations for growth and survival (Verdrager 1995). The risk of malaria both in terms of infection and clinical disease is higher in pregnant woman, likely due to the hormonal changes, immunological or other factors occurring during pregnancy. The adverse effects of *P. falciparum* are quite evident in primigravidae (Uneke 2007). The placenta is the most "privileged" site for parasite sequestration and multiplication. The local immune paresis towards malaria parasite remains obscure. This has a contribution towards the advent and intensification of resistance which have not been evaluated yet (Andrew R. Marks 2007). Pregnant woman and infants with low immunity are more prone to poorer treatment modalities. It has been found that pregnant woman has a detrimental effect towards treatment responses of antimalarials in low-transmission zones as compared with similar aged non-pregnant women from the same location (Takem and D'Alessandro 2013; McGready et al. 2001). Evidences also suggest that pregnant women are more susceptible to mosquitoes (Ansell et al. 2002). These observations clearly indicate that they might be an important contributor to Antimalarial drug resistance. Another example of immunocompromised cases are those of the African children suffering from malnourishment who had significantly poorer parasitological response to both chloroquine and SP treatment (Wolday et al. 1995).

In case of immunosuppressive subjects, the parasitological response towards treatment of acute malaria is significantly low. The co-infection of falciparum malaria along with HIV/AIDS is quite prevalent in high malaria transmission areas (Onyenekwe et al. 2007). Malaria is a serious problem of childhood, whereas HIV has higher mortality rates in infants and adults (Lallemant et al. 2010; Shapiro and Lockman 2010; Alemu et al. 2013). The increasing availability of antiretroviral drugs accelerates the life-span of HIV-infected patients, thus creating a situation where two infections will coincide more often. During pregnancy the HIV coinfection is associated with greater reduction in birth weight than that associated with malaria infection alone (Ayisi et al. 2003; Uneke et al. 2009). The common therapeutic practice to combat this problem is to use IPT and SP combinatorial medication in monthly basis to achieve improvement in birth weight (Wolday et al. 1995).

Parasitological response towards treatment of acute malaria among HIV-sero positive individuals is much poor than compared with HIV-negative patients (Chirenda et al. 2000; Grimwade et al. 2003). This clearly indicates that the immunosuppression is associated with HIV infection, can compromise the effect of antimalarial immunity in reducing the selection and spread of antimalarial drug resistance. Interestingly, antimalarials such as Trimethoprim-sulfamethoxazole, an antifol-sulfonamide combination is widely given to HIV/AIDS infected patients as prophylaxis against opportunistic infections. However, it is still elusive, whether this combinatorial drug has any role in emergence of antifolate resistance or not. The prevalence of these ailments can pose a tremendous threat to existing and future antimalarial drugs, thus giving a way for drug resistance (Onyenekwe et al. 2007).

The current phenomenon of cross resistance between chemically related drugs can lead to increased parasitological resistance. Amodiaquine and Chloroquine are both fall under4- aminoquinolines and cross-resistance between these two drugs is well known (Basco 1991; Hall et al. 1975). Resistance tomefloquine may also lead to development of resistance to quinine and halofantrine. The widespread use of sulfadoxine/pyrimethamine for the treatment of malaria may lead to increased parasitological resistance to other antifolate combination drugs (Watkins et al. 1997). Continuous mutations in DHFR domain leads to development of high levels of SP resistance which may compromise the useful lifespan of latest antifolate combination drugs such as chlorproguanil/dapsone (LapDap) even before they are brought into use (Karema et al. 2010).

P. falciparum has a point mutation rate of 10^{-6} to 10^{-9} per generation (Rathod and Govinda Rao 1997). It is estimated that an infected individual can have 10^{10} to 10^{13} parasites in the bloodstream. There is an approximation of 5×10^{8} infections per year (Rathod and Govinda Rao 1997). These clearly denote that at any particular time point triple -mutant parasite can arise anywhere in the world. Looking at these sheer numbers gives us a strong insight about how powerful these mutations can be; which has a major contribution towards generation of resistance. The variation in copy number is yet another fascinating feature of malarial strains and this can be seen easily in culture conditions. The plasmodium genome is quite unstable as there are frequent amplification and deletion events taking place at a time. This causes another greater challenge for antimalarials to work on as well as it poses a threat towards drug resistance.

The theory of genetic plasticity among some South-East Asian strains of falciparum parasites is the underlying adaptability feature which allows parasites to develop rapid resistance to any new drug (Rathod and Govinda Rao 1997). For example, one can assume of a parasite with duplication of a DHFR gene in which one copy confers resistance to one drug whereas the second confers resistance to the other drug.

There is a blurred image regarding prospective relationship between transmission intensity and development of resistance against anti-malarial drugs. Apparently there is presence of more genetically distinct clones per individuals in areas of more intense malarial transmission than in areas of lower transmission (Babiker and Walliker 1997). The relationship between transmission intensity and parasite genetic structure is yet another complex subject which also attributes towards developing resistance (Paul et al. 1998; Hastings and Mackinnon 1998). Although there are number of factors such as domain mutations, selectivity and drug pressure, prevalence immunological conditions, population movement which leads to development of resistance against antimalarial drugs but it is quite clear that reducing the intensity of transmission will likely facilitate increase in useful lifespan of drugs (Mackinnon and Hastings 1998; Molyneux et al. 1999).

15.5.2 Programmatic Influences

Programmatic influences on development of antimalarial drug resistance include pharmacokinetic and pharmacodynamic properties of the drug or combination of drugs. Overall drug pressure, poor compliance or in appropriate dosingregimens of drug intake are part of such influence (Wernsdorfer 1994). Antimalarial drug pharmacokinetics favors the emergence and spread of the disease. They are usually altered, due to an expanded apparent volume of distribution (mefloquine, quinine, atovaquone, and proguanil) which results in lower drug levels for the given dose (White 2004). Mismatched pharmacokinetics can also play a role in facilitating the development of resistance. The combination of sulfadoxine-pyrimethamine is a one of the examples of mismatched pharmacokinetics. The elimination half-life of sulfadoxine is between100 and 200 hours whereas that of pyrimethamine is between 80 and 100 hours, thus leaving sulfadoxine unprotected for a longer period of time in the patient's body (Hastings and Watkins 2006). In Thailand the apparent use of mefloquine-sulfadoxine-pyrimethamine (MSP) combination is even more dangerous as mefloquine has an elimination half-life of approximately 336 to 432 hours which makes it difficult for early elimination from the body (Boudreau et al. 1991).

The higher dependence on available probable treatments can also facilitate the development of antimalarial drug resistance. There are various studies which suggest that rate of resistance is higher in urban areas than rural areas, where access to and use of drug is greater. Overall drug pressure is exerted by operational research programs which utilizes mass drug administration and thus has the greatest impact on development of resistance example Federal government Malaria intervention program using Coartem, Eko free malaria programme (Wernsdorfer 1994). In

Africa, the costs of antimalarial combination therapies are over ten times more expensive than those of the traditional drugs currently available as monotherapy in the market. This also leads to inappropriate and inadequate drug intake which further leads to resistance (Arrow et al. 2004). Confusion over proper dosing regimen has also came into picture. The application of presumptive treatment for malaria has a higher potential for facilitating resistance by increasing the number of people who are treated unnecessarily with antimalarial drugs. This will further start exerting selective pressure on the circulating parasite population (Mwangi et al. 2005).

Concurrent treatment with other drugs can increase the chances of treatment failure and may contribute to advent of drug resistance. There is an increase chance of treatment failure among pregnant woman who are routinely administered with folate supplements for treatment of anemia (CDC, unpublished data, 1998).

Atovaquone-proguanil combination therapy is highly efficacious against P.falciparum, including strains that are resistant to chloroquine and mefloquine, with cure rates of 94–100%(Brockman et al. 2000). Current contra-indications include hypersensitivity to atovaquone or proguanil or presence of renal insufficiency. Similarly, concurrent illness may have an influence, as was mentioned previously with regard to malnourishment.

The quality of drug has also been implicated in feeble treatment and drug resistance. Poor manufacturing practices, deterioration of drugs due to improper handling and storage, intentional counterfeiting causes drugs not to contain adequate quantities of the active ingredients. It has been found that about 37% and 40% of samples of chloroquine and antibiotics available in Nigeria and Thailand, had substandard content of active ingredients (Maitland and Newton 2005).

15.6 The Future Prospective: Prevention of Drug Resistance

The number of assumptions defines the future of antimalarial drug resistance and cumulative efforts to combat it. Firstly, the requirement of antimalarials will persist long into the future. There is no strategy in existence or in development which can completely eradicate or prevent malaria infection (Breman 2001). Secondly, the continuous usage of present drugs itself create chances of resistance development (Plowe 2003).*P. falciparum* has developed resistance to almost all the antimalarial drugs available in the modern world. It is highly probable that soon the parasite will develop resistance to any broadly used drug. Thirdly, the pipeline for development of new drugs is almost chocked as it appears to be slower than development of parasitological resistance (Ridley et al. 1997). Affordability of drugs is the fourth major consideration for any strategic planning to control the drug resistance, especially in Africa (Foster and Phillips 1998; Goodman et al. 1999).

There are various interventions intended to prevent drug resistance by generally focusing on reducing overall drug pressure through more restrictive usage of drugs, prescribing proper dosages by clinicians, improving the mode of transport for the drugs and also by follow-up practices along with patient compliance (Seppala et al. 1997; Bauchner et al. 1999). Improving the diagnosis of malaria is one of the best approaches to decrease in antimalarial drug resistance. Although there are programs such as IMCI which aims to boost clinical diagnosis through well designed algorithms but unfortunately it has failed to achieve its efficiency in areas of relatively low malaria risk. Administration of drugs based on directly observed therapy (DOT) acts as a way to assure high degree of conformity. Utilization of single-dose DOT is very advantageous than using drugs having long half-lives. A recent clinical strategy that has gained much attention is the combination therapy. The drug combinations are made in such a way so that, they are intrinsically less likely to promote resistance or have properties which prohibits the development or spread of resistant parasites (Konotey-Ahulu 1999; White 1987). For the treatment of uncomplicated *Plasmodium falciparum* malaria, Artemisinin combination therapy (ACT) are the latest recommended first-line treatment. The efficacy of ACTs can further be enhanced by implementation of multiple first line therapies (MFTs) (Peters 1990; Boni et al. 2008).

Novel ideas include repurposing of drugs for eradication of malaria. In human erythrocytes, ferrochelatase, one of the enzymes play a key role in haem biosynthesis (Zumla et al. 2016); (Smith et al. 2015). Certain iron chelating agent, such as desferrioxamine (DFO) which is a strong ferrochelatase inhibitor when used in combination with quinine enhances parasite clearance and speeds recovery (Gordeuk et al. 2006; Smith and Meremikwu 2003). In other application excess of TNFα production promotes fulfillment of malaria parasite (Kinra and Dutta 2013). By using anti-TNFα drugs like adalimumab, etanercept could reduce parasite growth in a dose-dependent manner (Kwiatkowski et al. 1993). Recombinant protein could also be used for eradication of malaria. Anti-inflammatory drugs, the innate defense regulator (IDR) peptides such as IDR-1018 when used in combination with standard first-line antimalarial namely SP and CQ could decrease pathogen viability (Achtman et al. 2012). Recent identification of apicoplast in malarial parasites has generated a huge impact (Gleeson 2000). It is reported that they are involved in various functions in Apicomplexa such as protein synthesis, haeme synthesis, lipid metabolism and regulation of electron transport chain (Hackstein et al. 1995; Roy et al. 1999). Application of apicoplast towards multiple functions makes it suitable target for anti-malarial (Sadhukhan and Mukherjee 2016).

15.7 Conclusion

The recent emergence of drug resistance towards artemisinin derivatives, often termed as the last bastion in chemotherapeutical treatment against malaria rings an alarm bell stating there is a need to have new antimalarials or antimalarial drug policies. This might hopefully diminish the advent and spread of resistance (Fig. 15.3). Being unable to achieve so would certainly lead to a catastrophe where elimination of malaria would just be a dream and the number of malaria related morbidity and mortality would soon be higher.

To avoid such tragic set back there is urgent need to investigate the combination therapy based on fixed dose formulations, robustness of policy involved and

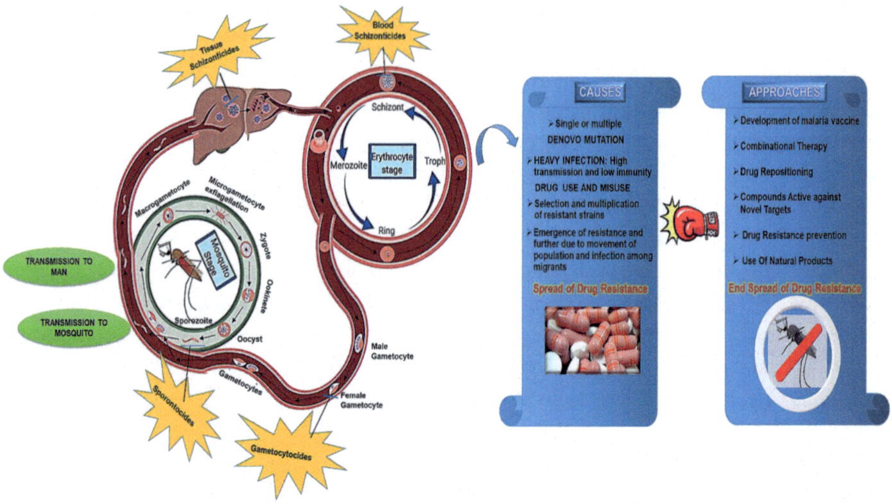

Fig. 15.3 Factors contributing towards onset of drug resistance in life cycle of malaria parasite and various ways to combat it

cost- effectiveness. Improvement towards access to and use of definitive diagnosis-based treatment might be encouraged. Anti- malarial vaccines that have transmission-blocking effect and could be used in combination with drugs active against blood-stage parasites would be best possible armament in targeting drug resistant parasites. We must explore new ways to increase effectiveness of drug regulatory systems and ability to control introduction of new antimalarial within endemic countries. Setting up proper health care infrastructure leading to higher quality services and their prompt availability would surely help us to fight against antimalarial drug resistance.

The current tools including indoor residual spraying, insecticide treated bed nets and effective case management have benefited in reduction of malaria to some extent. The implementation of these along with above mentioned interventions can surely par down the spread of drug resistance. All these without sustained political and financial support are not at all possible. Together, these efforts will eventually reduce global malaria burden and serve a significant benefit to global public health.

Acknowledgements AC thankfully acknowledge the financial support received from Shiv Nadar Foundation. VS is a recipient of SERB-NPDF from Department of Science and Technology (DST), Government of India, New Delhi. SS is a recipient of IYBA, Department of Biotechnology (DBT), Government of India, New Delhi. Authors acknowledge the support from Jawaharlal Nehru University.

References

Achan J, Talisuna AO, Erhart A et al (2011) Quinine, an old anti-malarial drug in a modern world: role in the treatment of malaria. Malar J 10

Achtman AH, Pilat S, Law CW et al (2012) Effective adjunctive therapy by an innate defense regulatory peptide in a preclinical model of severe malaria. Sci Transl Med 4. https://doi.org/10.1126/scitranslmed.3003515

Adjei GO, Goka BQ, Binka F, Al Kurtzhals J (2009) Artemether-lumefantrine: an oral antimalarial for uncomplicated malaria in children. Expert Rev Anti-Infect Ther 7:669–681

Alemu A, Shiferaw Y, Addis Z, et al (2013) Effect of malaria on HIV/AIDS transmission and progression. Parasites and vectors

Ansell J, Hamilton KA, Pinder M et al (2002) Short-range attractiveness of pregnant women to Anopheles gambiae mosquitoes. Trans R Soc Trop Med Hyg 96:113–116. https://doi.org/10.1016/S0035-9203(02)90271-3

Anvikar AR, Sharma B, Shahi BH et al (2012) Artesunate-amodiaquine fixed dose combination for the treatment of Plasmodium falciparum malaria in India. Malar J 11:97. https://doi.org/10.1186/1475-2875-11-97

Armstrong D J C (1999) Infectious diseases. In: Epidemiology and infection. Harcourt International 124(3):603–605

Arrow KJ, Panosian C, Gelband H (2004) Saving lives, buying time: economics of malaria drugs in an age of resistance

Ayisi JG, van Eijk AM, ter Kuile FO et al (2003) The effect of dual infection with HIV and malaria on pregnancy outcome in western Kenya. AIDS 17:585–594. https://doi.org/10.1097/01.aids.0000042977.95433.37

Babiker HA, Walliker D (1997) Current views on the population structure of plasmodium falciparum: implications for control. Parasitol Today. https://doi.org/10.1016/S0169-4758(97)01075-2

Basco LK (1991) Inefficacy of amodiaquine against chloroquine-resistant malaria. Lancet 338:1460

Bauchner H, Pelton SI, Klein JO (1999) Parents, physicians and antibiotic use. Pediatrics 103:395–401. https://doi.org/10.1086/250095

Belkaid Y, Hand TW (2014) Role of the microbiota in immunity and inflammation. Cell 157:121–141

Bloland PB, Lackritz EM, Kazembe PN et al (1993) Beyond chloroquine: implications of drug resistance for evaluating malaria therapy efficacy and treatment policy in Africa. J Infect Dis 167:932–937. https://doi.org/10.2307/30112358

Boni MF, Smith DL, Laxminarayan R (2008) Benefits of using multiple first-line therapies against malaria. Proc Natl Acad Sci 105:14216–14221. https://doi.org/10.1073/pnas.0804628105

Boudreau EF, Pang LW, Chaikummao S, Witayarut C (1991) Comparison of mefloquine, chloroquine plus pyrimethamine-sulfadoxine (Fansidar), and chloroquine as malarial prophylaxis in eastern Thailand. Southeast-Asian-J-Trop-Med-Public-Health

Breman JG (2001) The ears of the hippopotamus: manifestations, determinants, and estimates of the malaria burden. Am J Trop Med Hyg 64(1_suppl):1–11

Brockman A, Price RN, Van Vugt M et al (2000) Plasmodium falciparum antimalarial drug susceptibility on the north-western border of Thailand during five years of extensive use of artesunate-mefloquine. Trans R Soc Trop Med Hyg 94:537–544. https://doi.org/10.1016/S0035-9203(00)90080-4

Brocks DR, Mehvar R (2003) Stereoselectivity in the pharmacodynamics and pharmacokinetics of the chiral antimalarial drugs. Clin Pharmacokinet 42:1359–1382

Brossi A, Yeh HJC, Chrzanowska M et al (1988) Colchicine and its analogues: recent findings. Med Res Rev 8:77–94

Bryson HM, Goa KL (1992) Halofantrine. A review of its antimalarial activity, pharmacokinetic properties and therapeutic potential. Drugs 43:236–258. https://doi.org/10.2165/00003495-199243020-00009

Bunnag D, Poltera AA, Viravan C et al (1992) Plasmodicidal effect of desferrioxamine B in human vivax or falciparum malaria from Thailand. Acta Trop 52:59–67. https://doi.org/10.1016/0001-706X(92)90007-K

Chirenda J, Siziya S, Tshimanga M (2000) Association of HIV infection with the development of severe and complicated malaria cases at a rural hospital in Zimbabwe. Cent Afr J Med 46. https://doi.org/10.4314/cajm.v46i1.8514

Cui L, Su XZ (2009) Discovery, mechanisms of action and combination therapy of artemisinin. Expert Rev Anti-Infect Ther 7:999–1013

Cui L, Mharakurwa S, Ndiaye D et al (2015) Antimalarial drug resistance: literature review and activities and findings of the ICEMR network. Am J Trop Med Hyg 93:57–68

Cullen KA, Mace KE, Arguin PM (2016) Malaria surveillance — United States, 2013. MMWR Surveill Summ 65:1–22. https://doi.org/10.15585/mmwr.ss6502a1

Davidson DE, Ager AL, Brown JL, et al (1981) New tissue schizontocidal antimalarial drugs. Bull World Health Organ

Davies J, Davies D (2010) Origins and evolution of antibiotic resistance. Microbiol Mol Biol Rev 74:417–433. https://doi.org/10.1128/MMBR.00016-10

Demain AL, Sanchez S (2009) Microbial drug discovery: 80 years of progress. J Antibiot (Tokyo) 62:5–16. https://doi.org/10.1038/ja.2008.16

Eastman RT, Fidock DA (2009) Artemisinin-based combination therapies: A vital tool in efforts to eliminate malaria. Nat Rev Microbiol 7:864–874

Eckstein-Ludwig U, Webb RJ, Van Goethem IDA et al (2003) Artemisinins target the SERCA of Plasmodium falciparum. Nature 424:957–961. https://doi.org/10.1038/nature01813

Farooq U, Mahajan RC (2004) Drug resistance in malaria. J Vector Borne Dis

Foster S, Phillips M (1998) Economics and its contribution to the fight against malaria. In: Annals of Tropical Medicine and Parasitology

Frenk J (2006) Disease control priorities in developing countries. Salud Publica Mex 48. https://doi.org/10.1590/S0036-36342006000600011

Gaillard T, Madamet M, Pradines B (2015) Tetracyclines in malaria. Malar J 14:445

Galappaththy GNL, Tharyan P, Kirubakaran R (2013) Primaquine for preventing relapse in people with Plasmodium vivax malaria treated with chloroquine. Cochrane Database Syst Rev 2013(10)

Gero AM, Upston JM (1992) Altered membrane permeability: a new approach to malaria chemotherapy. Parasitol Today 8:283–286. https://doi.org/10.1016/0169-4758(92)90147-T

Gleeson MT (2000) The plastid in Apicomplexa: what use is it? Int J Parasitol 30(10):1053–1070

Gogtay N, Kannan S, Thatte Urmila M et al (2013) Artemisinin-based combination therapy for treating uncomplicated Plasmodium vivax malaria. Cochrane Database Syst Rev. https://doi.org/10.1002/14651858.CD008492.pub3

Goodman CA, Coleman PG, Mills AJ (1999) Cost-effectiveness of malaria control in sub-Saharan Africa. Lancet 354:378–385. https://doi.org/10.1016/S0140-6736(99)02141-8

Gordeuk VR, Onojobi G, Schneider MF et al (2006) The association of serum ferritin and transferrin receptor concentrations with mortality in women with human immunodeficiency virus infection. Haematologica 91(6):739–743

Grimwade K, French N, Mbatha DD et al (2003) Childhood malaria in a region of unstable transmission and high human immunodeficiency virus prevalence. Pediatr Infect Dis J 22:1057–1063. https://doi.org/10.1097/01.inf.0000101188.95433.60

Hackstein JHP, Mackenstedt U, Mehlhorn H et al (1995) Parasitic apicomplexans harbor a chlorophyll a-D1 complex, the potential target for therapeutic triazines. Parasitol Res 81(3):207. https://doi.org/10.1007/BF00937111

Hall AP, Segal HE, Pearlman EJ et al (1975) Amodiaquine resistant falciparum malaria in Thailand. Am J Trop Med Hyg 24:575–580. https://doi.org/10.4269/ajtmh.1975.24.575

Hastings IM, Mackinnon MJ (1998) The emergence of drug-resistant malaria. Parasitology 117:411–417. https://doi.org/10.1017/S0031182098003291

Hastings IM, Watkins WM (2006) Tolerance is the key to understanding antimalarial drug resistance. Trends Parasitol 22:71–77

Hemingway J, Shretta R, Wells TNC et al (2016) Tools and strategies for malaria control and elimination: what do we need to achieve a grand convergence in malaria? PLoS Biol 14:e1002380. https://doi.org/10.1371/journal.pbio.1002380
Hershko C (1989) Biological models for studying iron chelating drugs. Baillieres Clin Haematol 3:293–321. https://doi.org/10.1016/S0950-3536(89)80019-8
Hershko C, Peto TE (1988) Deferoxamine inhibition of malaria is independent of host iron status. J Exp Med 168:375–387. https://doi.org/10.1084/jem.168.1.375
Hotez PJ, Kamath A (2009) Neglected tropical diseases in sub-Saharan Africa: review of their prevalence, distribution, and disease burden. PLoS Negl Trop Dis 3:e412
Huijben S, Paaijmans KP (2018) Putting evolution in elimination: winning our ongoing battle with evolving malaria mosquitoes and parasites. Evol Appl 11:415–430
Hyde JE (2007) Drug-resistant malaria - an insight. FEBS J 274:4688–4698
Karema C, Imwong M, Fanello CI et al (2010) Molecular correlates of high-level antifolate resistance in Rwandan children with Plasmodium falciparum malaria. Antimicrob Agents Chemother 54:477–483. https://doi.org/10.1128/AAC.00498-09
Kay K, Hodel EM, Hastings IM (2014) Improving the role and contribution of pharmacokinetic analyses in antimalarial drug clinical trials. Antimicrob Agents Chemother 58:5643–5649
Kinra P, Dutta V (2013) Serum TNF alpha levels: a prognostic marker for assessment of severity of malaria. Trop Biomed 30(4):645–653
Konotey-Ahulu FID (1999) Averting a malaria disaster [12]. Lancet 354:258
Krogstad DJ, Schlesinger PH (1987) The basis of antimalarial action: non-weak base effects of chloroquine on acid vesicle pH. Am J Trop Med Hyg 36:213–220. https://doi.org/10.4269/ajtmh.1987.36.213
Kumar A, Valecha N, Jain T, Dash AP (2007) Burden of malaria in India: retrospective and prospective view. Am J Trop Med Hyg 77:69–78. doi:77/6_Suppl/69 [pii]
Kwiatkowski D, Molyneux ME, Curtis N et al (1993) Anti-tnf therapy inhibits fever in cerebral malaria. QJM 86(2):91–98. https://doi.org/10.1093/oxfordjournals.qjmed.a068783
Lallemant C, Halembokaka G, Baty G et al (2010) Impact of HIV/Aids on child mortality before the highly active antiretroviral therapy era: A study in Pointe-Noire, republic of Congo. J Trop Med 2010:1–6. https://doi.org/10.1155/2010/897176
Lell B, Kremsner PG (2002) Clindamycin as an antimalarial drug : review of clinical trials. Antimicrob Agents Chemother 46:2315–2320. https://doi.org/10.1128/AAC.46.8.2315
Lin JT, Juliano JJ, Wongsrichanalai C (2010) Drug-resistant malaria: the era of ACT. Curr Infect Dis Rep 12:165–173
Mackinnon MJ, Hastings IM (1998) The evolution of multiple drug resistance in malaria parasites. Trans R Soc Trop Med Hyg 92:188–195
Maitland K, Newton CRJC (2005) Acidosis of severe falciparum malaria: heading for a shock? Trends Parasitol 21:11–16
Andrew R. Marks (2007) Science In Medicine
McGready R, Cho T, Samuel, et al (2001) Randomized comparison of quinine-clindamycin versus artesunate in the treatment of falciparum malaria in pregnancy. Trans R Soc Trop Med Hyg 95(6): 651–656 doi: https://doi.org/10.1016/S0035-9203(01)90106-3
Molina-Cruz A, Barillas-Mury C (2014) The remarkable journey of adaptation of the Plasmodium falciparum malaria parasite to New World anopheline mosquitoes. Mem Inst Oswaldo Cruz 109:662–667. https://doi.org/10.1590/0074-0276130553
Molyneux DH, Floyd K, Barnish G, Fèvre EM (1999) Transmission control and drug resistance in malaria: A crucial interaction. Parasitol Today 15:238–240. https://doi.org/10.1016/S0169-4758(99)01453-2
Moody A (2002) Rapid diagnostic tests for malaria parasites. Clin Microbiol Rev 15:66–78. https://doi.org/10.1128/CMR.15.1.66-78.2002
Mwangi TW, Mohammed M, Dayo H et al (2005) Clinical algorithms for malaria diagnosis lack utility among people of different age groups. Trop Med Int Heal 10:530–536. https://doi.org/10.1111/j.1365-3156.2005.01439.x

Nixon GL, Moss DM, Shone AE et al (2013) Antimalarial pharmacology and therapeutics of atovaquone. J Antimicrob Chemother 68:977–985

Nosten F, White NJ (2007) Artemisinin-based combination treatment of falciparum malaria. Am J Trop Med Hyg. doi: 77/6_Suppl/181 [pii] 77:181–192

Nyunt MM, Adam I, Kayentao K et al (2010) Pharmacokinetics of sulfadoxine and pyrimethamine in intermittent preventive treatment of malaria in pregnancy. Clin Pharmacol Ther 87:226–234. https://doi.org/10.1038/clpt.2009.177

O'Meara WP, Mangeni JN, Steketee R, Greenwood B (2010) Changes in the burden of malaria in sub-Saharan Africa. Lancet Infect Dis 10:545–555

Obaldia N, Rossan RN, Cooper RD et al (1997) WR 238605, chloroquine, and their combinations as blood schizonticides against a chloroquine-resistant strain of Plasmodium vivax in Aotus monkeys. Am J Trop Med Hyg 56:508–510. https://doi.org/10.4269/ajtmh.1997.56.508

Onyenekwe CC, Ukibe N, Meludu SC et al (2007) Prevalence of malaria as coinfection in HIVinfected individuals in a malaria endemic area of southeastern Nigeria. J Vector Borne Dis

Palmer KJ, Holliday SM, Brogden RN (1993) Mefloquine. A review of its antimalarial activity, pharmacokinetic properties and therapeutic efficacy. Drugs 45:430–475. https://doi.org/10.2165/00003495-199345030-00009

Paul REL, Hackford I, Brockman A et al (1998) Transmission intensity and Plasmodium falciparum diversity on the northwestern border of Thailand. Am J Trop Med Hyg 58:195–203. https://doi.org/10.4269/ajtmh.1998.58.195

Peters W (1990) The prevention of antimalarial drug resistance. Pharmacol Ther 47:499–508. https://doi.org/10.1016/0163-7258(90)90067-C

Plowe CV (2003) Monitoring antimalarial drug resistance: making the most of the tools at hand. J Exp Biol 206(21):3745–3752

Ponts N, le Roch KG (2013) Chapter 100. Malaria

Rathod HT, Govinda Rao HS (1997) Integration of polynomials over N-dimensional linear polyhedra. Comput Struct 65:829–847. https://doi.org/10.1016/S0045-7949(97)00004-7

Ridley RG, Dorn A, Vippagunta SR, Vennerstrom JL (1997) Haematin (haem) polymerization and its inhibition by quinoline antimalarials. Ann Tropical Med Parasitol 91:559–566

Rieckmann KH, McNamara JV, Frischer H et al (1968) Gametocytocidal and sporontocidal effects of primaquine and of sulfadiazine with pyrimethamine in a chloroquine-resistant strain of Plasmodium falciparum. Bull World Health Organ 38(4):625–632

Rieckmann KH, Campbell GH, Sax LJ, Ema JE (1978) Drug sensitivity of plasmodium falciparum. An In-vitro Microtechnique Lancet 311:22–23. https://doi.org/10.1016/S0140-6736(78)90365-3

Riordan JR, Deuchars K, Kartner N et al (1985) Amplification of P-glycoprotein genes in multidrug-resistant mammalian cell lines. Nature 316:817–819. https://doi.org/10.1038/316817a0

Roy A, Cox RA, Williamson DH, Wilson RJM (1999) Protein synthesis in the plastid of Plasmodium falciparum. Protist 150:183–188. https://doi.org/10.1016/S1434-4610(99)70020-9

Sadhukhan GC, Mukherjee A (2016) Anti-malarial drug design by targeting Apicoplasts: new perspectives. J Pharmacopuncture 19:7–15. https://doi.org/10.3831/KPI.2016.19.001

Schapira A, France LTA, Averkiev L et al (1988) The plasmodium falciparum chloroquine in viva test: extended follow-up is more important than parasite counting. Trans R Soc Trop Med Hyg 82:39–43. https://doi.org/10.1016/0035-9203(88)90256-8

Seppälä H, Klaukka T, Vuopio-Varkila J et al (1997) The effect of changes in the consumption of macrolide antibiotics on erythromycin resistance in group A streptococci in Finland. N Engl J Med 337:441–446. https://doi.org/10.1056/NEJM199708143370701

Shapiro RL, Lockman S (2010) Mortality among HIV-exposed infants: the first and final frontier. Clin Infect Dis 50:445–447. https://doi.org/10.1086/649887

Shaukat AM, Gilliams EA, Kenefic LJ et al (2012) Clinical manifestations of new versus recrudescent malaria infections following anti-malarial drug treatment. Malar J 11:207. https://doi.org/10.1186/1475-2875-11-207

Smith HJ, Meremikwu MM (2003) Iron-chelating agents for treating malaria. In: Cochrane Database of Systematic Reviews

Smith CM, Jerkovic A, Puy H et al (2015) Red cells from ferrochelatase-deficient erythropoietic protoporphyria patients are resistant to growth of malarial parasites. Blood 125:534–541. https://doi.org/10.1182/blood-2014-04-567149

Soh PN, Witkowski B, Olagnier D et al (2009) In vitro and in vivo properties of ellagic acid in malaria treatment. Antimicrob Agents Chemother 53:1100–1106. https://doi.org/10.1128/AAC.01175-08

Srivastava IK, Rottenberg H, Vaidya AB (1997) Atovaquone, a broad Spectrum Antiparasitic drug, collapses mitochondrial membrane potential in a malarial parasite. J Biol Chem 272(7):3961–3966

Stover KR, King ST, Robinson J (2012) Artemether-Lumefantrine: an option for malaria. Ann Pharmacother 46:567–577. https://doi.org/10.1345/aph.1Q539

Sun W, Tanaka TQ, Magle CT et al (2014) Chemical signatures and new drug targets for gametocytocidal drug development. Sci Rep 4:1480–1482. https://doi.org/10.1038/srep03743

Takem EN, D'Alessandro U (2013) Malaria in pregnancy. Mediterr J Hematol Infect Dis 5:2013010. https://doi.org/10.4084/MJHID.2013.010

Uneke CJ (2007) Impact of placental Plasmodium falciparum malaria on pregnancy and perinatal outcome in sub-Saharan Africa: I: introduction to placental malaria. Yale J Biol Med 39:1175–1186. https://doi.org/10.1007/s00256-010-0904-6

Uneke CJ, Duhlinska DD, Ujam TN (2009) Effects of maternal Plasmodium falciparum malaria and HIV infection on birth weight in Southeastern Nigeria. Mcgill J Med

Vaidya Kuladeepa S (2012) Quantitative Buffy coat test and other diagnostic techniques for diagnosis of malaria:review of literature. Natl J Med Res

Van Eijk A, Terlouw DJ (2007) Azithromycin for treating uncomplicated malaria. Cochrane Database Syst Rev

Verdrager J (1995) Localized permanent epidemics: the genesis of chloroquine resistance in Plasmodium falciparum. Southeast Asian J Trop. Med. Public Health

Watkins WM, Mberu EK, Winstanley PA, Plowe CV (1997) The efficacy of antifolate antimalarial combinations in Africa: a predictive model based on pharmacodynamic and pharmacokinetic analyses. Parasitol Today 13:459–464. https://doi.org/10.1016/S0169-4758(97)01124-1

Wernsdorfer WH (1991) The development and spread of drug-resistant malaria. Parasitol Today 7:297–303

Wernsdorfer WH (1994) Epidemiology of drug resistance in malaria. Acta Trop 56:143–156. https://doi.org/10.1016/0001-706X(94)90060-4

Wernsdorfer WH, Payne D (1991) The dynamics of drug resistance in *Plasmodium falciparum*. Pharmacol Ther 50(1):95–121

White NJ (1987) Combination treatment for falciparum prophylaxis. Lancet 329:680–681

White NJ (2004) Antimalarial drug resistance. JClinInvest. https://doi.org/10.1172/JCI200421682.1084

Wilson BA, Garud NR, Feder AF et al (2016) The population genetics of drug resistance evolution in natural populations of viral, bacterial and eukaryotic pathogens. Mol Ecol 25:42–66

Wolday D, Kibreab T, Bukenya D, Kodes R (1995) Sensitivity of Plasmodium falciparum in vivo to chloroquine and pyrimethamine-sulfadoxine in Rwandan patients in a refugee camp in Zaire. Trans R Soc Trop Med Hyg 89:654–656. https://doi.org/10.1016/0035-9203(95)90431-X

Wongsrichanalai C, Sibley CH (2013) Fighting drug-resistant Plasmodium falciparum: the challenge of artemisinin resistance. Clin Microbiol Infect 19:908–916

World Health Organization (2015) Treatment of severe malaria. Guidel Treat Malar 85:556–557. https://doi.org/10.1016/0035-9203(91)90261-V

World Health Organization (2016) World malaria report 2016

Yingchoncharoen P, Kalinowski DS, Richardson DR (2016) Lipid-based drug delivery Systems in Cancer Therapy: what is available and what is yet to come. Pharmacol Rev 68:701–787. https://doi.org/10.1124/pr.115.012070

Zumla A, Rao M, Wallis RS et al (2016) Host-directed therapies for infectious diseases: current status, recent progress, and future prospects. Lancet Infect Dis 16:e47–e63. https://doi.org/10.1016/S1473-3099(16)00078-5

Zwang J, Ashley EA, Karema C et al (2009) Safety and efficacy of dihydroartemisinin-piperaquine in falciparum malaria: A prospective multi-Centre individual patient data analysis. PLoS One 4:e6358. https://doi.org/10.1371/journal.pone.0006358

Printed by Printforce, the Netherlands